Pollination and Agricultural Ecosystems

Pollination and Agricultural Ecosystems

Edited by Toby Hernandez

SYRAWOOD
PUBLISHING HOUSE

New York

Published by Syrawood Publishing House,
750 Third Avenue, 9th Floor,
New York, NY 10017, USA
www.syrawoodpublishinghouse.com

Pollination and Agricultural Ecosystems
Edited by Toby Hernandez

© 2022 Syrawood Publishing House

International Standard Book Number: 978-1-64740-068-2 (Hardback)

Cataloging-in-Publication Data

Pollination and agricultural ecosystems / edited by Toby Hernandez.
    p. cm.
Includes bibliographical references and index.
ISBN 978-1-64740-068-2
1. Pollination. 2. Agricultural ecology. 3. Plants--Reproduction. 4. Plant ecology. I. Hernandez, Toby.
QK926 .P65 2022
571.864 2--dc23

TABLE OF CONTENTS

Permissions

List of Contributors

Index

PREFACE

The world is advancing at a fast pace like never before. Therefore, the need is to keep up with the latest developments. This book was an idea that came to fruition when the specialists in the area realized the need to coordinate together and document essential themes in the subject. That's when I was requested to be the editor. Editing this book has been an honour as it brings together diverse authors researching on different streams of the field. The book collates essential materials contributed by veterans in the area which can be utilized by students and researchers alike.

Pollination is the process by which pollen is transferred from the male part of a plant to the female part. This enables fertilization and the production of seeds. Usually, pollination occurs within a species, but in cases where it occurs between two different species, a hybrid offspring is produced. Self-pollination occurs when pollen from the same plant reaches the stigma of a flower or ovule. Pollination is often aided by animals such as birds, insects and bats as well as by the wind, rain and water. Nearly 3/4th of world's supply of food is derived from plants that require pollination. The financial and ecological advantage of natural pollination by native pollinators in the case of agricultural crops can be witnessed in their improved quality and yield. However, there is a recent decline in pollinator populations owing to habitat destruction, pesticide use, climate change and parasitism or disease. Pollination management is thus vital to modern agriculture. Efforts to protect and enhance pollinators, and add new pollinators to monoculture situations, as in commercial fruit orchards, are important for better agricultural productivity. This book elucidates the concepts and innovative models around prospective developments with respect to utilizing pollination for improving the health of agricultural ecosystems. It consists of contributions made by international experts on pollination and pollinating agents. It will help new researchers by foregrounding their knowledge in this area.

Each chapter is a sole-standing publication that reflects each author's interpretation. Thus, the book displays a multi-facetted picture of our current understanding of application, resources and aspects of the field. I would like to thank the contributors of this book and my family for their endless support.

Editor

THE BREEDING SYSTEM AND EFFECTIVENESS OF INTRODUCED AND NATIVE POLLINATORS OF THE ENDANGERED TROPICAL TREE *GOETZEA ELEGANS* (SOLANACEAE)

Marcos A. Caraballo-Ortiz[1,2,3]* and Eugenio Santiago-Valentín[1,2,3]

[1]*Department of Biology, University of Puerto Rico, Río Piedras Campus, P.O. Box 70377, San Juan, Puerto Rico 00936–8377, USA*

[2]*Herbarium, Botanical Garden of the University of Puerto Rico, 1187 Calle Flamboyán, San Juan, Puerto Rico 00926, USA*

[3]*Center for Applied Tropical Ecology and Conservation, University of Puerto Rico, Río Piedras Campus, P.O. Box 23341, San Juan, Puerto Rico 00931–3341, USA*

Abstract—The impact of introduced species on native organisms is one of the main conservation concerns around the world. To fully understand the effect of introduced pollinators on native plants, it is important to know the reproductive biology of the focal species, especially its pollination biology. In this study we examined the breeding system of the endangered tree *Goetzea elegans* (Solanaceae), and compared pollination effectiveness of the two main floral visitors, *Coereba flaveola* (an avian nectarivore), and *Apis mellifera* (the introduced European Honeybee). We assessed the breeding system of *G. elegans* by applying several pollination treatments to flowers of cultivated trees to test fruit set, seed set, and seed viability. We also examined the pollination efficiency of *A. mellifera* and *C. flaveola*, and compared all the treatments with positive and negative controls. Our results indicate that the introduced honeybee *A. mellifera* is as efficient as the native bird *C. flaveola* in pollinating the flowers of *G. elegans*. This study also showed that *G. elegans* requires cross–pollination for fruit and seed set, and to obtain high seed viability rates. Despite the fact that many studies report exotic species as detrimental for native organisms, we document a case where an introduced insect has a beneficial impact on the reproductive biology of an endangered tropical tree.

Key words: Apis mellifera, Coereba flaveola, *geitonogamy, outcrossing, Puerto Rico, selfing*.

INTRODUCTION

The negative impacts of introduced species on ecosystems are a major issue in conservation biology (e.g. Callaway et al. 2004; Lugo 2005; Nogueira-Filho et al. 2009; Davis et al. 2010; Nunez et al. 2010). Introduced species can disrupt native plant-animal interactions, such as plant-pollinator mutualisms, which can have negative effects on reproductive success (Traveset & Richardson 2006; Aizen et al. 2008; Bartomeus et al. 2008; Padron et al. 2009). One of the most common introduced pollinators around the world is the honeybee, *Apis mellifera* L. (Apidae). Honeybees are native to Europe, Africa and the Middle East (Winston 1987), and they have been intentionally introduced in many countries to improve crop pollination and to produce honey (Hansen et al. 2002). The impacts of the introduction of *A. mellifera* on plant communities are a focal point of debate (Goulson 2003; Stanley et al. 2004; Moritz et al. 2005; Traveset & Richardson 2006; Kaiser-Bunbury & Müller 2009), with studies showing honeybees to be neutral or beneficial for the pollination of native plants (Gross 2001; Dupont et al. 2004;

Fumero-Cabán & Meléndez-Ackerman 2007), especially for dispersing pollen in fragmented habitats (Dick 2001; Dick et al. 2003). Conversely, other studies on the pollination of tropical plants show that *A. mellifera* has negative impacts on the fitness of native plants (Gross & Mackay 1998; do Carmo et al. 2004), on interactions with native pollinators by displacing them (Kato et al. 1999; Hansen et al. 2002; Goulson 2003), and they tend to be less efficient pollinators than native species, including other bees and birds (Westerkamp 1991; Freitas & Paxton 1998; Gross & Mackay 1998; Hansen et al. 2002).

The ecological disruption of introduced species on plant-pollinator mutualisms is expected to be more severe for organisms restricted to islands because island populations are usually smaller, have limited dispersal, and are believed to be inferior competitors than continental populations (Cronk & Fuller 1995; Whittaker 1998). In addition, islands often have depauperate pollinator communities which provide services to a wide array of plants (Olesen et al. 2002). However, rare plant species are often less likely to receive pollinator visits due to competition with more abundant floral resources of surrounding common plant species (Rymer et al. 2005). Pollen limitation and the resultant low seed set are particularly detrimental for rare tropical trees, as most of the species in the tropics are dioecious or self-incompatible (Bawa et al. 1985; Ward et al. 2005). The combination of pollen

---

*Corresponding author; current address: Department of Biology, Pennsylvania State University, 208 Mueller Laboratory, University Park, PA 16802, USA. email: marcoscaraballo@gmail.com

limitation and habitat destruction of tropical forests is believed to be especially harmful for rare self-incompatible trees in biodiversity hotspots. This is because in biodiversity hotspots, intense competition for pollinator services and high deforestation rates are prevalent (Vamosi et al. 2006).

The objectives of this study were to examine the impact of the introduced *A. mellifera* on the pollination system of the tropical tree *Goetzea elegans*, an endemic and rare tree found in the Caribbean Biodiversity Hotspot (Fig. 1a; Myers et al. 2000), to compare the pollinator effectiveness of *G. elegans* floral visitors, and to examine the *G. elegans* breeding system. Specifically, we hypothesized that (1) *A. mellifera* pollinates *G. elegans* less efficiently than its native pollinator, as found in previous studies on the impact of *A. mellifera* on pollination of native plants (Westerkamp 1991; Freitas & Paxton 1998; Hansen et al. 2002). We also hypothesized that (2) *G. elegans* is self-incompatible and requires outcrossing to set seeds, as this is the predominant reproductive strategy in tropical trees (Bawa et al. 1985, Nason & Hamrick 1997, Ward et al. 2005).

MATERIALS AND METHODS

Study species

Goetzea elegans Wydler is an endangered tree of the subfamily Goetzeoideae in the Solanaceae (sensu Olmstead et al. 2008) and only found in populations on the northwestern and the eastern region of the island of Puerto Rico (including Vieques Island; USFWS 1987; IUCN 2010 [C2a version 2.3]). Although *G. elegans* was once considered critically endangered, totalling around 50 individuals in three populations (USFWS 1987), new populations have been recently discovered. The total population is now estimated at around 820 individuals in 12 populations, most of them fragmented and restricted to ravines. As for most of the species designated as endangered, very little is known of its life history, such as breeding system, pollinators, dispersal agents and demographic characteristics. Preliminary studies by Santiago-Valentín (1995) suggest that *G. elegans* requires cross pollen to set fruits. However, additional experiments are needed to confirm these preliminary data. Although the populations of *G. elegans* have declined mainly due to deforestation and land-use changes (USFWS 1987), they may also be suffering the impact of introduced species.

In the wild, *G. elegans* trees can reach up to 18 m in height, and produces flowers and fruits throughout the year, with peak flowering occurring between February and July (Little et al. 1974; Santiago-Valentín 1995). The flowers are perfect with no fragrance perceptible to humans. The pale yellow corolla is funnel shaped, up to 2 cm long and 1.3 to 2 cm across (Fig. 1a). Six slender stamens are borne near the base of the corolla and are exserted. The pistil has a slender style with a bilobed stigma, and a hairy 2-celled ovary bearing few ovules. The fruits are orange drupes about 4 cm long covered by velvety hairs with a persistent calyx (Little et al. 1974). Fruit shapes are variable among trees (from round to pear–shaped), with shape being consistent within one tree. Fruits are commonly one-seeded, but fruits with two and up to nine seeds (more rarely) have been observed in the wild

(Santiago-Valentín 1995). The seeds are elliptic and about 0.7 cm in size (Little et al. 1974).

FIG. 1. Flower of *Goetzea elegans* at anthesis (a), receiving a visit from *Coereba flaveola* (b), and receiving a visit from *Apis mellifera* (c). Photo credits: W. Hernández Aguiar (a), T. A. Carlo Joglar (b), and M. A. Caraballo Ortiz (c).

The flowers of *G. elegans* are visited by both birds and insects. The most frequent flower visitor is the native nectarivorous bananaquit *Coereba flaveola* Bryant (Coerebidae; Fig. 1b; USFWS 1987; Santiago-Valentín 1995), which is considered the most common bird in Puerto Rico (Raffaele et al. 1998). *Coereba flaveola* has a reputation for robbing the nectar of many plant species (Kodric-Brown et al. 1984, Ricart 1992; Fumero-Cabán & Meléndez-Ackerman 2007), and although the culmen of *C. flaveola* appears to contact the reproductive structures of *G. elegans*, we lack experimental data confirming its role as a pollinator of this tree species. Another common visitor of *G. elegans* is the common honeybee *Apis mellifera* L. (Apidae; Fig. 1c; Santiago-Valentín 1995). This introduced bee has been observed foraging on flowers of many native species throughout the island in both managed and wild colonies. Both *C. flaveola* and *A. mellifera* are common visitors of *G. elegans* flowers in all of the wild populations of the species (M. A. Caraballo-Ortiz, pers. obs.). Aside from *C. flaveola* and *A. mellifera*, *G. elegans* flowers are infrequently visited by two native hummingbird species (*Anthracothorax dominicus aurulentus* Audebert & Vieillot and *Eulampis holosericeus holosericeus* L.; Santiago-Valentín 1995; M. A. Caraballo-Ortiz, pers. obs.). Although there are other common pollinator species in Puerto Rico (e.g. *Xylocopa* carpenter bees), they have not been observed interacting with *G. elegans* flowers.

Study site

Our study was conducted at the Botanical Garden of the University of Puerto Rico in San Juan, Puerto Rico (18°23'38.17" N, 66°03'46.35" W; 19 m asl), located within the subtropical moist forest zone (Holdridge system: Ewel & Whitmore 1973). All *G. elegans* plants used in this study were grown from seeds collected randomly from a wild population of about 50 trees at the municipality of Isabela, in northwestern Puerto Rico (Rafael Rivera-Martínez [Conservation Trust of Puerto Rico], pers. comm.). Experiments were performed from November 2003 to January 2005 with cultivated trees of *G. elegans* that had reached maturity. Under cultivation conditions, *G. elegans* begin first flowering during the first two or three years of age. Accessibility to an adequate number of trees and flower samples, as well as the fact that the two putative pollinators are common residents of the Botanical Garden, made the cultivated setting an ideal set-up for the pollination experiments.

Pollination treatments

We selected 20 cultivated adult trees of *G. elegans* to examine its breeding system. The trees were planted in 30 L pots, reaching a mean height of 2.5 m (S.E. ± 0.035 m) and all of them were positioned within an area of about 50 m². Each of the 20 experimental trees was subjected to 13 pollination treatments. Each pollination treatment was replicated three times (i.e. on three different flowers) on each tree (i.e. 39 flowers on each of the 20 trees: 780 flowers in total). The 13 pollination treatments were: (1) visits limited to *A. mellifera*, (2) visits limited to *C. flaveola*, (3) exposed flowers, (4) bagged flowers (see details of bagging technique below), (5) outcrossed hand pollination, (6) geitonogamous hand pollination, and (7) selfed (same flower) hand pollination. To exclude the possibility that experimental flowers were contaminated with their own pollen, treatments 1-6 were repeated with emasculated flowers (representing treatments 8-13, respectively). The following describes details of each treatment.

For the exposed pollination treatment, flowers were left open continuously to all visitors until abscission. In the bagged treatments, flowers were kept covered to exclude all pollinators. Flowers used to test pollination by animal vectors (*A. mellifera* and *C. flaveola*) were kept bagged and only exposed to allow a single visit per flower. Immediately after the vector visit, we re-bagged the flower and waited for fruit production. In the hand-pollinated treatments, flowers were bagged at all times except while performing manual pollinations and emasculations. For the outcrossing treatment, pollen was manually transferred between trees, and for the geitonogamy treatment, pollen was manually applied among flowers of the same tree. To examine within-flower selfing, pollen was manually transferred from anthers to the stigma of the same flower.

Hand-pollinations were performed using a fine brush with a long, pointed tip. The brush was rinsed thoroughly with water and dried between each pollen transfer to avoid contamination. Emasculations were performed one day prior to flower anthesis (and dehiscence of anthers) by removing all anthers in the flower bud with forceps. For all treatments involving bagged flowers, flower buds were covered one day prior to anthesis using a pellon cloth bag. Pellon is a lightweight, non-woven fabric used in the clothing industry that allows light and air to pass through. It has previously been used to cover flowers (Wyatt *et al.* 1992) without altering their development or longevity (Santiago-Valentín 1995). For each of the pollination treatments, fecundity was measured in three different ways: as fruit set, as seed set, and seed viability. Seedless fruits were classified as aborted and considered as zero in the fruit set analysis and not included in the seed set analysis. Seed viability tests were carried out at the Conservation Trust of Puerto Rico nurseries (located within the grounds of the Botanical Garden of the University of Puerto Rico in Río Piedras, Puerto Rico) by germinating the seeds in a humid peat moss bed with low sunlight conditions (to simulate natural germinating conditions for this species) and verifying normal shoot and root growth from January 2004 to April 2005. All seeds collected from an individual tree were planted in a block, and all the blocks were positioned at random along the peat moss bed.

Statistical analyses

We tested for differences in fruit set, seed set and seed viability among the 13 pollination treatments first by using the fraction of fruit set as the response variable. Fractions were calculated on a per-flower basis ($N = 3$ flowers per treatment per tree, thus, possible outcomes were 0, 0.33, 0.66 and 1). Seed set data were normalized using a logarithmic transformation, while seed viability data were normalized using an arcsine transformation. In statistical analyses, trees ($N = 20$ individuals) were treated as blocks where all treatments were represented and replicated. First we analyzed fecundity measures altogether using a multivariate analysis of variance (MANOVA; SAS Institute 2000) given that we measured multiple and correlated response variables on the same treatments. In the MANOVA, the responses were the percentage of fruit set per flower, seed number per flower and, percentage of seed viability per flower, while the factors were pollination treatments, individual, and the interaction between pollination treatments and individual. Following significance of this analysis (and as recommended by Scheiner 2001), we conducted post hoc univariate tests using two-way ANOVA to test the effects of pollination treatments (at 13 levels), individual, and the interaction terms using JMPIN (SAS Institute 2000). In the case of fruit set, we used a Kruskal-Wallis ANOVA (and post hoc pairwise Wilcoxon Signed Rank tests with Bonferroni correction) given that data structure violated ANOVA assumptions. Results are presented as mean ± standard deviation.

RESULTS

Pollinator visits

Visits by *A. mellifera* produced fruit on 28% of the unemasculated flowers, and 35% of the emasculated flowers (Fig. 2a). The average number of seeds recorded for *A. mellifera* visits was 0.7 ± 0.2 seeds in the unemasculated flowers, and 0.9 ± 0.2 seeds in the emasculated flowers (Fig. 2b). Lastly, seed viability in the unemasculated and in the emasculated flowers was 12% and 22%, respectively (Fig. 2c).

Following a similar pattern to *A. mellifera*, visits by *C. flaveola* produced fruit on 22% of the unemasculated flowers and 18% of the emasculated flowers (Fig. 2a). Seed set and seed viability for the unemasculated flowers visited by *C. flaveola* were 0.5 ± 0.2 seeds and 10% respectively, whereas the values for the emasculated flowers were 0.6 ± 0.2 seeds and 15%, respectively (Fig. 2b, c).

Although *A. mellifera* and *C. flaveola* were effective pollinators of *G. elegans*, in contrast to our expectations, both animals were statistically similar for all of the fecundity measures (Fig. 2). Also, we did not detect statistical differences between any of the unemasculated and emasculated pollination treatments for either *A. mellifera* and *C. flaveola* (Fig. 2).

Breeding system

The unemasculated exposed flowers showed 50% fruit set, while the emasculated exposed flowers yield a fruit set of 47% (Fig. 2a). For seed set and seed viability, the values for the exposed unemasculated flowers were 1.7 ± 0.3 seeds and 32% respectively, and for the emasculated flowers were 1.8 ± 0.3 seeds and 26%, respectively (Fig. 2b, c). On the other hand, the bagged treatment produced the lowest values for fruit set, with 3% for unemasculated flowers and 0% for emasculated flowers (Fig. 2a). Additionally, the unemasculated bagged flowers had a seed set of 0.2 ± 0.1 seeds, and a seed viability of 3% (Fig. 2b, c).

The highest fruit yield (70%) was recorded in the unemasculated outcrossed flowers, followed by its emasculated counterpart (57%; Fig. 2a). Similarly, the outcrossed pollination treatment obtained the highest seed set and seed viability for the unemasculated (2.7 ± 0.3 seeds and 52% viability) and for the emasculated (2.2 ± 0.4 seeds and 34% viability) flowers when compared to the other pollination treatments (Fig. 2b, c). In sharp contrast to the outcrossed treatment, the selfed treatment produced a fruit set of 7% and the lowest values for seed set and for seed viability,

with 0.1 ± 0.7 seeds and 2% viability (Fig. 2). The geitonogamy pollination treatment followed the same pattern as the selfed treatment, with a fruit set of 8%, a seed set of 0.2 ± 0.1 seeds, and a seed viability of 5% in the unemasculated flowers (Fig. 2). Likewise, emasculated flowers in the geitonogamy treatment produced a fruit set of 17%, a seed set of 0.4 ± 0.1 seeds, and a seed viability of 4% (Fig. 2).

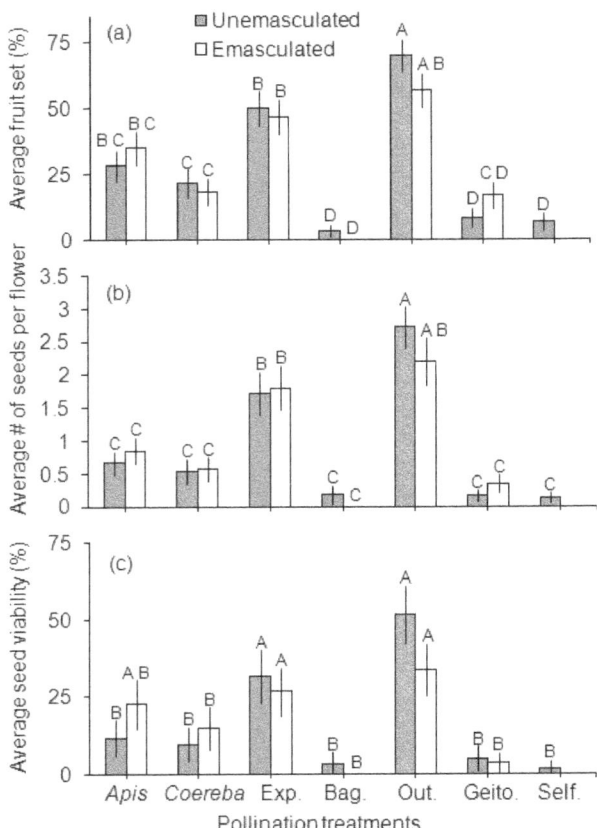

FIG. 2. Pollination treatments among unemasculated and emasculated flowers of *Goetzea elegans*. Columns show standard error bars, and different letters indicate statistical differences among treatments (α = 0.05). Pollination treatments include visits of *Apis mellifera* (*Apis*), visits of *Coereba flaveola* (*Coereba*), exposed flowers (Exp.), bagged flowers (Bag.), outcrossed pollination (Out.), geitonogamous pollination (Geito.), and selfed pollination (Self.). (a) Average percentage of fruit set (%). (b) Average number of seeds per flower. (c) Average percentage of seed viability (%).

While the bagged, selfed and geitonogamy treatments were not significantly different from each other, they were different from the exposed and outcrossed pollination treatments for all fecundity measures. The fruit set of the pollinator visitations was significantly different from the bagged, selfed and geitonogamy treatments (except for the emasculated geitonogamy treatment; $\chi^2 = 178$, df = 12, P = < 0.0001); however, seed set and seed viability values were not significantly different among pollination treatments (Fig. 2). Moreover, we did not detect significant differences among any of the unemasculated and emasculated pollination treatments for the breeding system (Fig. 2). The MANOVA test detected significant differences among the pollination treatments, the individual *G. elegans* tree, and the interaction between pollination treatments and individual (Tab. I). In the model, the pollination treatments were the most important factor explaining the variance (46%), followed by the individual (8%) and the interaction between pollination treatments and the individual (3%). After performing the two-way ANOVA tests, we found significant effects of pollination treatments ($F_{12,520} = 22.5$, P < 0.0001), individual ($F_{19,520} = 3.9$, P < 0.0001), and the interaction between pollination treatments and individual ($F_{228,520} = 1.4$, P = 0.002) on the seed set. Similarly, for the seed viability tests, we found significant effects of pollination treatments ($F_{12,520} = 17.1$, P < 0.0001), individual ($F_{19,520} = 3.1$, P < 0.0001), and the interaction between pollination treatments and individual ($F_{228,520} = 1.4$, P = 0.001).

Table I. Results for the multivariate analysis of variance (MANOVA) for the whole model, for pollination treatments, for individual trees of *Goetzea elegans*, and for the interaction between pollination treatments and individual trees.

Source	Pillai's trace	Approx. F	df	P
Whole model ($R^2 = 0.57$)	0.93	1.75	518, 1040	<0.0001
Pollination Treatments	0.31	7.88	24, 1040	<0.0001
Individual Tree	0.19	2.81	38, 1040	<0.0001
Pollination Treatments x Individual Tree	0.75	1.36	456, 1040	<0.0001

DISCUSSION

We found that both *A. mellifera* and *C. flaveola* were legitimate pollinators of *G. elegans*, with similar pollination effectiveness per visit. In addition, *G. elegans* is partially self-incompatible, requiring pollen outcrossing to produce a high fruit set. Outcrossing is the most common breeding system in tropical trees, although there are varying degrees of self-compatibility (Bawa et al. 1985; Doligez & Joly 1997; Nason & Hamrick 1997; Dick et al. 2003; Ward et al. 2005).

Pollinator visits

The legitimate pollination of *G. elegans* by *A. mellifera* corroborates results from some studies showing the beneficial role of the introduced honeybee as a pollinator of native plant species (Vaughton 1992; Dick 2001; Gross 2001; Dick et al. 2003; Fumero-Cabán & Meléndez-Ackerman 2007), and differs from those where *A. mellifera* has been found to be less effective than native nectarivorous birds (Vaughton 1996; Hansen et al. 2002). However, our results support the idea that the effectiveness of *A. mellifera* as pollinator is contingent on the structural attributes (size and shape) of flowers, as well as on the foraging behaviour of *A. mellifera* (Vaughton 1996). For example, when *A. mellifera* visited *G. elegans* flowers to collect pollen, it landed and crawled throughout the anthers and usually made contact with the nearby stigma, thus promoting pollination (Fig. 1b).

Conversely, when *A. mellifera* foraged for nectar, it landed on the inner surface of the corolla lobes from which it inserted its proboscis into the nectaries (located at the base of the ovary), which did not usually result in pollination (M. A. Caraballo-Ortiz, unpubl. data).

The pollination role of *C. flaveola* in *G. elegans* contrasts with their robbing behaviour reported in several studies on flower visitation (Kodric-Brown et al. 1984; Askins et al. 1987; Fumero-Cabán & Meléndez-Ackerman 2007). Ricart (1992) determined that robbing or legitimate pollination behaviour of *C. flaveola* was dependent on corolla length. The short size of the corolla tube in *G. elegans* (11.1 ± 1.6 mm in length and 10.2 ± 1.6 mm in width) provides easy access for *C. flaveola* to nectar (culmen length = 13.2 ± 0.7 mm), and the exserted stamens promote pollen deposition on the bird's forehead and consequently, pollen transport. Robbing by *C. flaveola* has been reported only in other plant species bearing longer corolla tubes (> 19 mm in length, unpublished data), where legitimate access to the flower nectaries by those birds with short bills is prevented by the disparity between bill and corolla length.

Breeding system

We detected a strong effect of inbreeding depression in the seed viability of the selfed treatments, suggesting that a single generation of selfing is sufficient to lower the germination vigour of a mainly outcrossing species such as *G. elegans*. By being partially self-incompatible, *G. elegans* decreases the deleterious effects of inbreeding and promotes genetic diversity in its populations (Charlesworth & Charlesworth 1979). Outcrossing plants generally have higher recombination rates, and thus, are more genetically variable and have higher levels of heterozygosity within populations than selfing plants (Hamrick et al. 1979; Loveless & Hamrick 1984; Hamrick & Godt 1989). Recombination is particularly important for plants that exist in small populations, as they often suffer from lower genetic diversity due to founder effects (Pfosser et al. 2005). Although outcrossing assures genetic variability of populations, a degree of self-compatibility enables reproduction in isolation when cross pollen limits seed set. In the Solanaceae family, there are several mechanisms through which the self-incompatibility system of a species can allow self-fertilization. These include changes in environmental conditions, the age of the flowers, and mutations in the *S*-alleles, among others (e.g., Levin 1996; Tsukamoto et al. 2003; Travers et al. 2004). It is possible that the partial self-compatibility in *G. elegans* evolved as a mutation that became fixed in the population, due to the fact that it increases the probability of reproduction in isolation or in fragmented habitats. As with *G. elegans*, other plants in Solanaceae have been found with partial self-incompatibility, such as the weed *Solanum carolinense* L., which has a plastic self-incompatible system (Travers et al. 2004).

The low fecundity exhibited by single visits of both *C. flaveola* and *A. mellifera* when compared to the exposed pollination treatment suggests that repeated visits to the flowers of *G. elegans*, which remained open an average of 3.4 days (S.D. ± 0.82), are necessary to increase chances of pollen deposition. However, the outcrossed pollination treatment achieved higher fecundity rates than the exposed treatment, indicating that a significant proportion of pollen deposition in exposed flowers might be the result of either heterospecific pollen or geitonogamy. Cultivated *G. elegans* trees used in this study usually displayed several dozens of flowers at once, which increases the probability of geitonogamous pollination. We predict that geitonogamy is frequent in larger and isolated wild trees, where thousands of flowers are presented to visitors at once.

The statistical similarity between the emasculated and unemasculated pollination treatments suggests that the pollen present in mature anthers does not interfere (i.e. blocking or clogging) with pollen deposition on the stigma, or that outcross pollen is more vigorous and fertilizes the ovaries faster than selfed pollen (Cruzan 1989; Aizen et al. 1990; Snow & Spira 1993). The fecundity of unemasculated bagged flowers was similar to the selfed treatment, demonstrating the ability of *G. elegans* flowers to set fruit in the absence of pollinators (i.e., autogamy). Autogamy was facilitated by the proximity of the anthers to the stigma due to the similar length of the reproductive whorls in most of the examined flowers. We observed, however, that a few trees presented a degree of herkogamy. Herkogamy - a disparity in the length of stamens and style of flowers - prevented those *G. elegans* trees from performing autogamy, and could be interpreted as an additional mechanism to promote outcrossing (Motten & Antonovics 1992). Finally, the lack of fruits in the emasculated bagged treatment suggests that *G. elegans* does not produce seeds by agamospermy, thus it requires pollen for sexual reproduction.

Conclusions

Unlike many introduced species, *A. mellifera* has a positive effect on the fecundity of the tropical endangered tree *G. elegans*. Besides being an additional pollinator, the ability of *A. mellifera* to connect isolated populations through pollen outcrossing in fragmented landscapes (Dick 2001; Dick et al. 2003) is of particular importance for rare and partially self-incompatible species such as *G. elegans*, since cross-pollination is critical for their reproduction. At present, the scattered distribution of *G. elegans* trees in the wild and the considerable geographic distance among populations (mean: 36.1 Km $\pm$ 55.3) could prevent optimal outcrossing, thus jeopardizing the continuous genetic diversity of the species. Currently, the main threats to *G. elegans* survival are habitat destruction and modification, such as deforestation for agriculture, selective logging for fence posts, and limestone quarrying, as these activities promote fragmentation of populations, decline of reproductive trees, and reproductive isolation of remnant trees (USFWS 1987). Although nearly all *G. elegans* populations are restricted to small canyons or ravines, which also shelter important populations of other rare and endangered plants and animals, none of these areas have legal designation for protection. However, propagation efforts and establishment of *ex situ* populations in protected areas has already been initiated (Rafael Rivera-Martínez [Conservation Trust of Puerto Rico], pers. comm.). The establishment of these *ex situ* populations, in conjunction with the conservation of extant wild populations and the effective pollination of wild trees by *C. flaveola* and *A.*

mellifera, are essential to maintain viable populations and to promote the conservation of *G. elegans*.

ACKNOWLEDGMENTS

The authors thank J. D. Ackerman, J. M. Wunderle, T. Giray, T. A. Carlo, R. Tremblay, E. Meléndez-Ackerman, C. Trejo, J. E. Aukema, A. W. Norris, R. Glenn, S. Yang, and A. Stephenson for their insightful comments and revisions on the manuscript. We also thank S. Flecha, W. Caraballo Castaing, L. Ortiz Ortiz, W. Caraballo Ortiz, R. Rivera Martínez, J. Rodríguez, J. Soltero, K. Umpierre, V. Mangual, S. Hernández, L. Rivera, E. Napoleón, C. Díaz, J. Colón, F. Ortiz, J. Rosario, C. Calo, B. Carrión, M. Miranda, N. M. Martínez, M. Robles, M. Vega, O. Díaz, A. Quiñones, M. Gómez, N. Cortés, D. Cuba, G. Bone, R. L. González, D. Torres, C. Viera, E. Vega, B. Martínez, A. M. Camacho and R. Rivera for providing materials and assistance with experimental work. We are grateful to the Botanical Garden of the University of Puerto Rico and to the Conservation Trust of Puerto Rico for permitting research at their facilities. This research was funded by National Science Foundation-Centers of Research Excellence in Science and Technology (CREST; HRD-0206200) through the Center for Applied Tropical Ecology and Conservation of the University of Puerto Rico (CATEC). This is scientific contribution number 002 of the Herbarium of the University of Puerto Rico Botanical Garden (UPR).

REFERENCES

Aizen MA, Searcy KB, Mulcahy DL (1990) Among- and within-flower comparisons of pollen tube growth following self– and cross–pollinations in *Dianthus chinensis* (Caryophyllaceae). American Journal of Botany 77:671–676

Aizen MA, Morales CL, Morales JM (2008) Invasive mutualists erode native pollination webs. PLoS Biology 6:e31

Askins RA, Ercolino KM, Waller JD (1987) Flower destruction and nectar depletion by avian nectar robbers on a tropical tree, *Cordia sebestena*. Journal of Field Ornithology 58:345–349

Bartomeus I, Vila M, Santamaria L (2008) Contrasting effects of invasive plants in plant-pollinator networks. Oecologia 155:761–770

Bawa KS, Perry DR, Beach JH (1985) Reproductive biology of tropical lowland rainforest trees. I. Sexual systems and incompatibility mechanisms. American Journal of Botany 72:331–345

Callaway RM, Thelen GC, Rodriguez A, Holben WE (2004) Soil biota and exotic plant invasion. Nature 427:731–733

Charlesworth D, Charlesworth B (1979) The evolutionary genetics of sexual systems in flowering plants. Proceedings of the Royal Society of London, Series B, Biological Sciences 205:513–530

Cronk QCB, Fuller JL (1995) Plant invaders. Chapman and Hall, London, UK, 256 pp

Cruzan MB (1989) Pollen tube attrition in *Erythronium grandiflorum*. American Journal of Botany 76:562–570

Davis NE, O'Dowd DJ, Mac Nally R, Green PT (2010) Invasive ants disrupt frugivory by endemic island birds. Biology Letters 6:85–88

Dick CW (2001) Genetic rescue of remnant tropical trees by an alien pollinator. Proceedings of the Royal Society of London, B, Biological Sciences 268:2391–2396

Dick CW, Etchelecu G, Austerlitz F (2003) Pollen dispersal of tropical trees (*Dizinia excelsa*: Fabaceae) by native insects and African honeybees in pristine and fragmented Amazonian rainforest. Molecular Ecology 12:753–764 do Carmo RM, Franceschinelli EV, da Silveira FA (2004) Introduced honeybees (*Apis mellifera*) reduce pollination success without affecting the floral resource taken by native pollinators. Biotropica 36:371–376

Doligez A, Joly HI (1997) Mating system of *Carapa procera* (Meliaceae) in the French Guiana tropical forest. American Journal of Botany 84:461–470

Dupont YL, Hansen DM, Valido A, Olesen JM (2004) Impact of introduced honeybees on native pollination interactions of the endemic *Echium wildpretii* (Boraginaceae) on Tenerife, Canary Islands. Biological Conservation 118:301–311

Ewel JJ, Whitmore JL (1973) The ecological life zones of Puerto Rico and U.S. Virgin Islands. U.S. Department of Agriculture, Forest Service Research Report ITF–18, Río Piedras, Puerto Rico, USA, 72 pp

Freitas BM, Paxton RJ (1998) A comparison of two pollinators: the introduced honeybee *Apis mellifera* and an indigenous bee *Centris tarsata* on cashew *Anacardium occidentale* in its native range of NE Brazil. Journal of Applied Ecology 35:109–121

Fumero-Cabán JJ, Meléndez-Ackerman EJ (2007) Relative pollination effectiveness of floral visitors of *Pitcairnia angustifolia* (Bromeliaceae). American Journal of Botany 94:419–424

Goulson D (2003) Effects of introduced bees on native ecosystems. Annual Review of Ecology, Evolution, and Systematics 34:1–26

Gross CL (2001) The effect of introduced honeybees on native bee visitation and fruit-set in *Dillwynia juniperina* (Fabaceae) in a fragmented ecosystem. Biological Conservation 102:89–95

Gross CL, Mackay D (1998) Honeybees reduce fitness in the pioneer shrub *Melastoma affine* (Melastomataceae). Biological Conservation 86:169–178

Hamrick JL, Linhart YB, Milton JB (1979) Relationships between life history characteristics and electrophoretically detectable genetic variation in plants. Annual Review of Ecology and Systematics 10:173–200

Hamrick JL, Godt MJ (1989) Allozyme diversity in plant species. In: Brown AHD, Clegg MT, Kahler AL, Weir BS (eds). Plant Population Genetics, Breeding and Germplasm Resources. Sinauer, Sunderland, Massachusetts, USA, pp. 43–63

Hansen DM, Olesen JM, Jones CG (2002) Trees, birds, and bees in Mauritius: exploitative competition between introduced honey bees and endemic nectarivorous birds? Journal of Biogeography 29:721–734

IUCN [International Union for Conservation of Nature] (2010) IUCN Red List of Threatened Species. Version 2010.4. [online] URL: http://www.iucnredlist.org, accessed on 03 December 2010

Kaiser-Bunbury CN, Müller CB (2009) Indirect interactions between invasive and native plants via pollinators. Naturwissenschaften 96:339–346

Kato M, Shibata A, Yasui T, Nagamasu H (1999) Impact of introduced honeybees, *Apis mellifera*, upon native bee communities in the Bonin (Ogasawara) Islands. Researches on Population Ecology 41:217–228

Kodric-Brown A, Brown JH, Byers GS, Gori DF (1984) Organization of a tropical island community of hummingbirds and flowers. Ecology 65:1358–1368

Levin DA (1996) The evolutionary significance of pseudo self-fertility. The American Naturalist 148:321–332

Little EL, Woodbury RO, Wadsworth FH (1974) Trees of Puerto Rico and the Virgin Islands. USDA Agriculture Handbook No. 449, US Forest Service, Washington, D.C., USA, 1024 pp

Loveless MD, Hamrick JL (1984) Ecological determinations of genetic structure in plant populations. Annual Review of Ecolog

and Systematics 15:65–95

Lugo AE (2005) The outcome of alien tree invasions in Puerto Rico. Frontiers in Ecology and the Environment 2:265–273

Moritz RFA, Hartel S, Neumann P (2005) Global invasions of the western honeybee (*Apis mellifera*) and the consequences for biodiversity. Ecoscience 12:289–301

Motten AF, Antonovics J (1992) Determinants of outcrossing rate in a predominantly self-fertilizing weed, *Datura stramonium* (Solanaceae). American Journal of Botany 79:419–427

Myers N, Mittermeier RA, Mittermeier CG, da Fonseca GAB, Kent J (2000) Biodiversity hotspots for conservation priorities. Nature 403: 853–858

Nason JD, Hamrick JL (1997) Reproductive and genetic consequences of forest fragmentation: two case studies of Neotropical canopy trees. Journal of Heredity 88:264–276

Nogueira-Filho SLG, Nogueira SSC, Fragoso JMV (2009) Ecological impacts of feral pigs in the Hawaiian Islands. Biodiversity and Conservation 18:3677–3683

Nunez MA, Bailey JK, Schweitzer JA (2010) Population, community, and ecosystem effects of exotic herbivores: a growing global concern. Biological Invasions 12:297–301

Olesen JM, Eskildsen LI, Venkatasamy S (2002) Invasion of pollination networks on oceanic islands: importance of invader complexes and endemic super generalists. Diversity and Distributions 8: 181–192

Olmstead RG, Bohs L, Migid HA, Santiago-Valentín E, García VF, Collier SM (2008) A Molecular Phylogeny of the Solanaceae. Taxon 57:1159–1181

Padron B, Traveset A, Biedenweg T, Diaz D, Nogales M, Jens M (2009) Impact of alien plant invaders on pollination networks in two archipelagos. PLoS One 4:e6275

Pfosser M, Jakubowsky G, Schlüter PM, Fer T, Kato H, Stuessy TF, Sun B-Y (2005) Evolution of *Dystaenia takesimana* (Apiaceae), endemic to Ullung Island, Korea. Plant Systematics and Evolution 256:159–170

Raffaele H, Wiley J, Garrido O, Keith A, Raffaele J (1998) A guide to the birds of the West Indies. Princeton University Press, Princeton, New Jersey, USA. 512 pp

Ricart CM (1992) Feeding ecology of nectar–feeding birds in the lower montane wet forest life zone, Maricao, Puerto Rico. Acta Científica 6:41–48

Rymer PD, Whelan RJ, Ayre DJ, Wetson PH, Russell KG (2005) Reproductive success and pollinator effectiveness differ in common and rare *Persoonia* species (Proteaceae). Biological Conservation 123:521–532

Santiago-Valentín E (1995) Reproductive and population ecology of *Goetzea elegans* Wydler (Solanaceae or Goetzeaceae). M.S. thesis, University of Puerto Rico, Mayagüez, Puerto Rico

SAS Institute (2000) JMPIN. Version 4.0.2, SAS Institute, Cary, North Carolina, USA

Scheiner SM (2001) MANOVA: multiple response variables and multispecies interactions. In Scheiner SM, Gurevitch J (eds). Design and analysis of ecological experiments (2nd edition). Oxford University Press, New York, USA, pp. 99–115

Snow AA, Spira TP (1993) Individual variation in the vigor of self pollen and selfed progeny in *Hibiscus moscheutos* (Malvaceae). American Journal of Botany 80:160–164

Stanley SS, DeGrandi-Hoffman G, Smith DR (2004) The African Honeybee: Factors contributing to a successful biological invasion. Annual Review of Entomology 49: 351–376

Travers SE, Mena-Alí JI, Stephenson AG (2004) Plasticity in the self-incompatibility of *Solanum carolinense*. Plant Species Biology

19:127–135

Traveset A, Richardson DM (2006) Biological invasions as disruptors of plant reproductive mutualisms. Trends in Ecology and Evolution 21:208–216

Tsukamoto T, Ando T, Takahashi K, Omori T, Watanabe H, Kokubun H, Marchesi E, Kao T-H (2003) Breakdown of self-incompatibility in a natural population of *Petunia axillaris* caused by loss of pollen function. Plant Physiology 131:1903–1912

USFWS [U.S. Fish & Wildlife Service] (1987) Beautiful *Goetzea* Recovery Plan. U.S. Fish and Wildlife Service, Atlanta, Georgia, USA, 35 pp

Vamosi JC, Knight TM, Steets JA, Mazer SJ, Burd M, Ashman T-L (2006) Pollination decays in biodiversity hotspots. Proceedings of the National Academy of Sciences of the United States of America 103: 956–961

Vaughton G (1992) Effectiveness of nectarivorous birds and honeybees as pollinators of *Banksia spinulosa* (Proteaceae). Australian Journal of Ecology 17:43–50

Vaughton G (1996) Pollination disruption by European honeybees in the Australian bird-pollinated shrub *Grevillea barklyana* (Proteaceae). Plant Systematics and Evolution 200:89–100

Ward M, Dick CW, Gribel R, Lowe AJ (2005) To self, or not to self... A review of outcrossing and pollen–mediated gene flow in Neotropical trees. Heredity 95:246–254

Westerkamp C (1991) Honeybees are poor pollinators– why? Plant Systematics and Evolution 177:71–75

Whittaker RJ (1998) Island biogeography: ecology, evolution, and conservation. Oxford University Press, Oxford, New York, USA, 304 pp

Winston ML (1987) The biology of the honey bee. Harvard University Press, Cambridge, UK, 294 pp

Wyatt R, Broyles SB, Derda GS (1992) Environmental influences on nectar production in milkweeds (*Asclepias syriaca* and *A. exaltata*). American Journal of Botany 79:636–642

THE ROLE OF ARTIFICIAL POLLINATION AND POLLEN EFFECT ON EAR DEVELOPMENT AND KERNEL STRUCTURE OF DIFFERENT MAIZE GENOTYPES

Fatih Kahrıman[1,*], Cem Ömer Egesel[2], Tuncay Aydın[1], Selinnur Subaşı[2]

[1]Çanakkale Onsekiz Mart University, Faculty of Agriculture, Department of Field Crops, 17020, Çanakkale, Turkey
[2]Çanakkale Onsekiz Mart University, Faculty of Agriculture, Department of Agricultural Biotechnology, 17020, Çanakkale, Turkey

Abstract—Pollen effect is important on several kernel traits in maize breeding and may vary under different pollination treatments. Our objectives in this study were i) to evaluate the effects of pollination treatments that are commonly used in maize breeding, on several ear and kernel traits, ii) to investigate if the genotypes so called "specialty corn" do have any different reaction to the pollen effect. A field trial was carried out at Dardanos Research and Application Center of Çanakkale Onsekiz Mart University, Turkey, in 2013. The experiment used a split plot design with three replicates. Four parents (three inbreds and one open pollinated landrace) were used as plant material. Three pollination treatments (open pollination, self-pollination and bulk pollination) were applied, and individual pollen effect of each parent on other parents was investigated. For this purpose, several ear and kernel traits (ear weight, kernel weight, kernel number, mean kernel weight) and biochemical features (protein, oil, carbohydrate and carotenoid content) were measured on harvested samples.

The results showed that pollination treatment affected the variation on all traits except for oil content ($P <$ 0.05). Self-pollination caused a significant reduction in kernel development. Pollen effect was found significant for most traits and this effect was evident on the related genotypes with open pollinated landrace. Results indicate that pollen effect is an important factor on kernel and ear development in small plot trials, where different types of maize are grown together.

Keywords: Protein, Oil, Zea mays, Pollination treatment

INTRODUCTION

"Pollen effect" refers to the changes in female parent's phenotype caused by the pollen source (Focke1881). It has been well studied in maize and shown to be the cause of important differences in various traits, such as oil content, protein content, fatty acid composition and embryo/seed ratio (Letchworth & Lambert 1998; Tsai & Tsai 1990; Weingartner et al. 2004; Dong 2007; Tanaka et al. 2009). The findings about the pollen effect were variable in the literature. Some studies concluded that pollen parent had no significant impact on protein content (Gilbert 1960; Letchworth & Lambert 1998), whereas, some others showed the lysine and tryptophan content in maize were affected by the pollen source (Pixley & Bjarnason 1994). It was found that pollen from different individual plants of the same variety might have an impact on kernel volume and weight (Bulant & Gallais 1998; Balestre et al. 2007). These investigations enlarged the area of study on pollen effect in maize. Several methods have been developed in order to take the advantage of pollen effect. The best known example is TopCross Blend (Dupont ®), used to elevate oil level of a high yielding hybrid (Thomison et al. 2002). Plus Hybrid

Effect is another application, designed to obtain high grain yield, where pollen effect and CMS (cytoplasmic male sterility) were collectively utilized (Weingartner 2002a; Weingartner 2002b).

Pollen must be under control in maize breeding studies and conventional seed production. To achieve this, artificial pollination methods are used in breeding, selfing being the most common one (Öz & Tuğay 2003). Selfing is accomplished by transferring the collected pollen from tassel to ear (pre-covered) in a controlled manner (Abdin et al. 1979). This method causes negative effects on many traits, known as inbreeding depression (Öz & Tuğay 2003). To avoid this, sib-pollination was introduced (Lindstorm 1939), which partly alleviated the ill effects of selfing. Another method, bulk pollination, is preferred in open pollinated varieties (OPV) to sustain the genetic constitution of OPV's, for which the other pollination methods have no use. In this method, pollen from a number of individual plants was bulked and then distributed to the same plants (Taba & Twumasi-Afriyie 2008). Natural pollination (open pollination) was also used in maize research in addition to the artificial pollination methods. Pollen contamination occurs in natural pollinated plants, resulting in undesired variations. Several researchers have studied the effects of pollination method on the kernel structure in maize. Open and self-pollinated genotypes (inbred and hybrids) were compared for protein, oil and starch content (Letchworth &

Lambert 1998; Schaefer & Bernardo 2013), fiber content (Şchiop et al. 2011) and starch properties (Krieger et al. 1998). Earlier studies exploring the effects of pollen and/or pollination treatment on the maize genotypes generally used a certain type of plant material, such as inbreds and hybrids. However, breeding nurseries may also contain genotypes with different characteristics, such as open pollinated varieties (OPVs). Use of plant material with different genetic composition such as OPVs, special types of inbreds or hybrids etc. in the studies on pollen and pollination effects could provide novel information. Additionally, evaluation of the traits rarely examined in previous studies, such as carotenoid content, could provide new findings about the pollen effect in maize.

The objectives of this study were: i) to evaluate the effects of pollination treatments that are commonly used in maize breeding, on several ear and kernel traits, ii) to investigate if the genotypes so called "specialty corn" do have any different reaction to the pollen effect.

MATERIALS AND METHODS

Plant material and trial organization

Four parental genotypes were used in this study, each possessing a different kernel quality (Tab. I). These genotypes were selected to study the pollen effect since they varied in terms of genetic, visual and biochemical features. Of these parents, IHO had high oil (~14%), low carotenoid and carbohydrate; Q2 had high carotenoid and low protein content; PR had high anthocyanin content; and OPV had normal values for all of the investigated traits. A 4 × 4 reciprocal full diallel mating design was generated by using these parental lines. A total of four parents and twelve hybrid combinations made up the plant material in this study. Except Q2 (dent), all of the parents were flint (Tab. I). To characterize the parents for flowering features, we collected

data on days to tasseling, days to anthesis and days to silking (number of the days from the sowing until 50% of number of plants in a plot had tassels, pollen shed, and silk emergence, respectively). Anthesis-silking interval (ASI) was also recorded as the number of days between 50% silking and 50% anthesis. Flowering data were collected only on open pollinated plots. Flowering features of the parental lines showed similar values except IHO, which was earlier for flowering as well as higher anthesis-silking interval than other genotypes. There was a good synchronization among the parental genotypes (Tab. I).

The field trial was carried out at the Çanakkale Onsekiz Mart University, Dardanos Research and Application Center in Çanakkale, Turkey. Experimental design was a split plot design with three replicates. Genotypes were assigned to main plots (8 rows), and pollination methods to subplots (2 rows). Planting was accomplished with a mechanical seed driller at a plant density of about 71,400 plants ha^{-1},on May 17th, 2013. The soil of the experimental area was clay-loam, with a pH of 7.8, containing 12.7% lime, 1.27%, organic matter content 37.8 kg ha^{-1} phosphorus, and 549.9 kg ha^{-1} potassium. A total of 170 kg ha^{-1} nitrogen was applied in two occasions (80 kg ha^{-1} at planting, and 90 kg ha^{-1} at pre-flowering), based on the soil analysis. Plots were drip-irrigated as needed. Weeds were mechanically controlled.

Pollination treatments

Three different pollination methods were applied in this study. As a natural pollination treatment, open pollination was used; while, self-pollination and bulk pollination were artificial treatments (Fig. I). To prevent pollen contamination among the genotypes, a controlled pollination method was used (Annonymous 2014). In the first step of this method, all of the ears on the plants were covered by shoot bags before the silk emerging to prevent pollen

TABLE I.   Plant materials used in the study.

	IHO	OPV	Q2	PR
Flowering Events (DT, DA, DS, ASI)	69,70,74, 4	74, 75, 76, I	75, 75, 76, I	72, 73, 75, 2
General Properties	High oil (14%), low carbohydrate, low carotenoid	Normal values for protein, oil and cabohydrate	Moderate oil, low protein, high carotenoid	High in anthocyanin; Normal values for protein, oil and cabohydrate
Source	NRPIC, USA	Trabzon, Turkey	NRPIC, USA	Çanakkale, Turkey
Hybrids made	IHO × Opaque-2 IHO × OPV IHO × PR	OPV × IHO OPV × Q2 OPV × PR	Q2 × IHO Q2 × OPV Q2 × PR	PR × IHO PR × OPV PR × Q2

NPRIC: North Central Plant Introduction Center. DT: Days to tasseling, DA: Days to anthesis, DS: Days to silking, ASI: Anthesis silking interval.

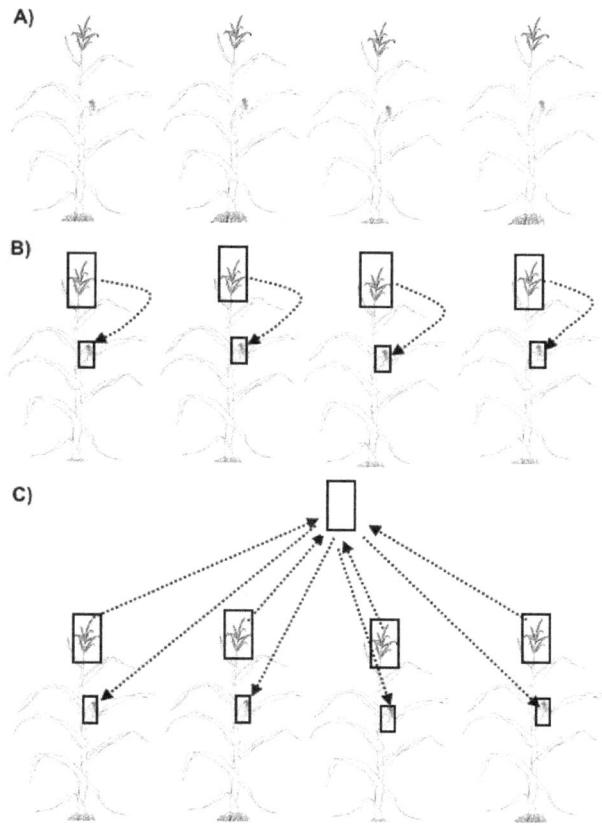

FIGURE 1. Graphical presentation of compared pollination methods: open pollination (A), self-pollination (B), and bulk pollination (C). Rectangular shapes indicate the pollination bags and shoot bags.

contamination. Emerging silk was truncated to get a satisfactory seed set. A tassel bag was carefully placed over the designated plants' tassels to collect pollen for the next day. Next morning (8:00-10:00 AM) the tassel bag was tapped to release pollen from the tassels. Then the tassel bag was brought onto the shoot exposing silk, and tapped gently so that the pollen was introduced to the fresh silk. Lastly, the flaps of the bag were tightly stapled against the stalk.

At least four plants were pollinated this way in each subplot. For self-pollination, the pollen collected from a tassel was used to pollinate the ear on the same plant. For bulk pollination, the pollen collected and bulked from at least four plants was distributed to the plants that provided this pollen bulk (Fig. 1). Unlike selfing, bulk pollination was made by pouring the pollen collected in the tassel bags onto the silks. The open pollinated plants were not bagged. To evaluate pollen effect of each genotype, all possible crossing combinations among the parental genotypes were made by artificial pollination. Harvest was done upon physiological maturity (black layer formation). All of the hand pollinated ears were harvested for the artificial pollination treatments, while four ears were sampled from the open pollinated plots ($N = 288$).

Observed traits

Data were collected on four ear traits and four biochemical traits to compare the pollination methods and to evaluate the pollen effect of parents. On the harvested ear samples, ear weight (g), kernel weight per ear (g), and kernel number per ear were measured. Mean kernel weight was determined by dividing kernel weight to kernel number for each sample. After collecting the data on ear traits, the kernels were grounded using a laboratory mill (Fritsch pulverisette 14) with 0.5 mm sieving. Grounded samples were analyzed to determine protein (%), oil (%), carbohydrate (%) and carotenoid contents (μg g^{-1}). Protein, oil and carbohydrate ratios were estimated with a Near Infrared Reflectance Spectroscopy (Spectrastar 2400D, Unity Scientific, USA). For this purpose, each sample was scanned within 1200-2400 nm interval, using the powder cup of the instrument. Carotenoid concentration was determined according to Rodriguez-Amaya & Kimura (2004). For this assay, two grams of sample weighed into a glassine tube. Then, 5mL of pure water added on the samples and incubated at 4 C° overnight. Samples were subjected to a series of pure acetone (15 mL two times) and acetone:hexane (25 mL one times) solutions. After each application, tubes were shaken (2 min) and liquid phase in tubes was collected. The upper phases of extracts that collected in 3 occasions were gathered in a glass funnel. About 300 mL cold water was added and the upper phase was transferred into a 25 mL flask. Three mL of extract was measured at 450 nm by using a quartz cuvette with 1 cm pathlenght in a pre-conditioned UV-VIS spectrophotometer (PG Instrument, England). Total carotenoid content (TCC) of each sample was determined by the following formula:

$$TCC\ (\mu g\ g^{-1}) = \frac{25\ x\ A1\ x\ 10^4}{2500\ x\ W}$$

where, $A1$ was the absorbance value of the sample at 450 nm and W was the sample weight.

Statistical analyses

Two different statistical methods were used to analyze the data. Firstly; analysis of variance was performed based on the following statistical model, using Proc GLM procedure of SAS (SAS Institute, 1999) to test the effect of the pollination methods on the parent traits.

$$y_{ijk} = \mu + \tau_i + \beta_j + (\tau\beta)_{ij} + \gamma_k + (\beta\gamma)_{jk} + \varepsilon_{ijk}$$

where; μ is the population mean, τ_i is replication effect, β_j is genotype effect (main plot), $(\tau\beta)_{ij}$ is whole plot error, γ_k is pollination treatment effect (sub-plot effect), $(\beta\gamma)_{jk}$ is interaction term between whole and sub-plot effect, ε_{ijk} is sub-plot error term. Pollination treatment was assigned to subplots and genotypes to main plots since we need a more accurate evaluation of the pollination effect. Type III error was used in ANOVA to calculate the significance levels of the sources of variation. LSD test was performed to compare the treatment means.

Second analysis of the data was performed to determine the effect of the male parent on the female parent. This was described as "pollen effect" and computed according to Bulant et al (2000) for each parental genotype. These computations give the deviation of hybrid values from the average of female parents. To compute the pollen effect of parental lines, samples from self-pollination treatment were used for all of the genotypes. Pollen effect of each parental genotype was equal to differences of least squares means in Proc MIXED procedure in SAS. REML based computations were made for each parental group, including one female parent and its related hybrids (e.g., IHO and the hybrids in which IHO was the female parent were collectively referred to as 'IHO and its Related Hybrids'). In each group, individual pollen effects of male parents over the female parents were compared by t test.

RESULTS AND DISCUSSION

Comparison of pollination treatments

Variance analysis revealed significant differences between the pollination treatments for ear and kernel development traits. Except the oil content, other biochemical constituents showed significant differences by the pollination treatment. Genotype × Treatment interaction was found significant for all biochemical features, while the genotype effect was significant for all of the evaluated traits (Tab. 2).

Open pollination produced higher ear weight, kernel number and total kernel weight compared to the artificial pollination methods. Selfing yielded lower values within the artificial methods for ear traits (Fig. 2). This variation may be due to the difference in the degree of pollen availability among the pollination treatments. In open pollination, great amount of pollen grain can reach the silk surface. In artificial pollination, however, the amount of viable pollen is limited. Thus, the amount of pollen available for the silks would be much more with the open and bulk pollination methods than it is with selfing. This is probably the main reason for getting more kernels from the open or bulk pollinated ears. This idea is supported by earlier research which showed

restricted pollination resulted in fewer seeds in maize (Borrás et al. 2002; Borrás et al. 2003). Nevertheless, the same researchers argued hand pollination would yield a better seed set compared to open pollination and restricted pollination. This is a contradiction with our results, probably due to the effect of synchronous pollination method used in those studies (Cárcova et al. 2000). The lower values for the mean kernel weight in open pollination could be attributed to the higher number of kernels per ear in open pollination. Borrás and Otegui (2001) reported a decrease in seed weight as the number of seeds increased on an ear. Bulant et al. (2000) argued that increased activity of ADPGPPase enzyme was associated with greater kernel size in the case of cross pollination. Our results from bulk pollination treatment agreed with this argument, whereas the results from open pollination did not. Having more kernels in open pollination may have masked the role of pollen effect in this case. Letchworth & Lambert (1998) reported that there was no significant difference for average mean kernel weight between open (30.7 mg) and self-pollination (30.9 mg) treatments. Similarly, we found no significant differences between these treatments for average of kernel weight (Fig. 2). A comparison within artificial pollination methods showed that kernel weight was slightly lower in selfing (pollen from the same plant) than in bulk pollination (pollen from different plants) (Fig. 2). In terms of kernel and ear development, the genotypes had similar reactions to different pollination treatments (Tab. 3), resulting in a non-significant G × Y interaction effect for ear traits (Tab. 2).

Artificial pollination had also significant effect on kernel biochemical structure. Protein, oil and carotenoid were found to be higher in self-pollination than the other methods. The lowest figures were obtained from open pollination treatment for these traits. The carbohydrate content significantly decreased with artificial pollination in our study. Oil content had no significant variation among the pollination treatments (Fig. 2). Our results for protein and carbohydrate contents were in agreement with Letchworth & Lambert (1998). Sulewska et al. (2014) also reported non-significant differences for oil content between

TABLE 2.   Means squares from the ANOVA for the investigated traits.

Source of Variation	df	Ear Weight	Kernel Weight	Kernel Number	Mean Kernel Weight
Replication (R)	2	79.2	434.7	13364.6	0.0001
Genotype (G)	3	20746.3**	15121.1**	171391.4**	0.0294**
Error I	6	2598.6	1403.8	8841.6	0.0023
Pollination (P)	2	12530.7*	9085.7*	117315.7**	0.0048
PxG	6	3003.2	2423.4	10871.9	0.0037
Error 2	12	1287.2	1035.9	5267.1	0.0019

Source of Variation	df	Protein Content	Oil Content	Carbohydrate Content	Carotenoid Content
Replication (R)	2	2.25	7.24	3.93	0.50
Genotype (G)	3	31.3**	70.1**	34.7**	22.9**
Error I	6	1.92	2.91	3.51*	1.52
Pollination (P)	2	5.13*	1.77	9.33**	12.1**
PxG	6	4.84*	4.72*	7.61**	52.2*
Error 2	12	1.28	2.78	1.63	1.56

df: Degrees of freedom. *, ** statistically significant at 0.05 and 0.01, respectively.

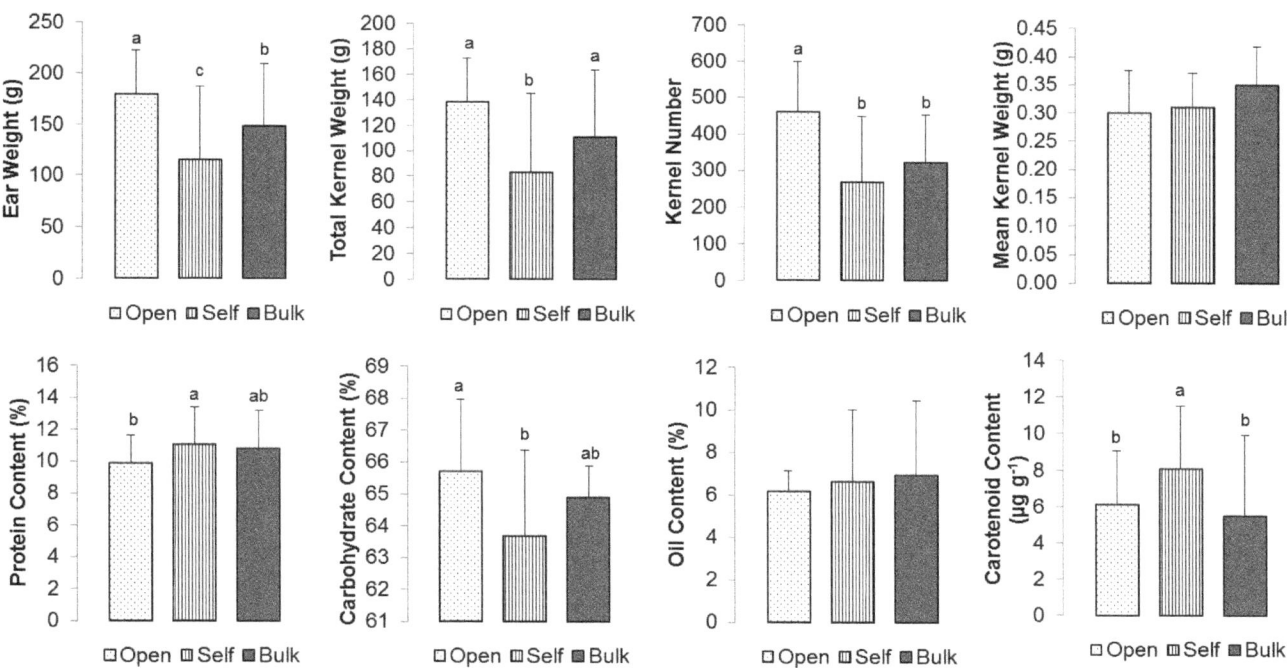

FIGURE 2. Differences among the pollination treatments (Open, Self, Bulk) for the investigated traits. Different letters among pollination treatments indicate the statistically significant differences at $P < 0.05$. Standard deviations are shown above the bars.

TABLE 3. Genotype means from the different pollination treatments for the investigated traits.

Trait	Treatment	IHO	PR	OPV	Q2
Ear Weight (g)	Open	147.2	231.6	155.1	185.7
	Self	103.2	156.4	31.6	170.1
	Bulk	114.1	236.3	111.1	133.3
Kernel Weight (g)	Open	114.9	178.2	112.3	147.5
	Self	77.9	116.2	6.7	132.1
	Bulk	82.4	186.2	78.6	97.1
Kernel Number	Open	477.2	572.3	261.8	536.0
	Self	308.6	361.0	20.7	391.0
	Bulk	316.3	493.5	185.5	298.7
Mean Kernel Weight (g)	Open	0.24	0.31	0.42	0.27
	Self	0.25	0.33	0.32	0.32
	Bulk	0.26	0.38	0.42	0.33
Protein Content (%)	Open	11.6 b	9.98 a	9.70 b	8.13 a
	Self	12.8 ab	9.48 a	13.33 a	8.94 a
	Bulk	13.7 a	11.58 a	10.28 ab	7.56 a
Oil Content (%)	Open	9.7 a	5.56 a	6.64 a	4.60 a
	Self	10.7 a	4.95 a	5.33 a	5.10 a
	Bulk	11.9 a	4.64 a	4.56 a	3.76 b
Carbohydrate Content (%)	Open	62.9 a	66.2 a	66.6 a	67.2 a
	Self	61.9 a	66.3 a	61.6 b	65.1 b
	Bulk	61.7 a	64.3 a	66.8 a	67.0 a
Carotenoid Content (mg g⁻¹)	Open	1.82 a	7.07 a	7.12 b	8.63 a
	Self	1.53 a	8.47 a	6.90 b	9.73 a
	Bulk	1.00 a	6.60 a	12.65 a	7.49 a

Note: Different letters in a column indicate significant differences at 0.05 alpha level.

open and self-pollination. The differences in chemical composition by the pollination methods may be a result of inbreeding effect. Inbreeding increases the homozygosity in maize, normally a cross-pollinated species. Combining recessive genes results in the expression of the respective phenotype that would be masked by the dominant alleles under heterozygote condition (Jalal et al. 2006). Our results suggest that selfing increased the frequency of favorable alleles for oil and protein content since self-pollination treatment resulted in elevated levels of these components. The differences in biochemical constituents among the pollination treatments are also related to kernel size. It is expected that higher kernel number per ear results in smaller kernels, thereby lower carbohydrate content per kernel. Negative relation between the carbohydrate and other constituents (Dado 1999) was also apparent in our data (Fig. 2). In contrary to our results, Schaefer & Bernardo (2013) obtained similar values from self and open pollination treatments for biochemical features. They concluded that there was no significant effect of pollination method on biochemical structure in maize kernel, based on their data collected on temperate inbreds. Our study includes more treatments and different specialty genotypes. The difference between the results of these studies indicates that the plant material may be a significant factor, as well as the pollination treatments tested, when investigating the pollen effect in maize. The genotypes used gave different responses to pollination treatments for kernel biochemical components while the effects on ear traits were similar (Tab. 3). As a result, we detected a significant G × T interaction effect on kernel biochemical properties. IHO had higher protein content in bulked samples, while selfed ears of OPV had higher levels. Genotypes had similar values for oil content in different pollination treatments, except the Q2 which had lower oil in bulk pollination compared to other treatments.

For carbohydrate content, IHO and PR had similar values for pollination treatments, while, Q2 and OPV had significantly lower values in self-pollination treatments. Carotenoid content significantly increased by the bulk pollination in OPV. Other genotypes had similar values for carotenoid content in samples with different pollination treatments. Based on these data, pollination treatment caused the variation of kernel quality traits and this situation was closely related to genetic specialties of the genotypes used.

Assessment of pollen effect

Results of variance analysis to assess the pollen effect by the genotypes were summarized in Tab. 4. Individual pollen effect varied by the genotypes for evaluated traits and it was found to be most significant for open pollinated landrace. In different studies, it was speculated that the pollen effect could be varied by year, genotype and climate conditions. There were also some studies reporting this effect was similar in different conditions for certain genotypes (Bulant et al. 2000).

Fig. 3 shows the differences between hybrids and their female parents for each variable. It was found that four crosses had significant differences for ear weight and total kernel weight, while three crosses did so for total kernel number and mean kernel weight (Fig. 3). OPV had significant increases for ear and kernel development when pollinated with other genotypes. IHO × OPV cross also had a significant pollen effect for ear weight and kernel weight per ear. Selfing in OPV resulted in a low ear weight, and cross pollination caused significant increases. Therefore, we concluded that pollen effect of the other genotypes on OPV was significant. In fact, inbreeding depression was reported to be higher in open pollinated landraces compared with the inbreds or hybrids (Öz & Tuğay 2003). Total kernel weight

TABLE 4.   F values for the investigated traits in the variance analysis by mixed model for the parental genotypes and their respective hybrids.

Trait	Source of Variation	Df	IHO and Related FIs	PR and Related FIs	OPV and Related FIs	Q2 and Related FIs
Ear Weight	Replication	2	2.31*	1.26	0.70	3.41
	Genotype	3	5.03*	0.89	17.6**	0.38
Kernel Weight	Replication	2	2.13	1.15	2.39	4.52
	Genotype	3	3.56	0.69	24.5**	0.39
Kernel Number	Replication	2	1.66	1.94	1.67	2.74
	Genotype	3	0.78	1.42	28.4**	0.30
Mean Kernel Weight	Replication	2	0.34	0.86	2.49	0.15
	Genotype	3	2.06	1.86	13.8**	0.51
Protein Content	Replication	2	0.00	1.89	0.43	0.92
	Genotype	3	4.51*	2.04	2.75	1.93
Oil Content	Replication	2	2.01	0.44	1.90	3.85
	Genotype	3	1.79	1.32	0.48	2.46
Carbohydrate Content	Replication	2	0.35	1.81	2.83	1.97
	Genotype	3	3.50	0.52	6.06*	3.53
Carotenoid Content	Replication	2	0.38	0.93	2.67	3.30
	Genotype	3	16.6**	11.2**	76.2**	47.3**

*,** statistically significant at 0.05 and 0.01, respectively.

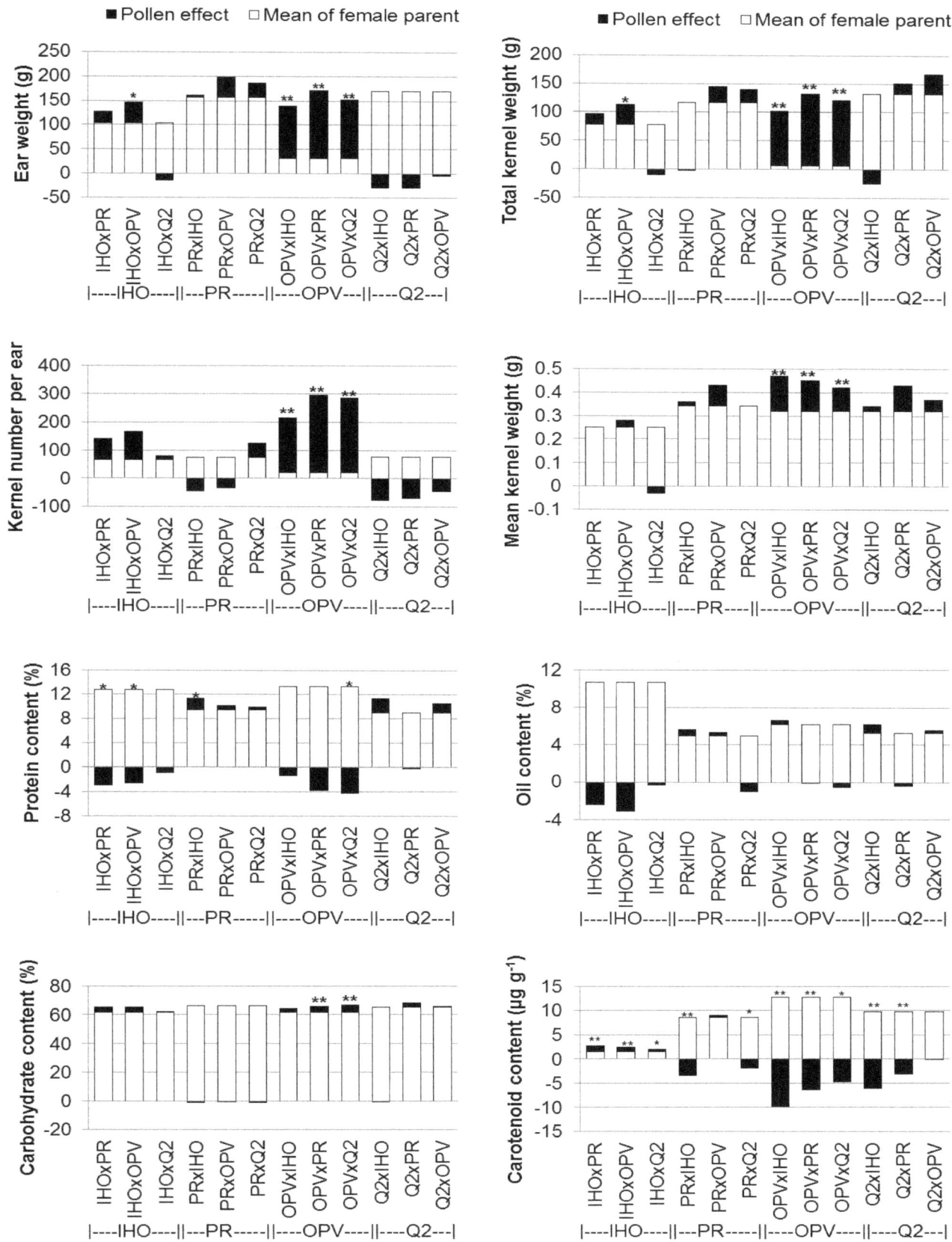

FIGURE 3.   Pollen effect of the genotypes on female parent.

*,** statistically significant at 0.05 and 0.01, respectively. Black portions of the bars represent the pollen effect (the difference between the pollen parent and the hybrid), while white portions indicate the mean value belong to the female parent.

in ear is a product of mean kernel weight and kernel number per ear. Increase in total kernel weight directly affects the ear weight. The increase in ear weight in OPV when pollinated with the other parents is a product of the increase in mean kernel weight and total kernel number per ear. It was reported that hybrids produced more kernels when pollinated by a different hybrid (Weingartner et al. 2002a; Weingartner et al. 2002b; Bozonovic et al. 2010). Pollen effect could not be determined when hybrids with similar kernel weight are crossed (Kannenberg & Hunter 1972). In our study, OPV hybrids had higher mean kernel weight than their parents, although OPV parent had similar values for mean kernel weight with those of pollen sources. Seka & Cross (1995) found that small kernel hybrids could bear bigger kernels when crossed with a large kernel hybrid. IHO has relatively small kernels within the set of genotypes used here. Interestingly, we did not observe any increase in kernel size when IHO was pollinated with the parents with larger kernel (Fig. 3). These results did not agree with earlier reports, where, unlike our study, mostly hybrids were tested. These results suggest that inbreds and landraces could display different responses to pollen source than the hybrids would. Moreover, carrying a special type endosperm (such as IHO) could be a factor on how the pollen effect would appear in a maize genotype.

Four of our crosses had significant differences for protein and carbohydrate content, while six crosses showed differences for carotenoid content (Fig. 3). Protein and carbohydrate contents were negatively affected by pollen source in IHO and OPV parents. These parents had higher protein and carbohydrate content than the others. In fact, pollination of these genotypes with lower parents resulted in decreases in protein and carbohydrate values. The low × high or high × low parent combinations gave similar results in earlier studies for protein and oil content (Curtis et al. 1956; Letchworth & Lambert 1998). The oil contents of crosses in our study were close to the pollen parent. However, the differences between the hybrids and their parents were not significant. The non-significant differences in our study may due to small number of genotypes ($N = 4$) to compare. All comparisons between crosses and parents for carotenoid concentration showed that pollen effect had a negative impact on this variable (Fig. 3). Vancetovic et al. (2014) found that antioxidant properties, including yellow pigments (i.e., carotenoids), were significantly affected the pollen parent. They concluded that when the hybrids with low pigment were pollinated by the high pigment hybrids, the pigment content increased. Our results did not agree their findings, probably because we used inbreds and OPV as female parent, while they used hybrids. Moreover, we observed that when the parent with low carotenoid content (IHO) was pollinated by the high carotenoid genotype, carotenoid content of the low parent had a very small increase. This indicates that maternal effects are more important than pollen effect for carotenoid content (Fig. 3).

In conclusion, results of this study showed that there were important differences among the pollination methods in terms of their effect on some kernel traits. Pollen effect played a significant role for ear and kernel development in

the investigated genotypes. This case was very distinct in open pollinated landrace. The change by pollen effect in carotenoid content was different than those in the other biochemical constituents. More accurate methods of chemical analysis (e.g., Reference Labs, HPLC methods) may provide a better picture for the biochemical constituents discussed here. Multi-year and multi-location trials would yield more informative results on pollen effect for these variables.

REFERENCES

Abdin Z, İdris M, Banta MGT (1979) A pollinating technique in maize. Pertenika 2:62-65.

Annonymous (2014) Controlled Pollination in Maize, Schnable Lab Protocols. [online] URL: http://schnablelab.plantgenomics. iastate.edu/resources/pollination/ (accessed December 2014).

Balestre M, Desouza JC, Dos Santos JB, Luders RR, Lima IJ (2007) Effect of self-pollination monitored by microsatellite markers on maize kernel weight. Crop Breeding and Applied Biotechnology 7:340-344.

Borrás L, Otegui ME (2001) Maize kernel weight response to post flowering source–sink ratio. Crop Science 41:1816–1822.

Borrás L, Cura JA, Otegui, ME (2002) Maize kernel composition and post flowering source-sink ratio. Crop Science 42:781-790.

Borrás L, Westgate ME, Otegui ME (2003) Control of kernel weight and kernel water relations by post flowering source-sink ratio. Annals of Botany 91:857-867.

Bozonovic S, Vancetovic J, Babic M, Filipovic M, Delic N (2010) The plus-hybrid effect on the grain yield of two ZP maize hybrids. Genetika 42:475-484.

Bulant C, Gallais A (1998) Xenia effects in maize with normal endosperm: I importance and stability. Crop Science 38:1517-1525.

Bulant C, Gallais A, Matthys-Rochon E, Priul JL (2000) Xenia in maize with normal endosperm: II kernel growth and enzyme activities grain filling. Crop Science 40:182-189.

Cárcova J, Uribelarrea M, Borrás L, Otegui ME, Westgate ME (2000) Synchronous pollination within and between ears improves kernel set in maize. Crop Science 40:1056-1061.

Curtis JJ, Brunson AM, Hubbard JE, Earle FR (1956) Effect of the pollen parent on oil content of the corn kernel. Agronomy Journal 48:551-555.

Dado RG (1999) Nutritional benefits of specialty corn grain hybrids in dairy diets. Journal of Animal Science 77:197-207.

Dong H (2007) Effect of high oil corn pollinator on kernel quality of common corn and their physiological and biochemical basis. Shandong Agricultural University, Master Thesis.

Gilbert N (1960) Xenia in quantitative characters. Journal of Genetics 57: 327-328.

Jalal A, Rahman H, Khan MS, Maqbool K, Khan S (2006) Inbreeding depression for reproductive and yield related traits in S1 lines of maize (Zea mays L.). Songklanakarin Journal of Science Technology 28:1170-1173.

Kannenberg LW, Hunter RB (1972) Yielding ability and competitive influence in hybrid mixtures of maize. Crop Science 12:274-277.

Krieger KM, Pollak LM, Brumm TJ, White PJ (1998) Effects of pollination method and growing location on starch thermal properties of corn hybrids. Cereal Chemistry 75:656-659.

Letchworth MB, Lambert R.J (1998) Pollen parent effects on oil, protein, and starch concentration in maize kernels. Crop Science 38:363-367.

Lindstorm EW (1939) Analysis of modern maize breeding principles and methods, International Congress of Genetics, Proceedings 7:191-197.

Öz A, Tuğay ME (2003) Variation on some agronomic characters in selfing generations in corn (*Zea mays indendata* Sturt). Journal of Agricultural Faculty of Gaziosmanpasa University 20:123-132.

Pixley KV, Bjarnason MS (1994) Pollen-parent effects on protein quality and endosperm modification of quality protein maize. Crop Science 34:404-409.

Rodriguez-Amaya DB, Kimura M (2004) HarvestPlus handbook for carotenoid analysis. HarvestPlus Technical Monograph Series 2.IFPRI, Washington, D.C., and CIAT, Cali.

SAS Institute (1999) SAS V8 User Manual, Cary, NC.

Schaefer CM, Bernardo R (2013) Pollen control and spatial and temporal adjustment in evaluation of kernel composition of maize inbreds. Maydica 58:135-140.

Seka D, Cross HZ (1995) Xenia and maternal effects on maize kernel development. Crop Science 35:80-85.

Sulewska H, Adamczyk J., Gygert H, Rogacki J, Szymanska G, Smiatacz K, Panasiewicz K, Tomaszyk K (2014) A comparison of controlled self-pollination and open pollination results based on maize grain quality. Spanish Journal of Agricultural Research 12:492-500.

Şchiop T, Haş I, Haş V, Coste I, Racz C (2011) Gene actions, cytoplasmic actions and cytoplasmicnuclear interactions involved in the determination of fiber content in a series of isonuclear maize lines. Research Journal of Agricultural Science, 43:282-288.

Taba S, Twumasi-Arfiyie S (2008) Regeneration Guidelines: Maize. Crop specific regeneration guidelines (CD-ROM), In: Dulloo ME, Thormann I, Jorge MA, Hanson J (eds) CGIAR System-wide Genetic Resource Programme, Rome, Italy.

Tanaka W, Matese AI, Maddoni GA (2009) Pollen source effects on growth of kernel structures and embryo chemical compounds in maize. Annals of Botany 104:325-334.

Thomison PR, Geyer AB, Lotz LD, Siegrist HJ, Dobbels TL (2002) TopCross High-Oil Corn Production: Agronomic Performance. Agronomy Journal 94:290-299.

Tsai CL, Tsai CY (1990) Endosperm modified by cross-pollinating maize to induce changes in dry-matter and nitrogen accumulation. Crop Science 30: 804-808.

Vancetovic J, Zilic S, Bozinovic S, Micic DI (2014) Simulating of Top-Cross system for enhancement of antioxidants in maize grain. Spanish Journal of Agricultural Research 12:467-476.

Weingartner U, Camp K-H, Stamp P (2004) Impact of male sterility and xenia on grain quality traits of maize. European Journal of Agronomy 21:239-247.

Weingartner U, Kaeser O, Long M, Stamp P (2002a) Combining male sterility and xenia increases grain yield of maize hybrids. Crop Science 42:1848-1856.

Weingartner U, Prest TJ, Camp K-H, Stamp P (2002b) The plus-hybrid system: A method to increase grain yield by combined cytoplasmic male sterility and xenia. Maydica 47:127-134.

3

Priorities for Research and Development in the Management of Pollination Services for Agriculture in Africa

Barbara Gemmill-Herren*[1], Kwame Aidoo [2], Peter Kwapong[2], Dino Martins[3], Wanja Kinuthia[4], Mary Gikungu[4], Connal Eardley[5]

[1]Food and Agriculture Organization of the United Nations, Viale delle terme di Caracalla, Rome 00153, Italy
[2] University of Cape Coast, Department of Entomology and Wildlife, School of Biological Sciences, Cape Coast, Ghana
[3]Turkana Basin Institute, Stony Brook University, P.O. Box 24467 – 00502, Nairobi, Kenya
[4]National Museums of Kenya, Centre for Bee Biology and Pollination, Box 40658, 00100, Nairobi, Kenya
[5]Agricultural Research Council, Plant Protection Research Institute, P.O. Box 8783, Pretoria, 0001, South Africa / School of Biological and Conservation Sciences, University of Kwazulu Natal, Private Bag X01, Scottsville, Pietermaritzburg, 3209, South Africa

Abstract—It is increasingly recognized that a sustainable future for agriculture must build on ecosystem services. Pollination is an important ecosystem service in all agroecosystems. In much of Africa the main challenge is conserving pollinator biodiversity in traditionally "ecologically-intensive" agroecosystems that are changing to meet different demands for food security and poverty alleviation, rather than safeguarding pollination in transition from conventional agricultural systems, with a high reliance on purchased inputs, to "ecologically-intensive" agroecosystems using natural inputs provided by biodiversity. Priority issues for research and development in pollination services in Africa include, inter alia: quantification and documentation of pollination deficits and finding measures to address these; socio-economic valuation of pollinator-friendly practices; assessment of lethal and sub-lethal effects of farming methods, such as pesticide use, on crop pollinators; identification of habitat management practices that enhance synergies between pollinator lifecycles and crop growing patterns; and policy analysis in relation to drivers and trends in pollination services and management.

Keywords: agriculture in transition, ecosystem services, poverty alleviation, sustainable agriculture

Introduction

Numerous recent reviews of agricultural science and technology (FAO 2011, Royal Society 2009) call for increasing support for systems of food production that are based on "ecological intensification" - understood as a means of increasing agricultural outputs (food, fiber, agro-fuels and environmental services), while reducing the use and the need for external inputs (agrochemicals, fuel, and plastic), and capitalizing on ecological processes that support and regulate primary productivity in agro- ecosystems (Titonnell and Giller 2013). Pollination is of course a key ecological process supporting such productivity. In many regions of the world with high input, high output agricultural systems, the approach may be to restore such ecological processes while reducing external inputs. However, in much of Africa, the strategy may be quite different, to work within traditionally ecologically-intensive agroecosystems to meet changing demands for food security and poverty alleviation.

Both approaches are needed depending on the region and commodities farmed therein. Therefore the expectation is that Africa will have many unique challenges but will also be able to adapt practices introduced in other parts of the world.

Before the African situation can be compared with that in other regions certain basic questions need to be addressed. The purpose of this article is to highlight current priority issues for research and development in managing pollination services that will be applicable in Africa and gather data needed to assess the situation, compare it to global trends and set new priorities. The current priorities are: quantification and documentation of pollination deficits and development measures to address these; socio-economic valuation of pollinator-friendly practices; assessment of the negative impacts of farming methods, such as pesticide use on crop pollinators; identification of habitat management practices that enhance synergies between pollinators lifecycles and crop growing patterns; and policy analysis in relation to drivers and trends in pollination services and management.

Pollination Services to Agricultural Production

Pollination service and poverty alleviation

Recent stressors on the global economy, such as financial instability, soaring commodity prices, energy crises and climatic changes have had negative impacts on human livelihoods, reflected in the fact that the number of undernourished people in the world now exceeds one billion (Gilland 2002; Ash et al. 2010). There are considerable

*Corresponding author; email: Barbara.herren@fao.org

pressures on the world food production system to intensify and expand agricultural systems to feed the human population, expected to reach 9.3 billion in 2050 (Lee 2011). Over the last fifty years, agriculture has intensified through the use of high-input, high-yielding crop varieties and livestock systems that are recognised to carry environmental consequences and vulnerabilities to farmers, along with increased production levels (Sachs et al. 2010; Pretty et al. 2010). Within the current production challenge lies opportunities to build food production systems less vulnerable to shocks impacting on those least able to withstand them, and more resilient and responsive to the ecosystem processes that support productivity.

Measures that can build resilience include incorporating ecological linkages and biological processes in agricultural systems, and enhancing the contribution of these over the use of agrochemical inputs (Bommarco et al. 2013). Agricultural systems, by design or by the nature of their development, may be relatively conducive to sustaining biological processes. This is the case in much of Africa, where due to the extensive nature of most farming systems and the relatively low use of inputs, such as fertilisers, most farmers currently rely on natural pollination services, the presence of natural enemies that control pests, and natural means of restoring soil fertility such as through fallow (Styger and Fernandes 2006). While yields are typically low, there are strong possibilities to increase yields through biological processes (Pretty et al. 2005).

Transitioning from one agroecosystem to another, be it from either a high input or ecologically extensive system to an ecologically intensive system, implies an overall "package" of practices to replace previous systems. Yet farmers are rarely able to change to entirely new practices, nor are they inclined to, as they seek to reduce risks. Adopting new practices always entails some measure of trade-offs and decision making that weighs benefits versus costs, and often small but significant investments, for example in sourcing planting material that can sustain pollinators, natural enemies and/or build soil fertility. In some instances, ecologically intensive systems may favour lower yearly yields but longer sustained yields over time (Tilman et al. 2002). Benefits to farmers may accrue through better quality produce, less variability in yield over time and less input costs resulting in better overall profits, or higher incomes from access to specialised and restricted markets. Agricultural policy and support services can serve to structure those benefit and cost relationships, and access to investments that assist in transitions.

The availability of fruits and vegetables is a central component of measures that effectively deal with hunger and malnutrition. In subsistence diets of people below the poverty line there are often insufficient vitamins and minerals within limited caloric intake diets (Graham et al. 2007). Yet those calories contribute much more to health by providing essential micro-nutrients, than those from grains alone, which can lead to obesity and poor health (Ciati and Ruini 2012). Potential pollinator declines are likely to negatively impact the production of vitamin (e.g. vitamin C) and mineral (e.g. calcium and fluoride) rich crops like fruits and vegetables, leading to higher market prices, increasingly unbalanced diets and eventually health problems (Eilers et al. 2011). Globally, fruits and vegetables have never received the same price support and subsidies that grains have, and tend to be more expensive on a calorie-basis.

Horticulture, including fruit production, has been the fastest growing food sector worldwide, with production increasing from 495 million tonnes in 1970 to 1574 million tonnes in 2010 (318%) (FAOSTAT 2007). During the same period, the vegetable subsector alone grew at an impressive annual average rate of 3.8%, from 255 million tonnes in 1970 to 966 million tonnes in 2010 (FAOSTAT 2010). Taking Kenya as a specific, and perhaps leading, example within Africa. Horticulture is the fastest growing agricultural subsector in the country, ranked third in earnings from exports after tourism and tea (Gioè 2006). In recent years, Kenya has been the world's leading exporter of fresh green beans in terms of value (FAOSTAT 2010). Economic analyses have shown that horticultural production in Kenya is capable of producing substantially higher returns per hectare for farmers than staple food crop production; one crop of French beans can generate gross margins more than ten times greater than maize-bean intercropping (Gioè 2006). Of equal relevance is the contribution of pollinator-dependent crops to agricultural development. In Kenya as elsewhere it has often been the case that large-scale commercial growers have not been able to compete with smallholders who have lower labour costs and greater motivation to provide the careful husbandry that meets many of the quality standards required by exporters (Jaffee 2003).

Attempts at valuing the worldwide economic benefit of pollination services to crops have been published. A recent study estimated the pollination service to crop production at about USD 208 billion p.a. in 2005, or 9.5 percent of the total value of the world's agricultural food production (Gallai et al. 2009). This result is also reflected at smaller scales. For instance in Ghana , the overall contribution of pollination services to agricultural production is estimated at 11.1 % of the national agricultural production p.a. of circa US$ 7 million (Gallai and Vaissière 2009a, b) (Table 1). In addition to the overall economic importance of pollination services, the production value per unit farming area of insect pollinated crops is four times that of crops that do not need insect pollination (Gallai et al. 2009). Thus farmers can make more money and produce more nutritious foods if they cultivate high-value, often pollinator-dependent, crops. The value of pollination service is however higher than the current global estimates since the contribution of sugar-acid-ratios and fruit firmness which improve the shelf life thus fruit loss by 11% (Klatt et al. 2014).

Agricultural development programmes aiming at poverty reduction need to recognize the crucial role sufficient animal pollination plays in maintaining and increasing yields of horticultural crops and thus in improving human nutrition and food security.

TABLE I. Economic impact of insect pollination on the 2005 agricultural production used directly for human food in Ghana (from Gallai and Vaissiere 2009a).

Crop category following FAO	Average value per metric ton in US$	Total value of crop (TVC) Price * Production in US$	Economic value of insect pollinators (EVIP) =TVC*%yield dependent on pollination (D) in US$
Cereals	422	821,267,900	0
Fruits	55	190,191,024	5,895,398
Oilcrops	141	400,822,900	30,717,694
Pulse	687	10,307,100	0
Roots and Tubers	286	4,356,036,458	0
Spices	1940	138,127,909	6,142,868
Stimulant crops	994	756,426,216	710,888,934
Sugar crops	28	3,981,600	0
Treenuts	466	6,060,990	3,296,046
Vegetables	617	396,491,526	31,505,314
TOTAL		7,079,713,622	788,446,253

The role of pollination in crop production

Animal pollination is being increasingly recognised as an essential ecosystem service, whose sufficient provisioning leads to overall increased and stabilized crop production (Garibaldi et al. 2011), and therefore sustained income levels and food security. For instance, coffee (Roubik 2002), avocado and mango (Johannsmeier 2001) which are all important crops for farmers' income and livelihoods in Africa. These crops benefit from insect pollination in terms of number of fruits produced and/or weight of fruits/seeds.

Around three-quarters of food crops worldwide depend on animal pollinators (Klein et al. 2007), including primarily vitamin and mineral rich crops like fruits and vegetables. In Africa, pollinator dependent crops include numerous indigenous vegetables, such as African nightshades (*Solanum scabrum*), amaranths (*Amaranthus blitum*), spiderplant (*Cleome gynandra*), slenderleaf (*Crotalaria ochroleuca* and *Crotalaria brevidens*), African kale (*Brassica carinata*), jute mallow (*Corchorus olitorius)* and African eggplant (*Solanum macrocarpon* and *Solanum gilo*) (Abukutsa-Onyango et al. 2010). Also important are nuts (e.g. macadamia, Johannsmeier 2001) and wild fruits harvested from natural and semi-natural areas such as shea nuts *(Vitellaria paradoxa)*. The vast majority of pollinators for these vitamin rich crops are insects such as bees, moths, flies, wasps and beetles (Klein et al. 2007), underlying the importance of insects in securing crop pollination services. Pollination services also contribute to other aspects of crop production. For example, strawberry producers and pollination researchers in Kenya demonstrated differences in their crop yields depending on whether the fields were located near bee hives. Increased insect pollination/visitation resulted in more uniform and marketable strawberries (Asiko 2012); and runner bean with good exposure to honeybees produced fewer "sickle shaped" pods that horticultural exporters refuse to accept (Vaissière et al. 2010). Insect pollination can also influence ripening speed, as is the case for chilli peppers (Bruijn and Ravestijn 1990), which means farmers are able to secure higher, off season, prices for their crop.

PRIORITIES FOR RESEARCH AND DEVELOPMENT IN AFRICA

Agricultural paradigms and practices

As noted above, pollination is a biological process in an agricultural system which can intensify production through sustainable agricultural development. There is a critical need to develop and expand agricultural paradigms and practices that sustain and increase crop yield and quality on existing cultivated land, to meet demands for higher agricultural production by current and future populations. These practices must increasingly rely on the key ecosystem services provided by biodiversity, such as nutrient cycling, pest regulation and pollination that enable the healthy functioning of the agricultural ecosystem. These functions ensure sustainability of agriculture as it intensifies to meet growing demands for food production. For high-value fruits and vegetables, harnessing the pollination services provided by biodiversity, along with companion services, such as soil nutrient cycling processes and natural pest control, is key to sustaining or increasing yields (Bommarco et al. 2013).

Status of pollinators and pollination services in Africa

While there is no solid documentation on the status and trends of pollinators in the African continent, the overall global trends of demands for pollination against anticipated supply is relevant in an African context. The area covered by pollinator-dependent crops has increased by more than 300 percent during the past 50 years (Aizen et al. 2008; Aizen and Harder 2009). A rapidly increasing human population will reduce the amount of natural habitats through an increasing demand for food-producing areas, urbanization and other land-use practices, putting pressure on the ecosystem service delivered by wild pollinators. At the same time, the demand for pollination in agricultural production will increase in order to sustain food production. Current trends, linked to the increase in the horticultural sector, show vastly greater increases in pollinator-dependent crops in

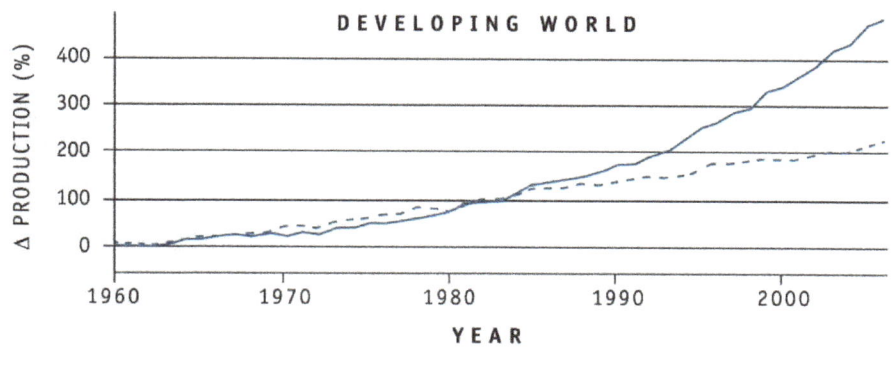

CROPS DEPENDENT UPON POLLINATION SERVICES
- - - - - CROPS NOT DEPENDENT UPON POLLINATION SERVICES

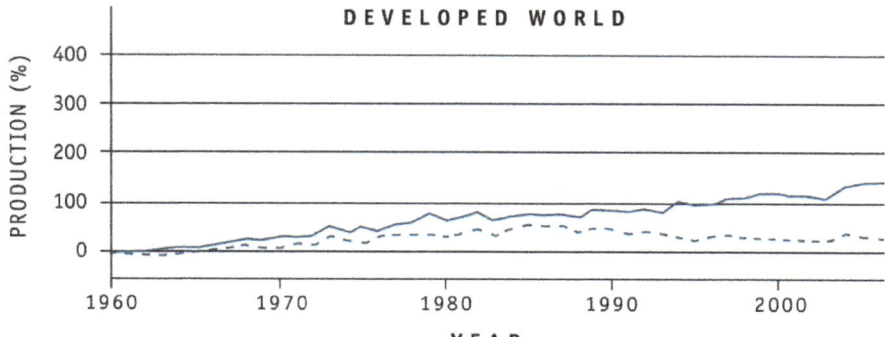

FIGURE I.   Temporal trends in total crop production from 1961 to 2006

developing regions of the world than in developed countries (Figure I).

Use of managed honeybee colonies for crop pollination has been virtually the sole practice to increase levels of pollination service to agricultural production (e.g. apples in South Africa, Johannsmeier 2001). As with other ecosystem-service practices, future recommendations should focus on past experiences and present information. Honeybees can easily be managed, and their populations increased and moved around to match flowering periods (Radar et al. 2009). However, they are not always the most optimal pollinators (Westerkamp 1991), and can be susceptible to disease and degradation of the natural environment (Johannsmeier 2001). Recent research has shown that interactions between wild pollinators and managed honeybees may lead to more effective pollination than either alone (Greenleaf and Kremen 2007; Carvalheiro et al. 2011). A Recent global meta-analysis has shown that wild pollinators pollinate crops more effectively than honey bees, enhancing fruit set by twice as much as equivalent levels of visitation by honey bees (Garibaldi et al. 2013). Strategic crop pollination recommendations need to be developed, drawn from the possible contributions, benefits and costs of wild and managed pollinators as appropriate to specific systems.

Status of research on crop pollination services

The status of research on African pollination biology was reviewed in 2004 (Rodger et al. 2004) at which point it was noted that relatively little work had been done on pollination biology in Africa. Much of the research which had been done was of an evolutionary nature, with comparatively little focus on either agriculture or conservation. In a number of instances, results of African studies, although not unique, contribute to our understanding of pollination biology. For example, higher levels of dioecy are found in some African plant groups and floras than in comparable areas elsewhere (van Wyk and Lowry 1988; Steiner 1987). Marula, for example, is harvested from wild trees but villagers must know that they must conserve male trees even though they do not bear fruit. Unique pollination syndromes involving, amongst others, oil-collecting bees (Steiner and Whitehead 1996; Steiner 1999), long proboscid flies (Goldblatt and Manning 1999; Manning and Goldblatt 1997) and monkey beetles (Picker and Midgley 1996; Goldblatt and Manning 2000) point to the highly diverse nature of the pollinating fauna in southern Africa's arid regions. On the other hand, high plant diversity and specialized pollination syndromes in some environments has been attributed to a paucity of pollinators, with plants competing intensely, and thus diversifying, to attract the relatively rare visitors (Johnson 1996; Johnson and Steiner 2003).

Surprisingly little work has been carried out in Africa on the pollination of a number of crops that were domesticated on the continent, such as coffee and okra. For several crops important to horticultural production in Africa (e.g. aubergines, tomatoes, peppers, papayas and passion fruits among others) honeybees are not effective pollinators, e.g., they do not buzz-pollinate, do not trip Lucerne flowers and in some areas avoid onion. The focus where honey bees are not effective pollinators must be on alternative, wild species of pollinators. Equally, for those crops pollinated both by honeybees and other pollinators, studying African crops in Africa offers an excellent focus for better understanding the interactions between the two groups.

Over the last ten years, since the African Pollinator Initiative was established (API 2003), an expanding focus on the role of pollination in natural and agricultural systems has been seen. A document taking stock of the state of knowledge of pollinators in agricultural production in Africa was produced in 2005 (API 2005). A resource book to introduce the concept and needs for pollination management to global policy makers and practitioners was drafted and coordinated by the API (Ahmed et al. 2006). Progress has been made on making taxonomic information on African bees accessible to end-users, with a key to the African genera of bees (Eardley et al. 2010). There are also numerous revisions of groups of bees occurring in Africa, such as *Scrapter* (Davies et al. 2005), *Melitta* (Eardley and Kuhlmann 2006), *Andrena* (Eardley 2006), *Capicola* (Michez et al. 2007) and *Ceratina* (Eardley and Daly 2007) and listed in Eardley et al. (2010).

African researchers contributed a review article to a special issue of the journal Apidologie in 2009 (Eardley et al. 2009) devoted to bee conservation; the review concluded that, although Africa contains seven biodiversity hotspots, the bee fauna appears rather moderate given the size of the continent (Eardley et al. 2009). Several factors were proposed for this pattern, an important one being the dearth of bee taxonomists working in Africa and difficulties in carrying out research in many regions. Anecdotal observations suggest a very large number of undescribed bee species on the continent. A number of serious threats to this diversity were also noted to exist, especially habitat destruction and degradation (Eardley et al. 2009). Bee diversity in these regions is likely to be important for both agriculture and maintaining indigenous ecosystems, but is under-appreciated and relatively poorly researched. Reliance on conserved areas such as national parks will not be sufficient to preserve bee diversity in Africa and Madagascar because national parks are more geared to conserve vertebrates than invertebrate and all ecosystems are not adequately protected. Changes to land use practices and development of industries that facilitate conservation will be essential (Eardley et al. 2009) for conserving of pollinating species.

Recent research has described the pollination needs and dynamics of a number of important horticultural crops in Africa. The role of native bees and natural habitats to the pollination of eggplant has been documented (Gemmill-Herren and Ochieng 2007). The contributions of a diversity of pollinators to smallholder agriculture in western Kenya, and their economic benefits have been recorded (Kasina et al. 2009a, b). An important contribution to the economic valuation of pollination services has been made by South African researchers (Allsop et al. 2008). Amongst non-conventional pollinators, the role of hawk moths in papaya

pollination has been shown to be of great significance (Martins and Johnson 2009), and in a more recent study hawk moths were found to also visit numerous indigenous plant species (Martins and Johnson 2013). An important study on cowpea pollination has documented gene-flow dynamics between cultivated and wild species (Pasquet et al. 2008). Nderitu and colleagues (2007, 2008) have detailed the detrimental effects of insecticides applied to sunflowers in Kenya on the diversity of bees and consequent seed yield. An interest in managing wild stingless bee species for both pollination and honey production has been published in recent articles (Cortopassi-Laurino et al. 2006; Kasina et al. 2009c). Interesting and unique syndromes of pollinator-plant interactions, such as strong associations of bees with grasses in Africa, have been reported (Bogdan 1962; Immelman and Eardley 2000; Gemmill and Martins 2004).

A reanalysis of the existing knowledge database on pollination in Africa, based on the original analysis of Rodger et al. (2004), has turned up a number of interesting insights (Figures 2a and b). First, and in part due to a Food and Agriculture Organization (FAO) coordinated project in Ghana, Kenya and South Africa that has supported national partners to database relevant studies, an additional 122 studies have been identified. Some of the studies come from more obscure sources as well as mainstream literature, and the facility to identify them now, as opposed to ten years ago, reflects the improvement in search functions and sharing of information over the Internet. Sixty-two of the new entries, however, are from new studies carried out between 2004 and 2013. The studies making the largest contribution (n=34) had an applied subject, always with an agricultural focus. The geographic location making the largest contribution over this period was East Africa (n= 39). Thus more attention needs to be given to pollination research in other areas of Africa.

Priorities for research and development

While considerable work has been undertaken to document the importance of pollinators to sustainable agriculture in Africa (i.e. studies mentioned above), work is still required to identify agricultural management practices that can increase the amount of pollination and thus yield of pollinator dependent crops. The next step will be to translate the knowledge base on pollination and other ecosystem services into a set of practices using biological processes that can be implemented to sustainably increase agricultural production in Africa. There is a strong need to understand how such ecosystem services can be enhanced and sustained, such that they provide a sustainable underpinning for production and livelihoods.

An important basis for such research is the recognition that considerable local knowledge already exists, making smallholder agriculture in Africa more efficient than large scale farms in some cases (Lele et al. 2010; Pinghali 2010). Interventions to sustain ecosystem services are likely to be highly site-specific and will need to be developed through a synthesis of existing, traditional knowledge and innovations by agricultural researchers. Specific development and research

A)

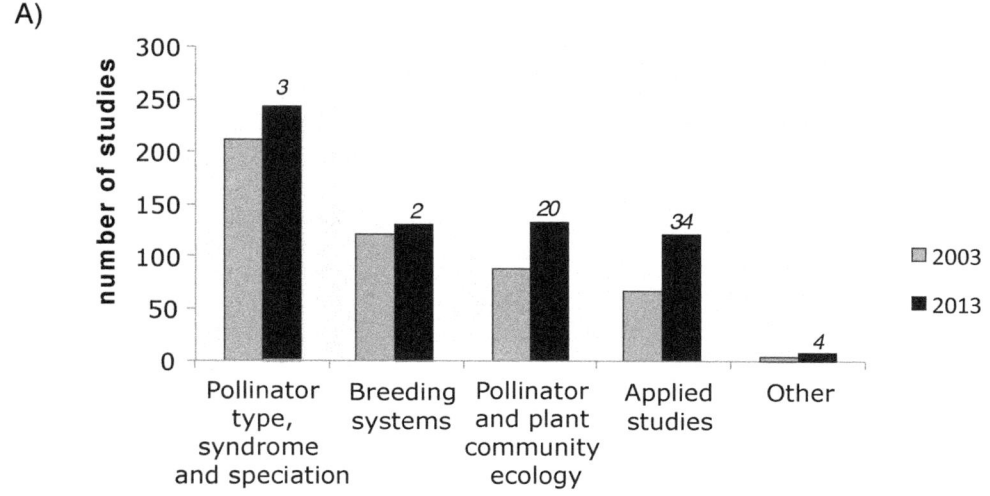

B)

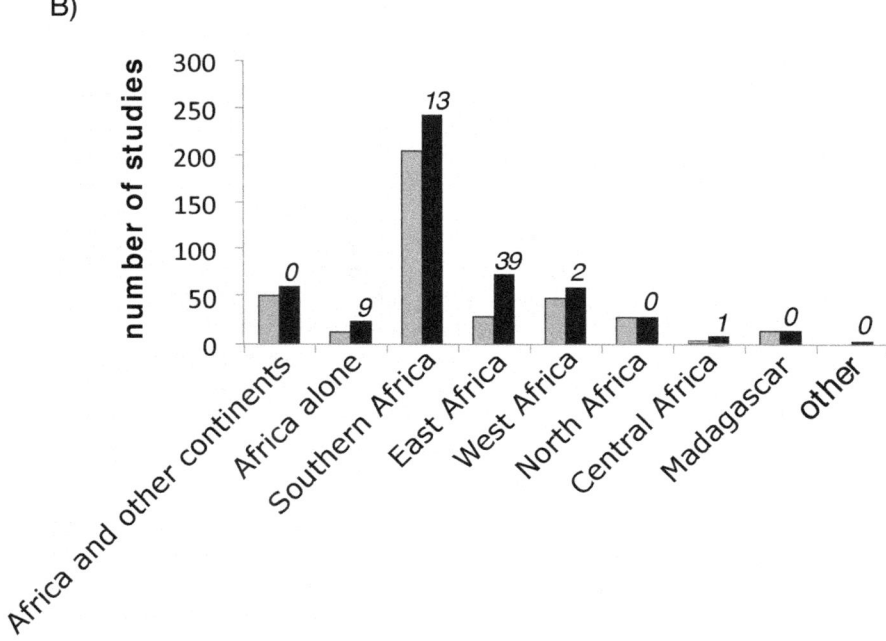

FIGURE 2. Classification of pollination studies carried out in Africa as identified in 2003, and in 2013. A total of 393 studies were identified in 2003, and 515 in 2013, with the increase in 2013 reflecting: (1) improved search ability and access to literature, including theses, (2) an FAO-coordinated project in Ghana, Kenya and South Africa that includes databasing the knowledge base and (3) 62 new studies published from 2004-2013. Numbers above the bars for 2013 results indicate the number of new studies in each class, from 2004-2013. A classification criterion is found in Rodgers, Balkwill and Gemmill (2004). A) Classified by subject matter. B) Classified by location of study

agendas that have been proposed as priority areas include the following:

First, there is need to develop a systematic assessment of the occurrence and consequences of pollinator declines to agricultural production over a range of crops all over the world. A protocol to assess pollination deficits has been developed through collaboration between FAO and the Institut National de la Recherche Agronomique (INRA) in France (Vaissière et al. 2010). The protocol should be applied to a range of cropping systems, both extensive and intensive, to detect and assess the extent to which insufficient pollination limits crop productivity across crops and across regions.

Second, the identification of habitat management practices that can best contribute to building up wild pollinator populations to service crop pollination needs. Farmers can supply pollinator foraging resources by encouraging the establishment of attractive indigenous plant

species that flower throughout the year, or increase nesting sites (e.g. by providing wooden bee nests or empty reeds for solitary bees), and applying conservation tillage to safeguard ground nesting bees (Gemmill-Herren, Azzu and Biesmeijer in prep). Documentation is needed on flowering plants species that can be used in hedgerows, fallows and natural habitats adjacent to the farms to provide a source of nectar, food, nesting opportunities and shelter for wild pollinators.

It is recognized that for farmers to implement pollinator-friendly practices, the benefits accrued from improved pollination service they receive must outweigh the costs of such practices. Participatory methods could assist farmers in recording costs and benefits of their practices. Many pollinator-friendly practices may involve minimal costs, such as to encourage (or not weed) selected wild flowering plants that are not pernicious weeds, near crops and document floral visitors over time. Other measures, including taking some crop land out of production to allow for native plant restorations, or reducing applications of pesticides, involve more complex understandings of costs and benefits. Research on the barriers and incentives for the uptake of ecological intensification practices, including pollinator-friendly measures, is also needed to support a transition to a more sustainable form of agriculture built on the enhancement of ecosystem services.

Third, agricultural pesticides may adversely affect pollinators in a variety of ways. For example, high concentrations can cause direct mortality while sub-lethal exposures of bees to pesticides during field applications can induce changes in individual bee activities and colony performance (Fisher and Moriarty 2011). There is little information available about the effects of pesticides on pollinators in Africa (van der Valk et al. 2013). Studies are needed to provide basic information on lethal and sub-lethal effects of selected insecticides commonly used by farmers, on social and solitary bees (Martins 2011). For example, work on pesticides used in fishing has been shown to be detrimental to dragonflies (Martins 2009) and similar patterns are expected for pollinators.

Forth, determining the lifecycles of wild pollinators (generation time, number of generations in a year, timing of reproduction) and assessing how they interact with crops during growing seasons, e.g., pollinator availability during off-season cultivation through irrigation and alternative food plants for pollinators when crops are not in flower. Development of pollinator management strategies including inventories of wild pollinators requires detailed understanding of their phenology, life history and distribution in relation to crop growing patterns. For this purpose, the taxonomic capacity to identify pollinators, at least to morphospecies, needs strengthening because pollinator may only visit certain crops types, and not all flower visitors are pollinators.

POLICY ANALYSIS

Pollination rarely receives much attention in the policy arena, but the farming category to which it substantially contributes, horticultural production, is growing in importance in Africa. The supply of fruits and vegetables remains little addressed component in fighting hunger and malnutrition. Multiple drivers affect, and will increasingly impact the future of horticultural production in Africa, and the contribution of pollination services to production. Pollinators cannot be protected in nature reserves alone, and so there is need to establish policies that guide pollinator protection in all terrestrial ecosystems including urban areas due to the prevailing threats such habitat modification, pollution and use of chemicals pose.

Population growth and agricultural intensification

African governments recognize the need to increase food production to provide food security for growing populations. Increasing production may have many possible paths: it can lead to larger, more consolidated farm operations producing for commercial markets. Where there are markets with value chains that dictate standards, such standards can shape agricultural practices. For example, in

some markets, fruit is only marketable if it is blemish-free, leading to increased use of pesticides for cosmetic purposes. In other instances, produce must not exceed limits of pesticide residues, leading to decreased use of pesticides. Alternatively, in consolidated operations, there is a growing recognition that smallholder farms may be significantly more productive than large farms in some situations (Pinghali 2010). With appropriate policies and support from the state, smallholder production sectors can be formidable engines for economic growth and poverty alleviation. Amongst these trends, small farm sizes and reduced pesticides usage are outcomes that are likely to benefit the provisioning of pollination service, and may be impacted by policy measures and voluntary standards.

Climate change

Climate change is expected to lead to fluctuating and frequently reducing, crop yields in Africa, as El-Niño-like events increase as is predicted in most climate models (Stige et al. 2006). Cultivating drought tolerant annual fruits and vegetables is one of the options that have been proposed for farmers adapting to climate change (Lobell et al. 2008). Horticultural crops such as melons (cucurbits) and vegetables generally produce over a shorter period than grain crops, and thus can fit into requirements for later start dates, when rains are delayed, or early harvest dates when rains are curtailed. Certainly, drought and consequent water deficits are likely to pose greater resource limitations, than pollination but both merit consideration under changing climates and changing crop options. Little is known about how pollinators may adapt their life history strategies when growing seasons are either shorter, or lengthened with irrigation, and research addressing this is needed. However, long-term irrigation is likely to disrupt biological processes of ground nesting bees especially in natural arid and semi-arid regions.

Farmers may select different crops, and thus change the distribution of crops in response to climate change, but wild pollinators cannot be so easily transplanted. Maintenance of reserves and habitat corridors between reserves, may enable some pollinators with long foraging ranges to readily disperse to new areas in response to different temperature and rainfall regimes.

Changes in diets

Economic development has been identified as one of the drivers of diet change in many economies (Caballero 2002). The pace of shifts in diets has recently accelerated (Popkin 1998; 2002), and these trends may have strong impacts on demands for pollinator-dependent fruits and vegetables. In many societies, as threats of famine recede, the consumption of fruits, vegetables, and animal protein increases, with starchy staples becoming a smaller percentage of food consumed (Drewnowski and Popkin 1997; Popkin 2002). As the consumption of processed foods, high in total fat, cholesterol, sugar and other refined carbohydrates and low in polyunsaturated fatty acids and fiber increases along with more sedentary lifestyles, dietary disorders such as obesity may also increase. But a third dietary trend is emerging, associated with desires to avoid dietary disease and promote health, this trend focuses on the importance of good nutrition for children and a return to fruits, vegetables, pulses and less processed food (Nestle 2006). This latter trend may be instituted and encouraged by consumers or by government policy. The impacts of this can be seen in countries such as Kenya, where products that incorporate flours from pulses into meal flours are increasingly popular for child nutrition, urban demands for fresh fruits and vegetables per capita are increasing, and soft drink providers are introducing products that incorporate fruits as a base (Gioè 2006; Gikungu pers. communication).

Local market access

Diversification into horticultural crops is becoming an avenue to poverty alleviation amongst many farmers around the world (Weinberger and Lumpkin 2005). For farmers to benefit from expansion into horticultural production, market access is critical. Fruits and vegetables are traditionally sold in regularly convened markets in centers of population. Although the presence of supermarkets in developing countries is a rapidly increasing phenomenon, horticultural produce tends to be largely locally sourced (McCullough et al. 2008). Supermarkets can still offer favourable access for farmers growing horticultural produce. In addition, urban consumers tend to be more concerned and conscious of how their fruits and vegetables have been grown (for example whether there may be pesticide residues in their food), thus there may be a potential for capturing price premiums for crops grown under sustainable practices (Onozaka et al. 2006).

An important aspect of market access for producers of pollinator-dependent crops, however, is that many of these crops (fruits in particular, but also vegetables) are highly susceptible to spoilage. It is increasingly recognized that food waste has tremendous impacts on both food security, farmer livelihoods, and natural resource use (Food Wastage Footprint and FAO 2013); thus more investment in efficient supply chains has multiple benefits. Recent publication may have found a way of ameliorating this problem by the increased shelf-life contributed by pollination in strawberry (Klatt et al. 2014) and by extrapolation other crops.

Globalization of trade

The profound changes in the nature of the international economy with the advance of globalized trade brings both threats and opportunities to agricultural production in Africa. Given the large and growing trade of horticultural products from Africa to Europe, it can be argued those countries that import horticultural crops from the continent are as vulnerable to pollinator losses in Africa as in their own countries (Gallai et al. 2009). Thus, pollination deficits may be matters for supra-national policy concerns.

CONCLUSION

Yields from agricultural production in Africa are amongst the lowest in the world (Tittonell and Giller 2013). There are strong expectations that intensifying farming systems will contribute to addressing issues of hunger and poverty throughout the continent. If productivity is not increased, the rate of natural habitat loss and degradation

may be accelerated, resulting in a further decrease in food security. A tremendous opportunity exists to instead use sustainable intensification methods to build on and enhance the natural wealth and ecosystem services underlying most African ecosystems, rather than to intensify through greater reliance on anthropogenic inputs. Pollination services and their contribution to the production of horticultural crops, may serve as an excellent flagship area of research and development in African agriculture as sustainable agricultural solutions are sought for the continent.

REFERENCES

Abukutsa-Onyango MO, Tushabomwe-Kazooba C, Mwai W, Onyango GM, Macha ES (2010) Diversity of African Indigenous Vegetables with Nutrition and Economic Potential in the Lake Victoria Region. Paper presented at the KARI Science Conference, Nairobi, Kenya 2010. [online] http://www.kari.org/conference/conference12/docs/DIVERSITY%20OF%20AFRICAN%20INDIGENOUS%20VEGETABLES%20WITH%20NUTRITION.pdf (last accessed 14.12.2013).

Ahmad F, Banne S, Castro M, Chavarria G, Clarke J, Collette L, Eardley C, Fonseca V, Freitas BM, French C, Gemmill B, Griswold T, Gross C, Kwapong P, Lundall-Magnuson E, Medellin R, Partap U, Potts SG, Roth D, Ruggiero M, Urban R, Willemse G (2006) Pollinators and Pollination: A resource book for policy and practice. Eardley C, Roth D, Clarke J, Buchmann S, Gemmill B (eds) African Pollinator Initiative, Pretoria.

Aizen MA, Garibaldi LA, Cunningham SA, Klein AM (2008) Long-term global trends in crop yield and production reveal no current pollination shortage but increasing pollinator dependency. Current Biology 18:1572-1575.

Aizen MA, Harder LD (2009) The global stock of domesticated honey bees is growing slower than agricultural demand for pollination. Current Biology 19:915-918.

Ali M (2008) Horticulture Revolution for the Poor: Nature, Challenges, and Opportunities. A background paper prepared for IAASTD, World Development Report/World Bank 2008.

Allsopp MH, de Lange WJ, Veldtman R (2008) Valuing Insect Pollination Services with Cost of Replacement. [online] PLoS ONE 3(9): e3128. doi:10.1371/journal.pone.0003128.

API (2003) Plan of Action of the African Pollinator Initiative. (reprinted 2008) FAO, Rome, Italy.

API (2005) Crops, Browse and Pollinators in Africa: an Initial Stocktaking. (reprinted 2008) FAO, Rome, Italy.

Asiko G (2012) Pollination of strawberry in Kenya, by stingless bees (Hymenoptera: Apinae, Meliponini) and the honeybees (Hymenoptera: Apinae) for improved fruit quality. PhD Thesis, University of Nairobi.

Ash C, Jasny BR, Malakoff DA, Sugden AM (2010) Feeding the future. Science 327(5967):797-797.

Bogdan AV (1962) Grass Pollination by bees in Kenya. Linnean Society of London 173:57-60.

Bommarco R, Kleijn D, Potts S (2013) Ecological intensification: harnessing ecosystem services for food security. Trends in Ecology and Evolution 28(4):230-238.

Bruijn JDe, Ravestijn WVan (1990) Capsicum fruits are better and heavier through bee pollination. Groenten + Fruit, Algemeen 46(25):48-49.

Byrne A, Fitzpatrick U (2009) Bee conservation policy at the global, regional and national levels. Apidologie 40:194–210.

Caballero B (2002) Nutrition Transition: Diet and Disease in the Developing World. Academic Press, New York.

Carvalheiro LG, Veldtman R, Shenkute AG, Tesfay GB, Pirk CWW, Donaldson JS, Nicolson SW (2011) Natural and within-farmland biodiversity enhances crop productivity. Ecology Letters 14(3):251-259. [online] DOI:10.1111/j.1461-0248.2010.01579.x.

Ciati RA, Ruini LA (2011) Double pyramid: healthy food for people, and sustainable for the planet. In: Burlingame B, Dernini S (eds) Sustainable Diets and Biodiversity. FAO, Rome, Italy.

Cortopassi-Laurino M, Imperatriz-Fonseca VL, Roubik DW, Dollin A, Heard T, Aguilar I, Venturieri GC, Eardley C, Nogueira-Neto P (2006) Global meliponiculture: Challenges and opportunities. Apidologie 37:275-292.

Davies GBP, Eardley CD, Brothers DJ (2005) Eight new species of *Scrapter* (Hymenoptera: Apoidea: Colletidae), with descriptions of *S. albifumus* and *S. amplispinatus* females and a major range extension of the genus. African Invertebrates 46:141-179.

Drewnowski A, Popkin BM (1997) The nutrition transition: new trends in the global diet. Nutrition Reviews 55 (2):31-43. [online] doi: 10.1111/j.1753-4887.1997.tb01593.

Eardley C (2006) The southern Africa species of *Andrena* Fabricius (Apoidea: Andrenidae). African Plant Protection 12:51-57.

Eardley C, Daly HV (2007) Bees of the genus *Ceratina* Latreille in southern Africa (Hymenoptera, Apoidea). Entomofauna 13(96):1-96.

Eardley CD, Gikungu M, Schwarz MP (2009) Bee conservation in Sub-Saharan Africa and Madagascar: diversity, status and threats. Apidologie 40:355–366.

Eardley C, Kuhlmann M (2006) Southern and East African *Melitta* Kirby (Apoidea: Melittidae). African Entomology 14(2):293-305.

Eardley C, Kuhlmann M, Pauly A (2010) The Bee Genera and Subgenera of sub-Saharan Africa. ABC Taxa 7:1-144.

Eilers EJ, Kremen C, Smith Greenleaf S, Garber AK, Klein A-M (2011) Contribution of pollinator-mediated crops to nutrients in the human food supply. [online] PLoS ONE 6(6): e21363. doi:10.1371/journal.pone.0021363.

FAO (2011). Save and grow: A policymaker's guide to sustainable intensification of smallholder crop production, Food and Agriculture Organization of the United Nations.

FAOSTAT. [online] www.fao.org/faostat 2007 (last accessed 14.12.2013).

FAOSTAT. [online] www.fao.org/faostat 2010 (last accessed 14.12.2013).

Fisher D, Moriarty T (2011) Pesticide risk assessment for pollinators: summary of a SETAC Pellston workshop. 15–21 January 2011. Pensacola, Florida, USA, Society for Environmental Toxicology and Chemistry (SETAC).

Food Wastage Footprint and FAO (2013) Food wastage footprint Impacts on natural resources. FAO, Rome.

Gallai N, Vaissière BE (2009a) Guidelines for the Economic Valuation of Pollination Services at a National Scale. FAO, Rome, Italy.

Gallai N, Vaissière BE (2009b) Tool for the Economic Valuation of Pollination Services at a National Scale. FAO, Rome, Italy.

Gallai N, Salles JM, Settele J, Vaissière BE (2009) Economic valuation of the vulnerability of world agriculture confronted with pollinator decline. Ecological Economics 68(3):810-821.

Garibaldi LA, Steffan-Dewenter I, Winfree, W,Aizen MA,

Bommarco R, Cunningham SA, Kremen C, Carvalheiro LG, Harder LD, Afik O, Bartomeus I, Benjamin F, Boreux V, Cariveau D, Chacoff NP, Dudenhöffer JH, Freitas BM, Ghazoul J, Greenleaf S, Hipólito J, Holzschuh A, Howlett B, Isaacs R, Javorek SK, Kennedy CM, Krewenka KM, Krishnan A, Mandelik Y, Mayfield MM, Motzke I, Munyuli T, Nault BA, Otieno M, Petersen J, Pisanty G, Potts SG, Rader R, Ricketts TH, Rundlöf M, Seymour CL, Schüepp C, Szentgyörgyi H, Taki H, Tscharntke T, Vergara CH, Viana BF, Wanger TC, Westphal C, Williams N, Klein AM (2013) Wild pollinators enhance fruit set of crops regardless of honey-bee abundance. Science 339:1608-1611.

Garibaldi LA, Aizen MA, Klein AM, Cunningham SA, Harder LA (2011) Global growth and stability of agricultural yield decrease with pollinator dependence. PNAS 108(14):5909-5914.

Gemmill B, Martins D (2004) Bees associated with grasses. Nature East Africa 34(2):24-30.

Gemmill-Herren B, Ochieng AO (2008) Role of native bees and natural habitats in eggplant (*Solanum melongena*) pollination in Kenya. Agriculture, Ecosystems and Environment 127:31–36.

Gioè M (2006) Can Horticultural Production Help African Smallholders to Escape Dependence on Export of Tropical Agricultural Commodities? Crossroads 6(2):16-65.

Gilland B (2002). World population and food supply: Can food production keep pace with population growth in the next half-century? Food Policy 27(1):47-63.

Goldblatt P, Manning JC (1999) The long proboscid fly pollination system in Gladiolus (Iridaceae). Annals of the Missouri Botanical Gardens 86:758–774.

Goldblatt P, Manning JC (eds) (2000) Cape Plants: A Conspectus of the Cape Flora of South Africa. National Botanical Institute of South Africa, Pretoria.

Graham RD, Welch RM, Saunders DA, Ortiz-Monasterio I, Bouis HE, Bonierbale M, De Haan S (2007) Nutritious subsistence food systems. Advances in Agronomy 92:1-74.

Greenleaf SS, Kremen C (2006) Wild bees enhance honey bees' pollination of hybrid sunflower. PNAS 103(37):13890-13895.

Hazell P, Wood S (2008) Drivers of change in global agriculture. Philosophical Transactions of the Royal Society B: Biological Sciences 363(1491):495-515.

Immelman K, Eardley CD (2000) Gathering of grass pollen by solitary bees (Halictidae: Lipotriches) in South Africa. Mitteilungen aus dem Museum für Naturkunde in Berlin, Zoologische Reihe 76(2):263-268.

Jaffee S. (2003) 'From challenge to opportunity: The transformation of the Kenyan fresh vegetable trade in the context of emerging food safety and other standards'. PREM Trade Unit. Washington, DC: World Bank.

Johannsmeier MF (ed) (2001) Beekeeping in South Africa. Plant Protection Research Institute Handbook 14:1-288.

Johnson SD (1996a) Pollination, adaptation and speciation in the Cape flora of South Africa. Taxon 45:59–66.

Johnson SD, Steiner KE (2003) Specialised pollination systems in southern Africa. South African Journal of Science 99:345–348.

Kasina M, Manfred K, Martius C, Wittmann D (2009a) Diversity and activity density of bees visiting crop flowers in Kakamega, Western Kenya. Journal of Apicultural Research and Bee World 48(2):134-139. DOI10.3896/IBRA.1.48.2.08.

Kasina JM, Mburu J, Kraemer M, Holm-Mueller K (2009b) Economic benefit of crop pollination by bees: A case of Kakamega small-holder farming in Western Kenya. Journal of Economic Entomology 102(2):467-473.

Kasina M, Manfred K, Martius C, Wittmann D (2009c) Stingless bees in Kenya, Bees for Development Journal 83. [online] DOI 10.3896/IBRA.1.48.2.07.

Klatt BK, Holzschuh A, Westphal C, Clough Y, Smit I, Pawelzik E, Tscharntke T (2014) Bee pollination improves crop quality, shelf life and commercial value. Proc. R. Soc. B 281: 20132440. http://dx.doi.org/10.1098/rspb.2013.2440.

Klein AM, Vaissière BE, Cane JH, Steffan-Dewenter I, Cunningham SA, Kremen C, Tscharntke T (2007) Importance of pollinators in changing landscapes for world crops. Proceedings of the Royal Society: Biological Sciences 274:303–313.

Lee R (2011) The outlook for population growth. Science 333(6042):569-573.

Lele U, Pretty J, Terry E, Trigo E, Klousiam M (2010) Transforming Agricultural Research for Development, GCARD Background Paper, Montpellier, France.

Lobell DB, Burke MB, Tebaldi C, Mastrandrea MD, Falcon WP, Naylor RL (2008) Prioritizing Climate Change Adaptation Needs for Food Security in 2030. Science 319:607-610.

Manning JC, Goldblatt P (1997) The *Moegistorhynchus longirostris* (Diptera: Nemestrinidae) pollination guild: long-tubed flowers and a specialised long-proboscid fly pollination system in southern Africa. Plant Systematics and Evolution 206:51–69.

Martins DJ (2009) Differences In Odonata Abundance And Diversity In Pesticide-Fished, Traditionally-Fished And Protected Areas In Lake Victoria, Eastern Africa (Anisoptera). Odonatologica 38(3):247-255.

Martins DJ (2011) Repercussions of pesticides (including carbofuran) on nontarget beneficial insects and use of insects in forensic analyses in Kenya. In: Richards N (ed) Carbofuran and Wildlife Poisoning: Global Perspectives and Forensic Approaches. John Wiley and Sons Ltd, Chichester.

Martins DJ, Johnson SD (2009) Distance and quality of natural habitat influence hawkmoth pollination of cultivated papaya. International Journal of Tropical Insect Science 29(3):114–123. [online] doi:10.1017/S1742758409990208.

Martins DJ, Johnson SD (2013) Interactions between hawkmoths and flowering plants in East Africa: polyphagy and evolutionary speciation in an ecological context. Biological Journal of Linnean Society, London. [online] DOI:10.1111/bij.12107, pp 1-15.

McCullough EB, Pinghali P, Stamoulis KG (eds) (2008) The Tranformation of Agri-Food Systems: Globalization, Supply Chains and Smallholder Farmers. FAO, Rome, Italy.

Michez D, Eardley C, Kuhlmann M, Patiny S (2007) Revision of the bee genus *Capicola* (Hymenoptera: Apoidea: Melittidae) distributed in the Southwest of Africa. European Journal of Entomology 104:311-340.

Nestle M (2006) What to Eat. New York: North Point Press (Farrar, Straus and Giroux):611. ISBN 978-0-86547-738-4.

Nderitu J, Kasina J, Nyamasyo G, Oronje ML (2007) Effect of insecticide application on sunflower (*Helianthus annuus* L.) pollination in eastern Kenya. World Journal of Agriculture Science 3(6):731-734.

Nderitu J, Kasina J, Nyamasyo G, Oronje ML (2008) Diversity of sunflower pollinators and their effect on seed yield in Makueni District, Eastern Kenya. Spanish Journal of Agricultural Research 2008 6(2):271-278.

Onozaka Y, Bunch D, Larson D (2006) What Exactly Are They Paying For? Explaining the Price Premium for Organic Fresh Produce. Agricultural and Resource Economics Update, University of California Giannini Foundation 9(6). July/August 2006.

Pasquet RS, Peltier A, Hufford MB, Oudine E, Saulnier J, Paul L, Knudseen JT, Herren HR, Gepts P (2008) Long-distance pollen flow assessment through evaluation of pollinator foraging range suggests trans gene escape distances. PNAS 105(36):13456-13461.

Picker MD, Midgley JJ (1996) Pollination by monkey beetles (Coleoptera: Scarabaeidae: Hopliini): flower and colour preferences. African Entomology 4:7–14.

Pinghali P (2010) [online] http://ilriclippings.wordpress.com/2010/10/15/prabhu-pingali-of-gates-foundation-remarks-at-the-borlaug-dialogueworld-food-prize-ceremony-in-iowa/ (last accessed 14.12.2013).

Popkin BM (1998) The nutrition transition and its health implications in lower-income countries. Public health nutrition 1(01):5-21.

Popkin BM (2002) The dynamics of the dietary transition in the developing world. The nutrition transition: diet and disease in the developing world pp. 111-128.

Pretty J, Noble AD, Bossio D, Dixon J, Hine RE, Penning de Vries FWT, Morison JIL (2005) Resource conserving agriculture increases yields in developing countries. Environmental Science and Technology 40(4):1114-1119.

Pretty J, Sutherland WJ, Ashby J, Auburn J, Baulcombe D, Bell M, Bentley J (2010) The top 100 questions of importance to the future of global agriculture. International journal of agricultural sustainability 8(4):219-236.

Rader R, Howlett BG, Cunningham SA, Westcott DA, Newstrom-Lloyd LE, Walker MK, Teulon DA, Edwards W (2009) Alternative pollinator taxa are equally efficient but not as effective as the honeybee in a mass flowering crop. Journal of Applied Ecology 46:1080–1087.

Rodger JG, Balkwill K, Gemmill B (2004) African pollination studies: where are the gaps? International Journal of Tropical Insect Science 24(1):5–28.

Roubik, DW (2002) The value of bees to the coffee harvest. Nature 417:708-708 [doi:10.1038/417708a].

Royal Society. (2009). Reaping the benefits: science and the sustainable intensification of global agriculture. RS Policy document 11/09. London: 86.

Sachs J, Remans R, Smukler S, Winowiecki L, Andelman SJ, Cassman KG, Castle D (2010) Monitoring the world's agriculture. Nature 466(7306):558-560.

Steiner KE (1987) Breeding systems in the Cape flora, pp. 22–51. In: Rebelo AG (ed) A Preliminary synthesis of Pollination Biology of the Cape Flora. Council for Scientific and Industrial Research, Pretoria.

Steiner KE (1999) A new species of *Diascia* (Scrophulariaceae) from the Eastern Cape (South Africa) with notes on other members of the genus in that region. South African Journal of Botany 65:223–231.

Steiner KE, Whitehead VB (1996) The consequences of specialisation for pollination in a rare South African shrub, *Ixianthes retzioides* (Scrophulariaceae). Plant Systematics and Evolution 201:131–138.

Stige LC, Stave I, Chan KS, Ciannelli L, Pettorelli N, Glantz M, Herren H, Stenseth NC (2006) The effect of climate variation on agro-pastoral production in Africa. PNAS 103(9):3049-3053.

Styger E, Fernades ECM (2006) Contributions of Managed Fallows to Soil Fertility Recovery. In: Uphoff N, Ball AS, Fernandes E, Herren H, Husson O, Laing M, Palm C, Pretty J, Sanchez P, Sanginga N, Thies J (eds) Biological Approaches to Sustainable Soil Systems. Taylor and Francis Group. Boca Raton.

Tilman D, Cassman KG, Matson PA, Naylor R, Polasky S (2002) Agricultural sustainability and intensive production practices. Nature 418(6898):671-677.

Tittonell P, Giller KE (2013) When yield gaps are poverty traps: The paradigm of ecological intensification in African smallholder agriculture. Field Crops Research 143:76–90.

Vaissière B, Frieitas B, Gemmill-Herren B (2010) A protocol to detect and assess pollination deficits: a handbook for its use. FAO, Rome.

Valk van der H, Koomen I, Nocelli RCF, Ribeiro MdeF, Freitas BM, Carvalho S, Kasina JM, Martins DJ, Mutiso M, Odhiambo C, Kinuthia W, Gikungu M, Ngaruiya P, Maina G, Kipyab P, Blacquière T, Steen van der J, Roessink I, Wassenberg J, Gemmill-Herren B (2012) Aspects determining the risk of pesticides to wild bees: risk profiles for focal crops on three continents. In: Oomen PA, Thompson H (eds). Hazards of Pesticides to Bees. Wageningen.

Weinberger K, Lumpkin TA (2005) Horticulture for Poverty Alleviation. The Unfunded Revolution'. AVRDC Working Paper Series, No.15 Taiwan: The World Vegetable Centre.

Westerkamp C (1991) Honeybees are poor pollinators – why? Plant Systematics and Evolution 177:71-75.

Wyk van AE, Lowrey TK (1988) Studies on the reproductive biology of Eugenia L (Myrtaceae) in southern Africa. Monographs in Systematic Botany, Missouri Botanical Garden 25:279–293.

4

NATIVE BEES POLLINATE TOMATO FLOWERS AND INCREASE FRUIT PRODUCTION

Carlos de Melo e Silva Neto, Flaviana Gomes Lima, Bruno Bastos Gonçalves, Leonardo Lima Bergamini, Barbara Araújo Ribeiro Bergamini, Marcos Antônio da Silva Elias and Edivani Villaron Franceschinelli*

Departamento de Botânica, Instituto de Ciências Biológicas, Universidade Federal de Goiás, 74001-970, Goiânia, GO, Brazil

Abstract—The tomato plant has a specific relationship with native pollinators because the form of its flowers is adapted to buzz pollination carried out by some pollen-gatherer bees that vibrate their indirect flight muscles to obtain that floral resource. The absence and the low density of these bees in tomato fields can lead to pollination deficits for crop. The aim of this study is to demonstrate that open tomato flowers, probably visited by native pollinator, have greater pollen load on their stigma than unvisited flowers. Another objective is to show that this great pollen load increases fruit production. We selected crops of the Italian tomato cultivar in areas of the State of Goiás, Brazil. Thirty seven plants of three crops each had one inflorescence bagged in the field. Bagged and non-bagged flowers had their stigmas collected and the amount of pollen on their surfaces was quantified. For the comparison of fruit production, we monitored bagged and not-bagged inflorescences and after 40 days, their fruits were counted, weighed, measured and had their seeds counted. The amount of pollen grains on the stigma of flowers available to pollinators was higher than that on the stigma of bagged flowers. On average, fruit production was larger in not-bagged inflorescences than in bagged inflorescences. In addition, not-bagged flowers produced heavier fruits than did bagged flowers. There was a significant difference in the number of seeds between treatments, with significantly more seeds in the non-bagged fruit. Our results show that native bees buzz-pollinate tomato flowers, increasing the pollen load on their stigma and consequently fruit production and quality.

Keywords: *Solanum lycopersicum L., pollen load, pollination deficit*

INTRODUCTION

Pollination is one of the most important ecological interactions and the first step for the sexual reproduction of most plant species (Murcia 1996). Pollination carried out by animals is considered an important ecosystem service with 35% of the plants cultivated in the world benefitting from this interaction (Klein et al. 2007). Bees are the main pollinators of most crops pollinated by animals (Free 1993; Delaplane & Mayer 2000; Klein et al. 2007). Many species of native bees contribute greatly to the pollination of crops such as coffee (*Coffea* spp.) (Klein et al. 2003; De Marco & Coelho 2004; Vergara & Fonseca-Buendía 2012), melon (*Citrullus lanatus*) (Winfree et al. 2007), tomato (*Solanum lycopersicum*) (Greenleaf & Kremen 2006a; Macias-Macias et al. 2009; Vergara & Fonseca-Buendía 2012), sunflower (*Helianthus annuus*) (Greenleaf & Kremen 2006b), canola (*Brassica* spp.) (Morandin & Winston 2005) and blueberries (*Vaccinium* spp.) (Kevan et al. 1983), among others.

The tomato plant belongs to the genus *Solanum* of the family Solanaceae. This plant, formerly in the genus *Lycopersicon*, originated from the Andean regions. Today, it is widely cultivated throughout the world and adapted into many cultivars (Olmstead & Palmer 1997; Chetelat et al. 2009). Wild tomatoes are self-incompatible and feature a close relationship with their pollinators for the formation of fruits (Chetelat et al. 2009). The cultivated plant is autogamous. However, one of the features of the genus is the poricidal opening of its anthers, which requires the agitation of the flowers by wind and/or the presence of pollinators that vibrate their indirect flight muscles for the release of pollen grains, even in cultivated varieties of tomatoes and especially in the still air of greenhouses (Kevan et al. 1991; Morandin et al. 2001a). Teppner (2005), while conducting studies on tomato plants in central Europe, observed that bees, such as *Bombus* and *Lasioglossum*, can be good pollinators of the flowers by vibrating their anthers easily. In respect of our study, we note that some families of bees from Brazil that perform buzz pollination are: Andrenidae, Apidae (except *Apis*), Colletidae, Halictidae, and Megachilidae (Harter et al. 2002).

Even though the importance of pollinators to tomato crops, especially in greenhouse production, is recognized, studies which demonstrate the direct relationship of pollinators to the pollen load on stigma and fruit production are scarce (Macias-Macias et al. 2009; Vergara & Fonseca-Buendía 2012). Tomato flowers in field crops in the State of Goiás, Brazil are visited by native bees, such as *Exomalopsis analis* (Apidae), *Augochloropsis* sp. (Halictidae) and *Centris tarsata* (Apidae) (Silva Neto et al., unpublished data). The visit frequency with which they visit the plants is high and apparently every flower is visited, sometimes more than once (Santos & Nascimento 2011; Silva Neto et al., unpublished data). This can be verified by the bruises on the anthers

*Corresponding author; email: edivanif@gmail.com

caused by the bees' mandibles as the grip the flower to buzz-pollinate (Morandin et al. 2001a; Silva Neto at al., unpublished data). Thus, native pollinators are assumed to be important to the pollen doses delivered to tomato flower stigmas and consequently to fruit production. To test this assumption, we proposed to quantify the difference between pollen doses transferred to the stigmas of open tomato flowers and those found on self-pollinated bagged flowers and then to quantify and compare fruit production in these two treatments. Here, we assume that almost all open non-bagged flowers were visited by native bees at least once as evidenced by the bruising on the anthers. Further, we checked for the main pollinator species of the tomato flowers in the study crops.

MATERIALS AND METHODS

Area of study

The State of Goiás is the largest producer of tomato in Brazil (CEASA/GO 2013) with large crops of industrial and fresh-market tomato. Our experiments were made on field crops located in the municipalities of Nerópolis and Goianápolis in Goiás (area of study: https://docs.google.com/file/d/0ByQNaWHmeZI5Ymp SQk4yZXZ6bTQ/edit). The study area consisted of conventional field-grown and irrigated tomato crops (variety Italian). Trials were made from March to November, 2012.

To determine the main pollinators, we observed and collected bees that buzz pollinated the flowers. Four rows of about 120 plants were surveyed in each crop for 30 minutes and the presence of pollinators was recorded and quantified. The number of visits per pollinator species was recorded. Flower visitors were collected for identification. When collection was not possible, the identification to genus was made in the field. Those data were collected during two separate days in each property to make for statistically applicable sampling results. Our filed studies took place between 09:00 and 12:00 (UTC/GMT – 3 hours), which is when previous data indicated peak floral visitation and greatest amount of pollen available for pollination in the anthers (Silva Neto, unpublished data).

Pollen load on stigmas

To compare the amount of pollen grains on stigmas of pollinated and not pollinated (bagged) flowers, 37 plants had flower buds from one inflorescence bagged in the field. After the opening and senescence of these bagged flowers, three of them were collected. At the same time, another three senescent but not-bagged flowers were collected. All were fixed in FAA 80% in the field and taken to the laboratory. There, the stigmas of these flowers were separated and softened in a solution of NaOH 9N for one hour, stained with acetic carmine and observed under an optical microscope. The pollen grains per stigma were counted in three visualization fields at 40 x magnification. The fields consisted of the two opposite ends of the stigma and its central part (Dafni et al. 2005).

Statistical comparison of pollen doses between bagged and non-bagged flowers used the Kolmogorov-Smirnov test and paired Student's t-test with 95% significance (Malagodi-Braga & Kleinert 2007; Montemor & Souza 2009).

Fruit production

To assess the effect of pollination on tomato production, the flower buds bagged in the previous experiment and other buds tagged and not bagged were monitored on 37 different individual plants. Fruits were collected and taken to the laboratory forty days after the opening of their flowers. The fruits were weighed, measured (diagonal diameter) and their seeds counted in a similar manner to that described by Kevan et al. (1991). We chose to use only the first fruit produced by inflorescences because the remaining fruit had not fully developed over the sampling period.

Statistical comparisons between the number, mass and amount of seeds developed from bagged and not-bagged flowers used Kolmogorov-Smirnov test and subsequently through the paired Student's t-test with 95% significance. The relationship between mass and seeds was determined by simple linear regression (Spears 1983; Malagodi-Braga & Kleinert 2007; Montemor & Souza 2009).

RESULTS

The species of bees observed in the study crops were *Exomalopsis analis* Spinola (the most common), *Centris tarsata* Smith, *Bombus morio* Swederus, *Eulaema nigrita* Lepeletier and *Epicharis* sp. In three hours of observations, those bees were seen 47 times visiting tomato flowers (Table I). The bees that performed buzz pollination approached the tomato flowers from the front, landing on the anther cones. They clung to the cone by their mandibles, vibrated the anthers and the pollen was expelled and adhered to the abdomen and other parts of their body. The same bees made circular motions on the anther, vibrating many times and over various anthers. When large amounts of pollen were deposited in their bodies, the bee stopped the vibration and cleaned themselves by collecting the pollen and putting it in its pollen basket.

On average, the stigma of not-bagged flowers had on average 114 pollen grains (t = 5.678; $P = 0.0001$) (Figure 1A and Table 2) more than on the stigma of bagged flowers.

TABLE I. Species of native bees that visit tomato flowers in three conventional plantations of the Italian variety in the State of Goiás – Brazil and their total number of visits in three hours of observation during the flowering peak.

Pollinator	Number of flower visits
Exomalopsis analis Spinola	47
Centris tarsata Smith	16
Epicharis sp.	2
Bombus morio Swederus	2
Eulaema nigrita Lepeletier	1

TABLE 2. Gains in tomato plant crops obtained from bagged and not bagged flowers treatments. Compared with paired Student's *t*-test with 95% significance (N: sample number; t: Student's t test; df: degree of freedom; p: statistical significance; %: Percentage gain of the "not bagged" treatment in relation to "bagged")

	Flowers	Average	N	t	df	P	%
Pollen load (No. of pollen grains)	Not bagged	182.51±86.68					
	Bagged	67.83±68.33	37	5.67	36	0.00	168.31
Fruit set	Not bagged	0.82±0.22					
	Bagged	0.50±0.22	34	9.55	33	0.00	64.48
Tomato paste (g)	Not bagged	70.69±20.40					
	Bagged	47.06±26.88	31	4.91	30	0.00	50.21
Tomato size (mm)	Not bagged	51.12±5.43					
	Bagged	46.59±8.14	27	2.66	26	0.01	9.72
Seeds (No.)	Not bagged	183.94±46.34					
	Bagged	59.63±38.54	36	12.37	35	0.00	208.46

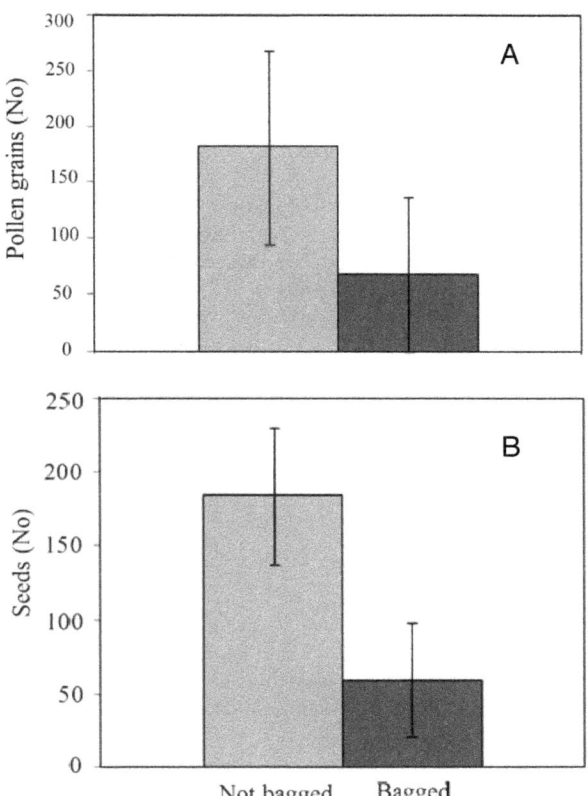

FIGURE 1. (A) Mean number (± SE) of pollen grains on the stigma of bagged and non bagged flowers. Compared with paired Student's *t*-test at 95% significance (t = 5.678; P = 0.0000). (B) Mean number (± SE) of seeds in the fruit of bagged and non bagged flowers. Compared with paired Student's *t*-test at 95% significance (t = 12.37; P = 0.000).

The fruit set was on average 64.48% (t = 9.55; P = 0.000) larger from not-bagged inflorescences than from bagged inflorescences (Table 2). In addition, not-bagged flowers produced 50.21% (t = 4.91; P = 0.000) heavier fruits than bagged flowers (Figure 2). The size of the fruits also was significantly difference between bagged and not-bagged treatments: Fruit from not-bagged flowers were 9.72% larger than fruit from bagged flowers (t = 2.66; P = 0.01).

Seed number increased 208.5% between treatments (t = 12.37; P = 0.000) (Figure 1B). The correlation between fruit mass and number of seeds was high (r² = 0.7047; r = 0.8395; P = 0.00003; y = 28.2011 + 0.3608*x) (Figure 3).

DISCUSSION

Our results show that the visits of native pollinators probably increase pollen doses transferred to the stigma of flowers. The buzz pollination behaviour probably contributed mainly to the deposition of self-pollen because the stigma surface is inside the anther cones of cultivated tomato varieties. On average, non-bagged flowers had 114 ± 68 more pollen grains than bagged flowers on the stigma. This difference is smaller than that found on greenhouses tomato crops with colonies of *Bombus impatiens* Cresson (Morandin et al. 2001a). Nevertheless, our results showed that the tomatoes from not-bagged flowers and probably visited by pollinators are larger, heavier and with more seeds compared to those of bagged flowers. The pollen dose added to the stigmas of tomato flowers should lead to an increase in the number of fertilized eggs and thus, an increase in the production of seeds in the fruits. Studies have shown that the number of seeds in development in tomato fruits influences the activity of the fw 2.2 gene, which is responsible for the production of stimuli for the ovary walls growth and fruit formation (Tanksley 2004; Paran & van der Knaap 2007).

It has been shown that in greenhouses with managed *Melipona quadrifasciata* bees, gains in fruit production reached 15% (Bispo dos Santos et al. 2009). With *B. impatiens*, gains reached 50% in fruit mass and up to the double in the number of seeds (Morandin et al. 2001b). Other studies showed similar results in greenhouses (Hogendoorn et al. 2006; Palma et al. 2008; Bispo dos Santos et al. 2009; Vergara & Fonseca-Buendía 2012). In open air (filed) cultivation in Mexico, Macia-Macia et

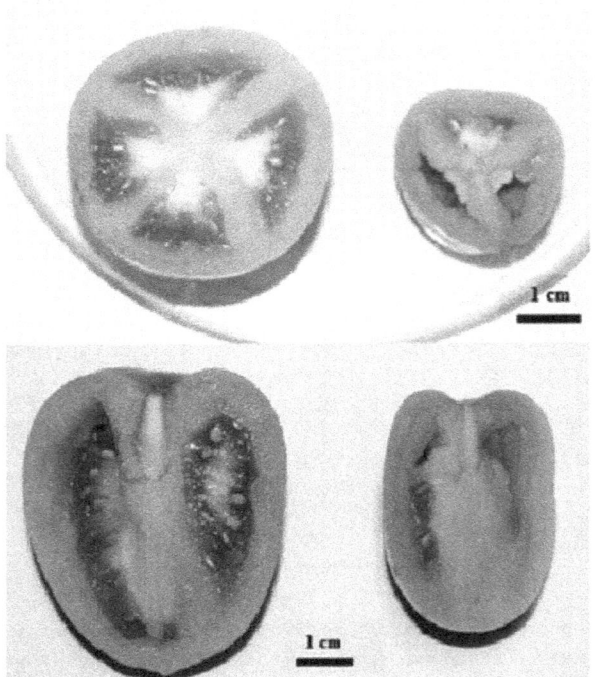

FIGURE 2. Tomatoes in cross (above) and longitudinal sections (below). The tomatoes on the left are from not bagged flowers and those on the right are from bagged flowers.

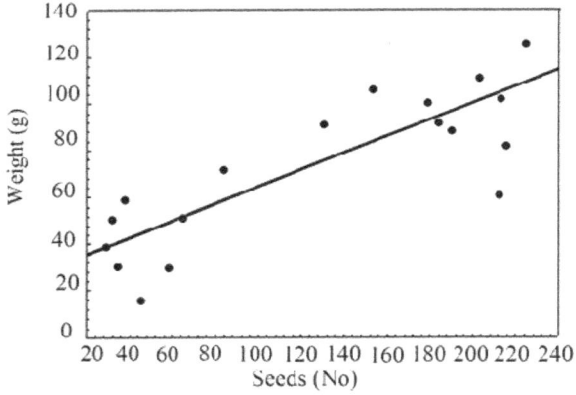

FIGURE 3. Relationship between seeds and tomato fruits mass carried out with simple linear regression (r^2 = 0.7047; r = 0.8395; P = 0.00003; y = 28.2011 + 0.3608*x).

production gains of 67.91% in mass and up to 208% in the number of seeds with native bee visits. Thus, the increase in the production of fruits seems to be greater in field-grown tomato crops than in greenhouses.

It has been suggested that a practical way to increase agricultural production of open grown crops (whose pollinators are native bees) is the conservation of wild or semi-managed vegetation areas around the perimeter of the areas of cultivation (Greenleaf & Kremen 2006a; Holzschuh et al. 2008; Winfree et al. 2007). Those areas of native vegetation are important sources for feeding and nesting to native bees (Kevan et al. 1990, Kevan, 1999). The influence of native areas to bee species may reduce the deficit of pollination in tomato crops is being analyzed by our research team in the State of Goiás where we have already identified 29 different native species visiting tomato flowers in 14 different crops in the same of our study reported herein. A further 17 species of bees were collected there in pan-traps.

Considering the relevance of native bees in nature and for food production, it is essential to understand their attributes, such as nesting, social behaviour or not, foraging behaviour (flight distance, type of food resource) and pollination (buzz pollination and other behaviours). Such studies should cover not only social bees, but also solitary and para-social bees, which have been shown to be important for pollination of many crops (for example, *Exomalopsis* in the case of tomato and pepper) (Raw 2000; Macias-Macias et al. 2009; Santos and Nascimento 2011; Burkart et al. 2011; Kremen et al. 2011; Giannini et al. 2012; van der Valk et al. 2013). Knowledge on the biology of those bees is of utmost importance in order to propose management and conservation strategies to the government and also to implement friendly practices by tomato producers and other pollinator-dependent crops in the State of Goiás and other areas where these species may occur. The use of pesticides for conventional tomato production negatively impacts native pollinators but the extent of that impact is not known for tomato production in Brazil and is part of our continuing research program.

ACKNOWLEDGEMENTS

The authors are especially thankful to Conselho Nacional de Desenvolvimento Científico e Tecnológico - CNPq (*National Council for Scientific and Technological Development*), FAO (Food and Agriculture Organization), FUNBIO (Brazilian Fund to Biodiversity), and to the Brazilian Environment Ministry (MMA) for their funding allocation. The authors also thank Coordenação de Aperfeiçoamento de Pessoal de Nível Superior - CAPES (*the government agency linked to the Brazilian Ministry of Education in charge of promoting high standards for post-graduate courses in Brazil*) for granting a Master's scholarship to the first author. We are grateful to two anonymous reviewers and Dr. Peter G. Kevan for their suggestions and comments that led to considerable improvements of the paper.

REFERENCES

Bezerra ELS, Machado IC (2003) Biologia floral e sistema de polinização de *Solanum stramonifolium* Jacq. (Solanaceae) em remanescente de mata atlântica, Pernambuco. Acta botanica brasiliense 17(2): 247-257.

Bispo dos Santos SA, Roselino AC, Hrncir M, Bego LR (2009) Pollination of tomatoes by the stingless bee *Melipona quadrifasciata* and the honey bee *Apis mellifera* (Hymenoptera, Apidae). Genetics and Molecular Research 8(2): 751-757.

Burkart A, Lunau K, Schlindwein C (2011) Comparative bioacoustical studies on flight and buzzing of neotropical bees. Journal of Pollination Ecology 6:118-124.

Centrais de Abastecimento - CEASA/GO – Histórico de Preços de Mercadorias e Produtos. URL: http://www.ceasa.goias.gov.br (accessed April 2013)

Chetelat RT, Pertuzé RA, Faúndez L, Graham EB, Jones CM (2009) Distribution, ecology and reproductive biology of wild tomatoes and related nightshades from the Atacama Desert region of northern Chile. Euphytica 167:77-93

Dafni A, Pacini E, Nepi M (2005) Pollen and stigma biology. In: Dafni, A., Kevan, P., Husband, B., editors. Practical Pollination Biology. Ontario: Enviroquest Ltd. pp 83-142.

Delaplane KS, Mayer DF (2000) Crop Pollination by Bees. Cambridge: Cabi.CABI

De Marco P, Coelho FM (2004) Services performed by the ecosystem: forest remnants influence agricultural cultures' pollination and production. Biodiversity and Conservation 13:1245-1255.

Free JB (1993). Insect Pollination of Crops. Academic Press London.

Giannini TC, Acosta AL, Garófalo CA, Saraiva AM, Alves dos Santos I, Imperatriz-Fonseca VL (2012) Pollination services at risk: bee habitats will decrease owing to climate change in Brazil. Ecological Modelling 244:127-131.

Greenleaf SS, Kremen C (2006a) Wild bee species increase tomato production and respond differently to surrounding land use in Northern California. Biological Conservation 13, 81-87.

Greenleaf SS, Kremen C (2006b) Wild bees enhance honey bees' pollination of hybrid sunflower. Proceedings of the National Academy of Sciences 103:13890–13895

Harter B, Leistikow C, Wilms W, Truylio B, Engels W (2002). Bees collecting pollen from flowers with poricidal anthers in a south Brazilian Araucaria forest: a community study. Journal of Apicultural Research 40 (1-2): 9:16.

Hogendoorn K, Gross CL, Sedgley M, Keller MA (2006). Increased tomato yield through pollination by native Australian Amegilla chlorocyanea (Hymenoptera: Anthophoridae). Journal of Economic Entomology 99(3):829-833.

Holzschuh A, Steffan-Dewenter I, Tscharntke T (2008). Agricultural landscapes with organic crops support higher pollinator diversity. Oikos 117: 354–361.

Kevan PG, Gadawski RM, Kevan SD, Gadawski SE (1983). Pollination of cranberries, Vaccinium macrocarpon, on cultivated marshes in Ontario. Proceedings of the Entomological Society of Ontario 114:45-53.

Kevan PG, Clark AE, Thomas VG (1990). Insect pollinators and sustainable agriculture. American Journal of Alternative Agriculture 5:13-22

Kevan PG, Straver WA, Offer M, Laverty TM (1991). Pollination of greenhouse tomatoes by bumblebees in Ontario. Proceedings of the Entomological Society of Ontario 122:15–19.

Kevan PG (1999). Pollinators as bioindicators of the state of the environment: species, activity and diversity. Agriculture, Ecosystems & Environment 74: 373-393.

Klein AM, Steffan-Dewenter I, Tscharntke T (2003). Fruit set of highland coffee increases with the diversity of pollinating bees. Proceeding Royal Society London B. 270:955–961.

Klein AM, Vaissière B, Cane JH, Steffan-Dewenter I, Cunningham SA, Kremen C, Tscharntke T (2007). Importance of crop pollinators in changing landscapes for world crops. Proceeding Royal Society London B, Biological Sciences 274, 303-313.

Kremen C, Ullman KS, Thorp RW (2011) Evaluating the Quality of Citizen-Scientist Data on Pollinator Communities. Conservation Biology 25: 607-617.

Macias-Macias O, Chuc J, Ancona-Xiu P, Cauich O, Quezada-Euán JJG (2009) Contribution of native bees and Africanized honey bees (Hymenoptera:Apoidea) to Solanaceae crop pollination in tropical México. Journal of Applied Entomology 133(6).

Malagodi-Braga KS, Kleinert AMP (2007) How bee behavior on strawberry flower (Fragaria ananassa Duchesne) can influence fruit development? Bioscience Journal 23(1): 76-81.

Montemor KA, Malerbo-Souza DT (2009) Biodiversidade de polinizadores e biologia floral em cultura de berinjela (Solanum melongena). Zootecnia Tropical 27: 97-103.

Morandin LA, Laverty TM, Kevan PG (2001a) Bumble bee (Hymenoptera: Apidae) activity and pollination levels in commercial tomato greenhouses. Journal of Economic Entomology 94(2): 462-467.

Morandin LA, Laverty TM, Kevan PG (2001b) Effect of bumble bee (Hymenoptera: Apidae) pollination intensity on the quality of greenhouse tomatoes. Journal of Economic Entomology 94(1): 172-179.

Morandin LA, Winston ML (2005) Wild bee abundance and seed production in conventional, organic, and genetically modified canola. Ecological Applications 15:871–881.

Murcia C (1996) Forest fragmentation and the pollination of neotropical plants. In: Schelhas, Greenberg R (Eds). Forest patches in tropical landscapes. Island Press, Washington, D.C., USA

Olmstead RG, Palmer JD (1997) Implications for the phylogeny, classification and biogeography of Solanum from cpDNA restriction site variation. Systematic Botany 22:19–29.

Palma G, Quezada-Euán JJG, Reyes-Oregel V, Meléndez V, Moo-Valle H (2008) Production of greenhouse tomatoes (Lycopersicon esculentum) using Nannotrigona perilampoides, Bombus impatiens and mechanical vibration (Hymenoptera: Apoidea). Journal of Applied Entomology 132: 79–85.

Paran I, van der Knaap E (2007) Genetic and molecular regulation of fruit and plant domestication traits in tomato and pepper. Journal of Experimental Botany 58:3841–3852

Raw A (2000) Foraging behaviour of wild bees at hot pepper flowers (Capsicum annuum) and its possible infuence on cross pollination. Annals of Botany 85: 487-492.

Santos AB, Nascimento F S (2011) Diversidade de visitantes florais e potenciais polinizadores de Solanum lycopersicum (Linnaeus) (Solanales: Solanaceae) em cultivos orgânicos e convencionais. Neotropical Biology and Conservation 6(3):162-169.

Spears EE (1983) A direct measure of pollinator effectiveness. Oecologia 57: 196-199.

Tanksley SD (2004) The genetic, developmental, and molecular basis of fruit size and shape variation in tomato. The Plant Cell 16:181-189.

Teppner H (2005) Pollinators of tomato, Solanum lycopersicum (Solanaceae), in. Central Europe. Phyton 45(2): 217.

Van der Valk H, Koomen I, Nocelli RCF, Ribeiro MF, Freitas BM, Carvalho SM, Kasina, Martins DJ, Maina G, Ngaruiya P, Gikungu M, Mutiso MN, Odhiambo C, Kinuthia W, Kipyab P, Blacquiera T, van der Steen J, Roessink I, Wassenberg J, Gemmill-Herren B (2013). Aspect determining the risk of pesticides to wild bees: Risk profiles for focal crops on three continents. Food and Agriculture Organization of the United Nations, Romes.

Vergara CH, Fonseca-Buendía P (2012) Pollination of greenhouse tomatoes by the Mexican bumblebee Bombus ephippiatus (Hymenoptera: Apidae). Journal Pollination Ecology 7:27–30.

Winfree R, Williams NM, Dushoff J, Kremen C (2007) Native bees provide insurance against ongoing honey bee losses. Ecology Letters 10 (11):1105–1113.

Winfree R, Williams NM, Gaines H, Ascher JS, Kremen C (2007) Wild bee pollinators provide the majority of crop visitation across land-use gradients in New Jersey and Pennsylvania, USA. Journal of Applied Ecology 45:793–802.

5

BIG BEES DO A BETTER JOB: INTRASPECIFIC SIZE VARIATION INFLUENCES POLLINATION EFFECTIVENESS

P.G. Willmer* & K. Finlayson

Sir Harold Mitchell Building, School of Biology, University of St Andrews, St Andrews KY16 9TH

Abstract—1. Bumblebees (*Bombus* spp.) are efficient pollinators of many flowering plants, yet the pollen deposition performance of individual bees has not been investigated. Worker bumblebees exhibit large intraspecific and intra-nest size variation, in contrast with other eusocial bees; and their size influences collection and deposition of pollen grains.

2. Laboratory studies with *B. terrestris* workers and *Vinca minor* flowers showed that pollination effectiveness PE, as measured from pollen grains deposited on stigmas in single visits (SVD), was significantly positively related to bee size; larger bees deposited more grains, while the smallest individuals, with proportionally shorter tongues, were unable to collect or deposit pollen in these flowers. Individuals did not increase their pollen deposition over time, so handling experience does not influence SVD in *Vinca minor*.

3. Field studies using *Geranium sanguineum* and *Echium vulgare*, and multiple visiting species, confirmed that individual size affects SVD. All bumblebee species showed positive SVD/size effects, though even the smallest individuals did deposit pollen. *Apis* with its limited size variation showed no such detectable effect when visiting *Geranium* flowers. Two abundant hoverfly species also showed size effects, particularly when feeding for nectar on *Echium*.

4. Mean size of foragers also varied diurnally, with larger individuals active earlier and later, so that pollination effectiveness varies through a day; flowers routinely pollinated by bees may best be served by early morning dehiscence and visits from larger individuals.

5. Thus, while there are well-documented species-level variations in pollination effectiveness, the fine-scale individual differences between foragers should also be taken into account when assessing the reproductive outputs of biotically-pollinated plants.

Keywords: *Pollinator, Pollen deposition, Bumblebee, Body size, Intraspecific variation*

INTRODUCTION

Bees are the most plentiful and successful of the pollinators, and depend entirely on flowering plants since they feed only on pollen and nectar throughout their lives. Eusocial bees are highly efficient pollen-gatherers, and bumblebees (*Bombus*) often collect significantly more pollen from anthers and deposit more pollen on stigmas than *Apis* honeybees (Willmer et al. 1994; Thomson & Goodell 2001; King, Ballantyne & Willmer 2013). *Bombus* have a substantial capacity to improve flower handling time by learning (e.g. Laverty 1994), and some species will learn from each other (Leadbeater & Chittka 2009; Dawson et al. 2013). However individual pollinator effectiveness (PE) variation within a species, whether for naïve foragers or after a learning period, has not been investigated.

Bumblebees show substantial inter-specific differences in worker size (Benton 2006), but unlike most other eusocial bees also demonstrate large intra-specific and intra-nest size variation (Plowright & Jay 1968; Peat et al. 2005); workers can exhibit a ten-fold size variation within a single nest

(Alford 1975) compared with less than two-fold variation in honeybees and stingless bees (Waddington et al. 1986; Roulston & Cane 2000). This is not genetically controlled as workers within a nest are normally full sisters, but instead probably stems from unequal larval feeding (Sutcliffe & Plowright 1988; Couvillon & Dornhaus 2009); Persson & Smith (2011) have shown that adult size of bumblebee workers is significantly related to the availability of floral resources. Intra-nest size variation dictates the polyethism found within the colony; larger bees are more likely to become foragers for the colony, and are more efficient as they can transfer more pollen and nectar to the colony (Morse 1978; Goulson et al. 2002). There is also a positive interspecific relation between worker size and foraging range in bees (Greenleaf et al. 2007) which may also hold within species and should mean that larger workers gain access to a wider range of floral resources. Larger individuals are also better able to broaden their diet when stored food supplies require it, when compared with smaller nest mates (Fontaine et al. 2008). However there has been little investigation of the pollen-depositing abilities of individuals, and hence the influence of visitor size variation on plant pollination.

Here we investigate whether individual flower foragers, especially bumblebees, vary in pollen deposition performance according to their body size, as well as aspects of their

---

*Corresponding author: pgw@st-and.ac.uk

behaviour and their foraging experience. A variety of techniques have been employed to quantify pollinator success (or effectiveness, or efficiency; see Willmer 2011), from visit number, frequency or duration to pollen carried or deposited, or eventual seed-set. However the number of pollen grains deposited on a stigma from a single visit is the most robust measure of pollinator effectiveness (PE) (Ne'eman et al. 2010), for a particular plant species and visitor pairing. Therefore we measured Single Visit Deposition (SVD), recently demonstrated by King, Ballantyne & Willmer (2013) as a reliable and practical method of distinguishing true pollinators from mere visitors. In this study SVD is used to address the specific pollination ability of individual visitors within a species.

MATERIALS AND METHODS

A. Laboratory Studies

Bombus terrestris was used from a single commercial nest box (Syngenta Bioline, The Netherlands), containing 50-80 worker bees, whose activity out of the nest (in an arena with test flowers) could be controlled by doors. The nest was in a glasshouse with exposure to both natural sunlight and additional overhead lights, at a temperature of $22 \pm 3°C$. During testing, bees had access into a wooden arena (100cm x 100 cm x 30 cm) with a plexiglass lid and side-access doors. In this enclosed space a worker could forage on the plants provided. The bees were fed artificial nectar in the nest, but could be isolated from this supply overnight to encourage foraging in the arena. No additional pollen pellets were provided in the nest (so avoiding heterospecific deposition onto flower stigmas), although withholding pollen does deprive workers of dietary protein and may decrease nest longevity (Smeets & Duchateau 2003). Bees that entered the arena were tagged with queen-marker discs on the dorsal thorax to distinguish individuals, and the thoracic inter-tegular width was recorded with digital calipers (LTL Linear Tools). Disc placement did not influence flight ability, and no bee was tested on the same day as a disc was applied, to reduce possible stress effects on behaviour.

Vinca minor (Apocynaceae) was the test flower, being native to temperate Europe (Fjell 1983) and pollinated by several insect genera, including *Bombus* spp. (Horwood 1919). Within the corolla the reproductive anatomy is unusual (Fig. 1), as first described by Darwin (1861). The pistil bears a horizontal stigmatic "wheel", with the hairy anthers above. Only the concave ventral wheel surface is receptive, and its sticky exterior rim prevents self-pollen from above reaching the stigma. Through this use of spatial herkogamy, the flower avoids self-fertilization and requires crossing by insect visitors (Fryxell 1957). Bumblebees can reliably accomplish this while probing for nectar, since pollen from other plants, borne on the tongue, reaches the underside of the stigma wheel as the tongue is withdrawn. The proboscis, now sticky with stigmatic secretions, then also collects the fresh dry pollen from the brush on top of the style, so that there can be a near-complete exchange of

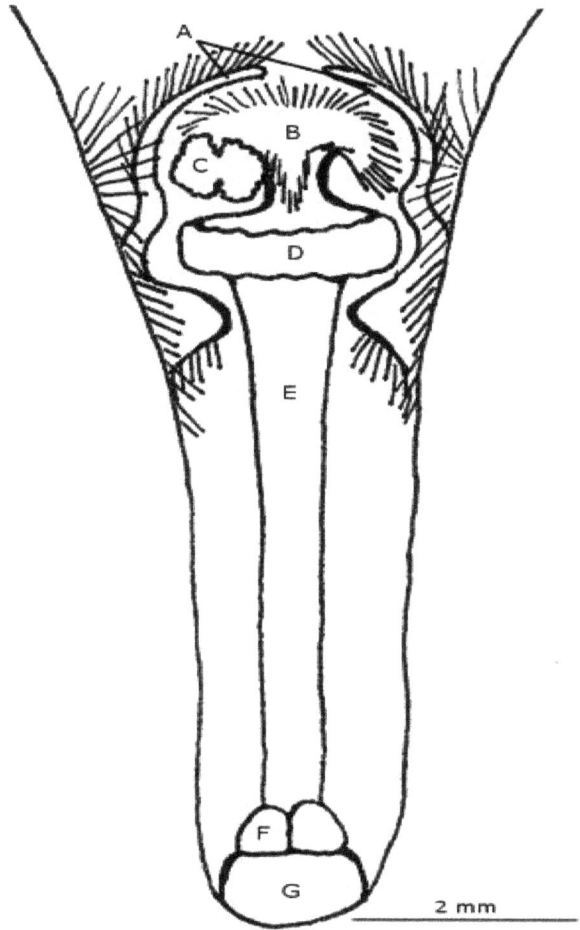

Figure 1. Internal anatomy of a *Vinca* flower. A - anthers; B - filament brush; C - pollen grains; D - stigmatic 'wheel'; E - style; F - ovaries; G - receptacle bearing nectaries.

pollen grains during nectar-feeding. Hence stigma wheels only reveal pollen sourced from another flower, and confounding variables from self-fertilization or the deposition of multiple flowers' pollen upon a single stigma are largely avoided.

Potted plants with unopened, virgin blossoms were purchased from local nurseries and kept in the closed greenhouse free from extraneous insect visitors, with two cultivars used interchangeably throughout the experiment (*Vinca minor* 'Atropurpurea' and *Vinca minor* Ralph Shugart'; similar in flower size and with identical reproductive anatomy). Each plant provided 30-50 flowers over 2-4 weeks.

The experiments took place in January-April 2013, from early morning to mid-afternoon, in accordance with maximum natural sunlight within the glasshouse and peak bee activity. Testing sessions lasted 1-4 hours, depending on the activity levels of the bees and their willingness to exit the hive to forage (varying mainly with outside weather and light levels).

Tagged bees were allowed free access for 24 hours into the arena containing one potted *Vinca minor* plant, as a familiarisation period. Thereafter, a single *V. minor* plant with virgin flowers was placed in the arena, and the door to the hive opened until a single bee entered. If the bee did not visit a flower within ten minutes, it was recaptured, tagged if necessary, and returned to the nest, as longer times in the arena rarely led to any visitation, irrespective of bee size. When a bee did visit a flower, it was closely observed for tongue extension (indicating nectar-feeding) and for grooming behaviours (tongue-wiping after leaving a flower). Each bee was normally allowed to visit three flowers per trial, and each flower was removed after a single visit.

Flowers removed from the plant after a single visit (234 in total) were assessed for pollen deposition (Single Visit Deposition, SVD) as detailed in King, Ballantyne & Willmer (2013). Briefly, the stigma was removed with clean forceps and stored in a plastic cell-culture array (24 cells, TPP test plate) kept covered and cool. The number of adherent pollen grains on each stigma was assessed with a dissecting microscope (x40), counting only the grains located on the receptive underside of the stigma 'wheel'. Unvisited control flowers were taken periodically to ensure the flowers were not experiencing self-fertilization.

B. Field Studies

Two plants were chosen for field work, based on their availability at a field site in NE Fife, Scotland (NO 3719) and on the ease of recognition of their pollen. *Echium vulgare* has distinctive purple/blue pollen and is characterised as a bumblebee-pollinated flower (Rademaker et al. 1999). Flowers produced were entirely hermaphrodite (though some gynodioecy may occur at other sites (Klinkhamer et al. 1991), and strongly protandrous, with spatially separated anthers and style. They were pink when opening (usually before midday) but turned mauve/blue during day 1 and deep blue on day 2, before wilting by day 3. Day 1 flowers were functionally male, and individual bees visiting the youngest flowers could pick up thousands of pollen grains (~7,000 per visit, Rademaker et al. 1999), so that pollen was substantially depleted from anthers by midday of day 1 whenever weather conditions permitted

regular visitation. Day 2 blue flowers were female phase with receptive stigmas, and were used for this study. *Geranium sanguineum* grew in the same site, having large orange pollen grains with distinctive reticulate sculpturing. The population contained entirely hermaphrodite flowers, which were almost completely protandrous and lasted 3-5 days. They were male on day 1 and usually part of day 2, pollen dehiscing within 1-2 hours of bud opening and often available for 24-30 hours thereafter. The stigmatic lobes opened on day 2 (occasionally day 3) after virtually all the pollen had been shed from that flower, so avoiding within-flower selfing. Only flowers with open stigmatic lobes, on days 2 and 3, were used for SVD analysis. *Echium* observations occurred in June-August 2012 and 2013, and *Geranium* in July 2013, covering all daylight hours of suitable weather conditions (between 0620 and 2030 on different days).

Buds were enclosed in fine net the evening before they would open, and then exposed at varying times the following day (or on subsequent days for *G. sanguineum*). The first visitor was noted (identification, nectar and/or pollen feeding, visit duration); then the flower was removed and its stigma picked. Methodology for measuring SVD was as above, but conspecific pollen grains on the stigma were sufficiently distinctive to be scored with a 20x lens in the field. At intervals fresh flowers were exposed and their stigmas counted immediately, with no visitation, to act as controls for self-pollen moved during netting and handling.

All visitors were identified to species as far as possible in the field, with uncertain insects caught for later checking. Within a species, each was scored as large, medium or small by eye. A proportion of all visitors already assessed in each size category were captured through the season for accurate measurement of individual size (inter-tegular thorax breadth, using calipers as above). Table 1 shows the measured size ranges for the various bee species that had been assigned to each size class, and mean size for the species; only one case of overlap between size classes occurred (a 'medium' *B. terrestris* with a size actually in the small range), so that the 3-level size scoring was accepted as appropriate.

TABLE I.   Measured size ranges (as inter-tegular thoracic width, mm) of bees and hoverflies at the field site in each size class, with the 3 ranges tailored for each species; only one case of overlap (*) occurred between assigned size class and subsequently measured thorax width. (Numbers in brackets for *Apis* were each for only one individual).

	Large	Medium	Small	Mean
Apis	(4.7)	3.3-4.3	(3.0)	3.6
B. terrestris	6.5-7.3	4.2-6.3	3.5-4.4*	5.0
B. lapidarius	5.6-6.2	4.6-5.5	3.5-4.5	4.5
B. lucorum	5.5-6.0	4.2-5.2	<4.0	4.8
B. pascuorum	5.5-6.5	4.5-5.3	3.1-4.2	4.3
B. pratorum	5.0-5.8	4.2-4.9	3.3-4.0	4.1
Episyrphus balteatus	2.6-2.9		2.2-2.5	
Platycheirus albimanus	2.2-2.4		1.9-2.1	

C. Statistical Analysis

Pollen grain counts and bee measurements were normally distributed. Pearson's correlations were therefore carried out to compare bee size and SVD, and t-tests to compare SVD following different behaviours. ANOVAs compared multiple species in relation to body size, and SVD/size effects within each species. Linear regression was used to test effects on SVD of learning over time. Tests were applied using either Minitab v.17 or SPSS v.21, the latter particularly for non-linear relations tested with simple and quadratic regression. Data are shown as means ± 1 SE, and vertical bars on Figures also show ± 1 SE.

RESULTS

A. Laboratory Studies with captive bumblebees.

In total 45 individual *B. terrestris* were tagged, with 26 participating in foraging activities. Thorax widths were 2.5-7.4 mm (mean 4.9 ± 0.2 mm); the range and mean are larger than some reported values (del Castillo & Fairbairn 2012; Persson & Smith 2013), but Peat et al. (2005) did record larger bees of this species in Scotland relative to English colonies, and commercial bee nests may often produce a wider size range of foragers than wild nests.

Pollen grain deposition overall (for 234 visits) varied from 0 to 300 grains per flower, with zero deposition on the first flower visit and means of 39.8 ± 2.6 and 38.0 ± 4.5 grains (difference NS) on the second and third visits (after pollen had been acquired from the previous flower(s)) within a trial. Mean SVD was therefore calculated from flowers 2 and 3. Fourteen of the test bees emerged and foraged multiple times. For example bee 8, with a thorax width of 6.9 mm, engaged in 14 flower-visiting trials over 7 days, depositing 0-130 grains per visit, but it did not get significantly better or worse over time in the trials (Fig. 2; linear regression, df = 1,12, F = 0.599, P = 0.45). The same was true of 13 other bees that each made between 2

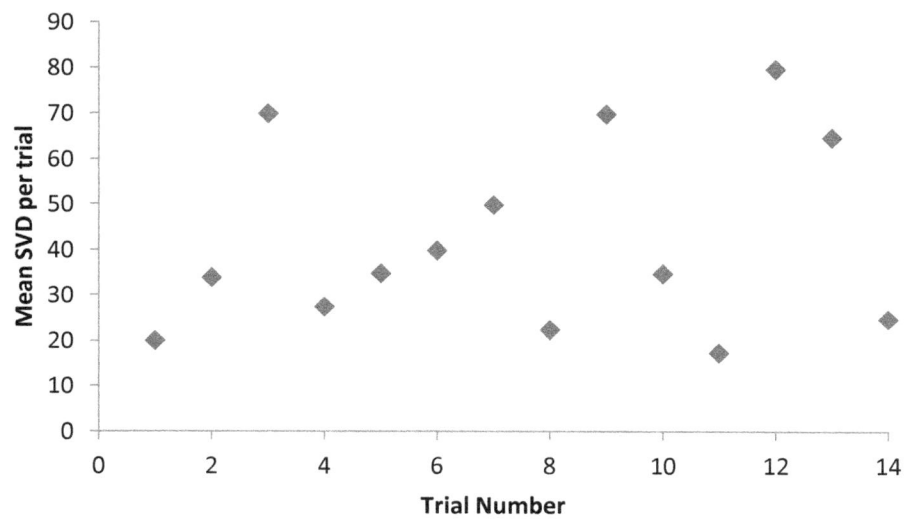

FIGURE 2. The performance of one bee on *Vinca* flowers (bee #8, thorax width 6.9 mm) for 14 trials across 7 days. Mean SVD range 20-80 grains, with no significant trend over time (linear regression, df=1,12, F = 0.599, P= 0.45).

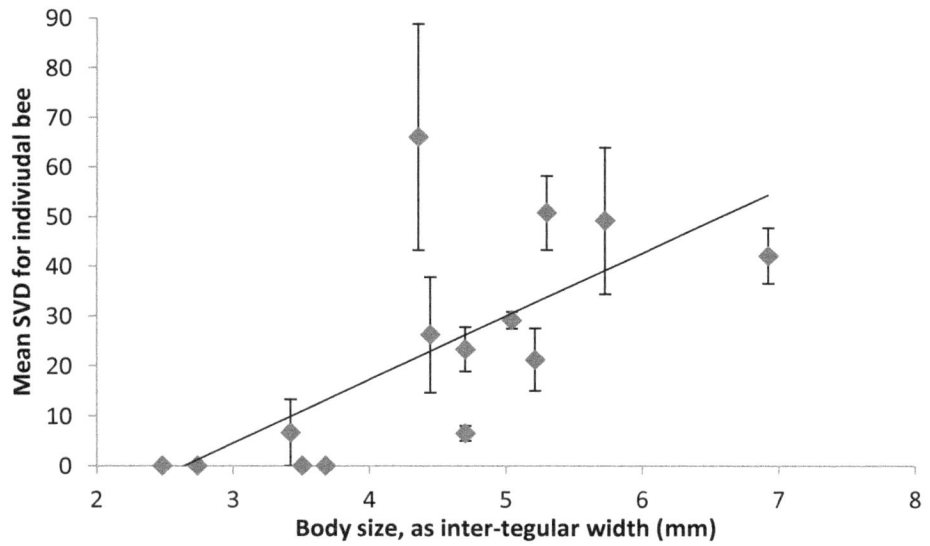

FIGURE 3. Mean SVD (± SE) in relation to body size for individual *Bombus terrestris* working *Vinca minor* flowers in laboratory conditions (for the 14 individuals where more than one flower visit was recorded). Pearson correlation line shown (r = 0.692, P= 0.006).

and 9 visits, thus there was thus no evidence of improved handling or learning over time with this flower, and time of trials was excluded from further analysis.

However larger bees generally deposited more pollen, and Fig. 3 shows the significant correlation between bee size (thorax width) and mean SVD (r = 0.692, P = 0.006). The smallest bees produced zero pollen deposition; these bees were commoner later in the trials (date versus thorax width, r = -0.52, P < 0.0001), perhaps because pollen in the nest was becoming depleted and the later-emerging brood were less well fed, and/or because smaller bumblebees are more resistant to starvation (Couvillon & Dornhaus 2010).

There was also a significant effect of individual grooming ('tongue-wiping', where some bees scraped pollen from their proboscis after withdrawal from the flowers) on pollen deposition and hence PE. When no wiping occurred, mean SVD was 93.3 ± 10.3, whereas in bees that tongue-wiped mean SVD on the next flower visited was 39.8 ±7.2. The very small bees that deposited no pollen never had pollen on their faces and thus never showed tongue-wiping behaviour; when they were removed from the analysis grooming behaviour (above a threshold body size) did significantly reduce SVD (t = 4.28, df = 1,61, P < 0.001).

B. Field analyses with multiple visitors

1. Echium vulgare

SVD results for a range of visitors to *E. vulgare*, summed across all dates and times, are shown in Table 2 arranged by body size (mean control SVD = 0.4 ± 0.2 pollen grains, too low to merit subtraction from the experimental data). Deposition of grains per stigma was similar to the range of 1-10 previously recorded in studies using manipulated dead bees as carriers (Rademaker et al. 1997). The great majority of visits (94%) were made by *Bombus* species, all purely nectar-feeding. For all five species of bumblebee mean SVD was greater in larger individuals, with around 3- to 6-fold differences between the large and small size categories. The

effect was significant for 4 of the 5 species, and also strongly significant for all bumblebees combined (ANOVA, df= 2,194, F = 27.15, P < 0.001). Several genera of hoverflies were also occasional visitors, especially in autumn 2012, but nearly always foraged only for pollen, feeding at the protruding anthers and depositing no pollen on stigmas. However *Eristalis pertinax*, *Episyrphus balteatus* and *Platycheirus* spp. also made a few nectar-collecting visits, and SVD values for the latter two (commonest) hoverflies are also given in the Table, split into two size categories; although size differences were smaller than for bumblebees, there were still significantly greater SVD values for larger individuals for *Platycheirus*, and for all hoverfly species combined (see Table). The only other visitors observed were *Pieris rapae* butterflies on two occasions, taking nectar and depositing 3 and 10 pollen grains. But as predicted from previous studies (Rademaker et al. 1999) bumblebees were by far the most important pollen-depositing visitors.

SVD also showed variation through time, summed for all bee visitors (Fig. 4). Mean pollen deposition was greatest in early- to mid-morning (0800-1000), and fairly constant at all other times. Peak SVD coincided with the observed peak of anther dehiscence in the majority of newly-opened flowers. However there were also variations in the mean size of recorded visitors through a day (Fig. 5), as expected from known thermal effects on insects in relation to their size (Willmer 1983; Willmer & Stone 2005). Larger individuals were more likely to be active before 1100h and after 1700h, with a preponderance of individuals in the small size category between midday and 1600h giving a curvilinear relationship (simple regression, quadratic term significant t_{195}= 2.19, P = 0.03). Thus larger bumblebees were mainly responsible for visitation during the peak period of pollen presentation and deposition in this plant species. Inevitably SVD did not increase in the evening when larger bumblebees were active again, because by then the pollen was substantially depleted from flowers of *Echium*.

TABLE 2.   Mean SVD for large, medium and small individual bumblebees (± SE, n in parentheses) visiting *Echium vulgare*, and for two size categories of the two commonest hoverflies. Details of ANOVA tests are also shown.

	Large	Medium	Small	All	F	P
Bees						
B. terrestris	13.8 ± 4.7 (5)	8.0 ± 1.7 (19)	4.0 ± 2.5 (4)	8.5 (28)	2.05	ns
B. lapidarius	12.4 ± 3.7 (5)	5.8 ± 1.1 (14)	5.0 ± 1.6 (4)	7.1 (23)	3.98	0.035
B. lucorum	15.8 ± 4.4 (5)	5.4 ± 1.6 (8)	2.7 ± 1.1 (7)	6.8 (20)	4.83	0.022
B. pascuorum	7.9 ± 3.0 (8)	5.2 ± 0.7 (40)	2.4 ± 0.7 (14)	4.9 (62)	3.67	0.031
B. pratorum	11.0 ± 5.8 (2)	4.9 ± 0.9 (25)	2.6 ± 0.4 (35)	3.9 (62)	7.53	0.001
All *Bombus*	11.8 ± 1.7 (25)	5.6 ± 0.5 (106)	2.8 ± 0.3 (64)		27.15	<0.001
Hoverflies	Larger		Smaller			
Episyrphus balteatus	6.4 ± 1.6 (11)		1.2 ± 0.7 (4)		2.21	ns
Platycheirus albimanus	9.0 ± 6.2 (3)		1.2 ± 0.4 (8)		4.84	0.055
All hoverflies	6.9 ± 1.6 (14)		1.2 ± 0.4 (12)		7.41	0.012

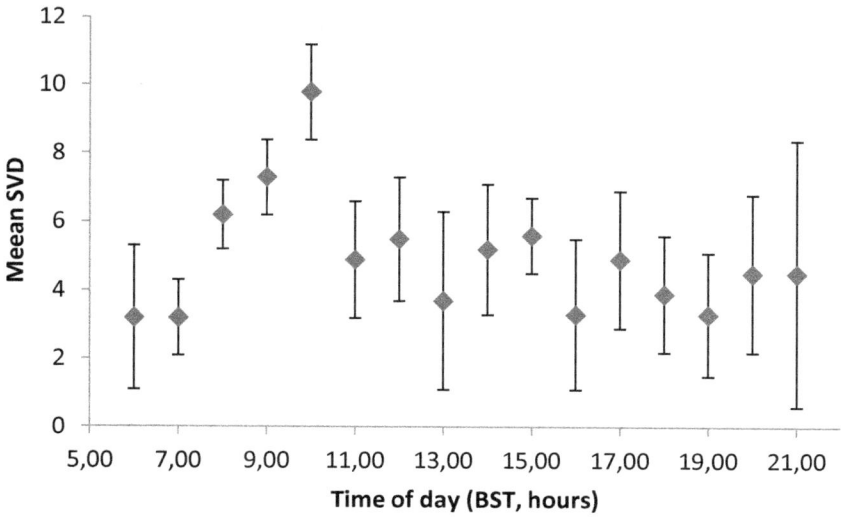

FIGURE 4. Mean SVD (± SE) for all visitor species to *Echium vulgare*, against time of day.

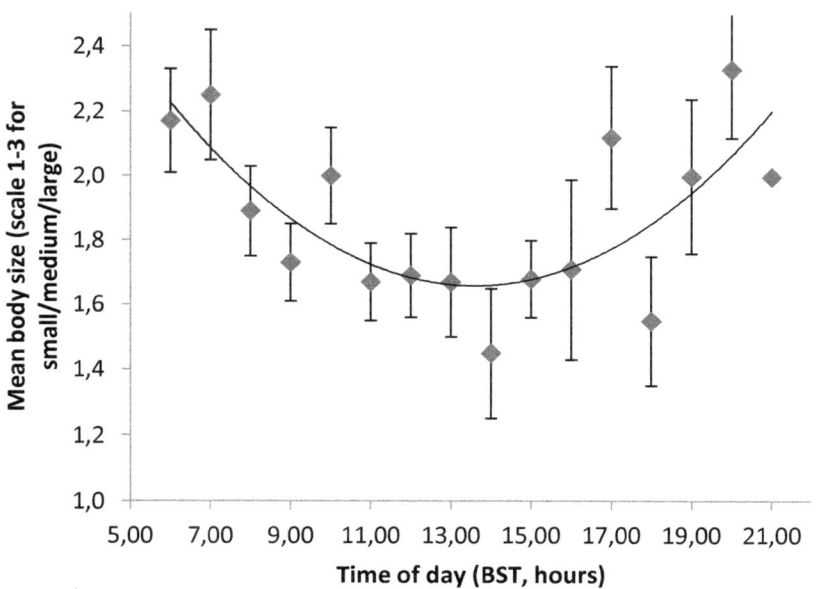

FIGURE 5. Mean body size (± SE) against time of day for all visitor species to *Echium vulgare*. (Best-fit polynomial is shown; simple regression, quadratic term significant t_{195} = 2.19, P = 0.03).

2. *Geranium sanguineum*

Data were only recorded at times when the flower was in the female phase, and are summed across flowers of age 2-5 days (predominantly day 2). Table 3 shows the mean SVD for all visitors, in three size categories. Rather few large individuals were recorded, but for *Bombus* larger bees still deposited more pollen than smaller ones, with the differences greatest for the abundant *B. terrestris*. For all *Bombus* combined size had a significant positive effect on SVD (ANOVA P = 0.041, see Table). Virtually no size variation occurred for the honeybees, and their mean SVD value was lower than the mean for *Bombus* species. Many hoverflies also visited the flowers but only for pollen when they were in the male phase, and occasional butterflies visited for nectar; but no records of SVD above the control level were recorded for visitors other than bees.

The diurnal pattern of SVD (Fig. 6) is low initially (10-30 grains per visit) until about 0800h, but then fairly constant (40-70 grains) through the remaining daylight hours, consistent with the observed presence of pollen in male-phase flowers over most of the daylight hours in the early life (day 1-2) of a flower. Fig. 7 shows the mean body size of visitors against time of day; here the pattern of larger visitors early and late is missing, with no significant trend (simple regression, quadratic term NS, t_{184} = 1.86, P = 0.07).

DISCUSSION

A) Effects of body size on SVD

Both in laboratory studies and in the field, intraspecific variation in body dimensions strongly influenced pollen

TABLE 3. Mean SVD for large, medium and small individual bees (± SE, n in parentheses) visiting *Geranium sanguineum*. Details of ANOVA tests are also shown.

	Large	Medium	Small	All	F	P
Apis				34.4 ± 3.4 (46)		
B. terrestris	69.4 ± 12.4 (5)	65.7 ± 10.4 (22)	36.2 ± 10.0 (19)	53.9 ± 6.5 (19)	2.79	ns
B. lapidarius		44.0 (2)		44.0 (2)		
B. lucorum		48.9 ± 7.8 (10)		48.9 ± 7.8 (10)		
B. pascuorum		37.0 ± 4.0 (30)	31.0 ± 7.6 (7)	35.8 ± 3.3 (37)	0.49	ns
B. pratorum		38.0 ± 5.5 (11)	29.5 ± 7.6 (8)	34.4 ± 5.1 (19)	0.65	ns
All *Bombus*	69.4 ± 12.4 (5)	47.4 ± 4.2 (75)	33.5 ± 3.9 (34)		3.29	0.041

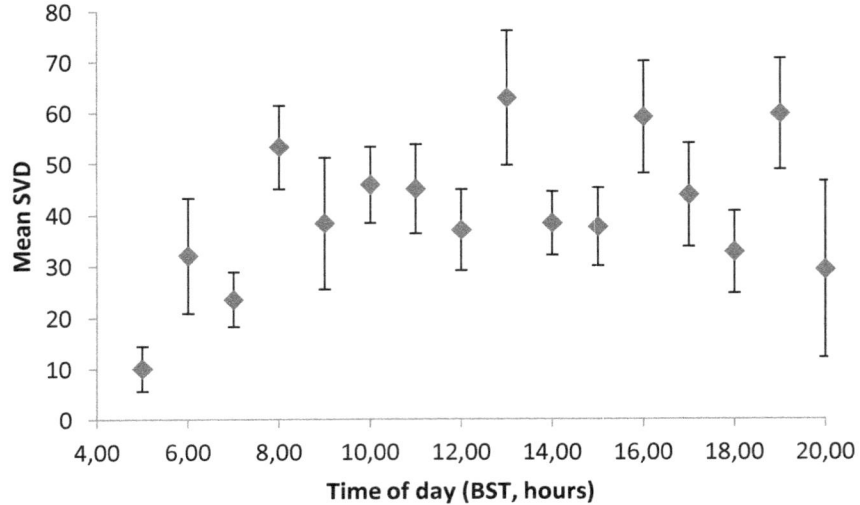

FIGURE 6. Mean SVD (± SE) for all visitor species to *Geranium sanguineum*, against time of day.

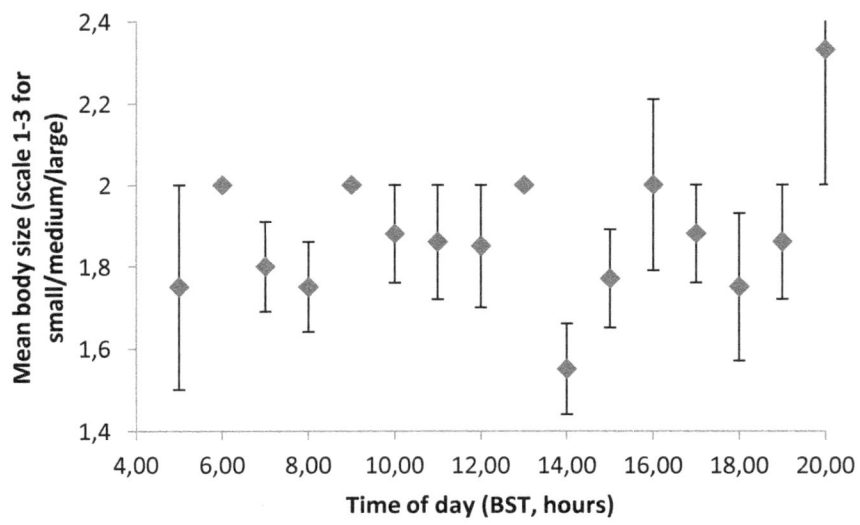

FIGURE 7. Mean body size (±SE) against time of day, for all visitor species to *Geranium sanguineum*, with no signifianct trend. (Simple regression, quadratic term not significant, $t_{184} = 1.86$, $P = 0.07$).

deposition by flower-visiting social bees, particularly for bumblebees where size variation is substantial. It is not surprising to find larger visitors being more efficient pollinators on reasonably open ('generalist') flower types, as documented by Sahli & Connor (2007) for *Raphanus* when considering interspecific body-mass differences and seed-set. Better pollination by larger visitors has also been explicitly reported for other *Geranium* species (Kandori 2002). For more complex flowers such as keel-type Fabaceae large body size is generally reported as advantageous for 'tripping' the flowers, although Stout (2000) showed that small and medium bees tripped *Cytisus* flowers better than very large queen bumblebees. But an intraspecific difference in pollinator effectiveness for a flower offering reasonably open access to visitors has not previously been demonstrated.

In eusocial bees, the survival of the colony depends upon the division of tasks amongst workers (polyethism). In most genera it is progressive ageing that determines the tasks a worker performs (Seeley 1982), notably in honeybees and stingless bees where workers are of fairly uniform size. However bumblebees do not employ age-determined polyethism, and do exhibit large intraspecific and intra-nest worker size variation (Goulson 2010); this results in alloethism, the performance of different behaviours and tasks as dictated by size (Goulson et al. 2002). In *Bombus terrestris*, alloethism leads to the largest bees becoming foragers, and switching to foraging behaviours earlier than smaller counterparts (Pouvreau 1989). Larger bumblebee foragers have longer tongues, resulting in more nectar collection (Peat et al. 2005; Peat & Goulson 2005), larger more sensitive eyes and brains (Macuda et al. 2001; Mares et al. 2005) and greater antennal sensitivity (Spaethe et al. 2007), leading to better learning and memory (Worden et al. 2005). They also show better thermoregulatory efficiency (Bishop & Armbruster 1999), so being more efficient foragers in colder weather (Heinrich & Heinrich 1983; but see Peat & Goulson 2005), where their size also positively influences flight ability (Kapustjanskij et al. 2007); and they may have better competitive capacity associated with improved access to resources (Inoue & Yokoyama 2006). Overall, they contribute more nectar and pollen per unit time to the colony than smaller foragers (Spaethe & Weidenmüller 2002) consistent with the view that worker size is largely determined by foraging-related non-reproductive factors (whilst size in males and queens is more strongly linked to selection on reproductive functions (del Castillo & Fairbairn 2011)).

Goulson et al. (2002) observed a linear relationship between forage mass collected and thorax width in bumblebees. From our study, individual bees of greater size (Fig. 3, Tables 2 and 3) were capable of depositing a higher mean number of pollen grains, whether on the underside of the *Vinca* stigma wheel or on the simpler protruding stigmas of *Echium* and *Geranium* flowers. This was not due to any variation in visit duration, which was unrelated to bee size but did increase later in the day for *Echium* as both pollen and nectar became scarcer (pers. obs.). Instead it may be largely attributable to their larger surfaces (of ventral body, or of tongue length) so that they may contact more anthers and gather more pollen; larger bees can evidently carry larger

pollen loads in their scopa and ungroomed pollen loads on the rest of the body should be similarly greater. Larger individual foragers are therefore potentially beneficial to the plant, as long as their depositions are not exceeding the maximum pollen grains required per flower to fertilise all ovules and are therefore not causing 'pollen clogging'.

For *Vinca minor* flowers the smallest bees failed to deposit pollen at all. A visiting nectar-seeking bumblebee must have a tongue able to extend the length of the corolla (mean 9.3mm) to access the basal nectaries. Thus a *Vinca* flower must be co-adapted with relatively long-tongued visitors. Goulson et al. (2002) found a simple proportional relationship between overall body size and tongue length in bumblebees, whilst Morse (1977) and Harder (1982) also showed that bees with larger wings, linked to a larger body, had longer tongues. Thus larger foraging bees with longer tongues are advantageous to *Vinca minor*, being able to contact the critical points within the flower to acquire and deposit pollen grains. The plant thereby has a potentially increased success in cross-fertilization from greater pollen deposition.

While small bumblebees proved unsuccessful at pollinating *Vinca minor*, and less effective on *Echium* and *Geranium*, they may of course function as effective foragers at other species of flowers. Individual bees tend to specialize on certain flowers (Heinrich 1979; Cane & Sipes 2006) and may feed on flowers appropriate to their body size and tongue length, so that smaller bees do prefer to forage at flowers with shallower corollae where they may more easily access nectar (e.g. Peat, Tucker & Goulson 2005). Size variation within the nest thus contributes to exploitation of a wider resource range and can be advantageous to the bee colony as well as to the plants that they visit. A *Bombus* colony with a large size range of individuals may therefore be more successful in food-gathering overall. It would be interesting to explore influences on this size range, other than the obvious weather and food constraints; for example pesticide exposure can influence the size of *Bombus* workers (Baron et al. 2014), and competition from honeybees may also have an effect (Butz Huryn 1997; Thomson 2004), both of these being potentially deleterious to colony success.

B) The influence of grooming behaviour

All Hymenoptera, including bumblebees, perform grooming behaviours to maintain their condition and remove foreign contaminants. The cleaning of the head, with the forelegs performing a scraping action, allows cleansing of mouthparts and antennae (Jander 1976). With *Vinca* a bee would occasionally perform 'tongue wiping' behaviour before moving on to another flower, which strikingly influenced pollen deposition and often reduced SVD to zero resulting in a total loss of cross-fertilization. Post-visit grooming is common in bees, with pollen being packaged into the scopa; but specific grooming behaviours between flower visits were observed only rarely for both *Echium* and *Geranium* flowers (less than one in 20 and one in 35 visits respectively) so in these cases did not affect mean SVD values.

C) The performance of foragers over time

Bumblebees normally show decreased handling time on flowers with increased practice (Goulson 2010), the effect increasing markedly with morphologically more complex flowers (Laverty 1994), although the direct effects may be mainly short-term (Durisko et al. 2011). They may take three times as many visits to learn the motor skills needed for effective pollen collection compared with learning how to feed on nectar (Raine & Chittka 2007). Hence a similar improvement in flower handling might be expected in our trials with individual bees, and would need to be taken into account in assessing effectiveness. However there was no significant increase in pollen deposition with prolonged exposure to *Vinca minor* flowers, presumably because of the unusual morphology involved. A visiting bee did not need to learn how to 'handle' the flower beyond the process of inserting its proboscis, and the inherent structure of the flower controlled pollen deposition onto the proboscis via its complex spatial arrangement of pollen and stigma.

Field trials did not involve individually marked bees, so no direct comparisons of SVD over time could be made. However for both flower species several visibly distinctive bees (based on wing-wear patterns and/or hair loss on dorsal thorax) did return multiple times to the flower patches within a day, but with no indication of improving SVD. This lack of effect is presumably attributable to the rather simple flower forms, where visitation for nectar extraction was a straightforward process of relatively constant and short duration (2.52 ± 0.06 s for *Echium* and 3.10 ± 0.11 s for *Geranium*). In more complex flowers handling time should indeed decrease, but whether this leads to an increase or a decrease in SVD is hard to predict: we previously found little or no relation between visit duration and SVD across 13 plant species (King, Ballantyne & Willmer 2013, though longer visits on particularly nectar-rich flowers may improve pollen deposition in some cases (e.g. Thomson & Plowright 1980). A bee that learnt to extract nectar more quickly but was not engaged in deliberate pollen collection might show decreasing SVD; one that was specifically gathering both nectar and pollen might become more efficient at pollen collection and show increased SVD even on shorter visits. Bumblebees are well known to specialise in either pollen or nectar trips on different flowers (Gonzalez et al. 1995; Goulson 2010), different days or different weather conditions (Peat & Goulson 2005). From the plant's point of view, then, the changes in pollinator performance could potentially be either positive or negative in effect.

D) Diurnal effects on size of visitors and resultant SVD

Visitor size at individual flowers varies on a diurnal basis, and there will also be seasonal effects and between-year effects at the level of individual flowers and between flowering communities (explicitly documented for *Geranium*, spp. by Kandori (2002)). Since visitor patterns vary in this way, we cannot accurately give a single SVD 'pollinator effectiveness' value for a given visitor to a given plant species, but must record the situation for a particular time and place.

This variation should also have significant effects for the plant's reproduction, and particular plants may be able to exploit it to improve their own cross-pollination. *Echium* flowers mainly dehisce in the early morning, and have pollen available primarily in the first half of their first day of opening; and across many plant species this pattern has been widely assumed to relate to attracting bees as pollinators (Shelly & Villalobos 2000; Castellanos et al. 2006; Willmer 2011) and so to be indicative of a bee-pollination syndrome. In practice, it may specifically be related to attracting the largest and most efficient bumblebees as pollinators. In contrast *Geranium* flowers are longer-lasting and in this study had pollen available for at least 24 hours, visitors collecting it with equal efficiency at any hour of the day; this pattern should be better suited to a more generalist flower, with open or bowl-shaped anatomy, targeted by many visitors (including those with shorter tongues), and able to achieve reproductive success from the services of many of them.

Conclusions

There is a significant positive correlation between body size and average pollen deposition for individual bees, and the smallest individuals may be unable to collect or deposit pollen onto the stigma of certain flowers. Individual body size, as well as specific behaviours on the flower and after leaving the flower (such as grooming), can result in significant variation in pollen deposition, so that it is inappropriate to give values of SVD for a particular visitor species on a particular flower without taking these factors into account. This is particularly true for visitors such as bumblebees, where intra-nest size variation is substantial at any one time and can be variable through a season (though with no consistent pattern; see Goulson & Sparrow 2009). This has numerous implications for colony and plant success. Bumblebee nests might benefit from producing larger workers, as these will be more efficient foragers and may visit a range of flowers inaccessible to smaller workers; however this must be offset against greater initial investment and the metabolic costs of larger individuals. Flowering plants also benefit from being visited by larger bumblebees, as more pollen grains may be collected during a single visit, and more may then be deposited at the next flower or next few flowers (useful so long as the number deposited does not exceed the number needed for full fertilisation of ovules). Potentially (as long as bees are moving between plants) this should improve cross-fertilization for the plant. Small forager bees, however, may be less efficient for both the colony and flowering plants. They are unable to collect as much nectar or pollen (Goulson et al. 2002), and may be too small to pollinate certain flowers at all.

Our results have marked implications for pollination biology. Generally in the current literature visiting organisms are 'ranked' on a species-level (or even a generic or family level) for their suitability for a given flower or flower type, and their true effectiveness as pollinators is rarely assessed. King, Ballantyne & Willmer (2013) showed that pollination ability and efficiency varies more within functional group, family or genus than commonly supposed; the results here argue that we must additionally consider individual-level variance in PE. This also accords with the findings of Tur et al. (2014) who showed that pollinator network structure can be affected by individual (within-species) variation associated with differing levels of generalization and specialization in flower visiting behaviours, although those authors did not link the effect to size differences.

From a plant's-eye-view, it is most beneficial to have visitors (even within species) that are appropriately size to maximise efficient pollen transfer. Flowers employ many methods to attract pollinators of the appropriate type or species (as described by pollination syndromes; see Willmer 2011) but may also modify their architecture and/or by the timing and pattern of reward presentation to draw in the 'best-fit' individuals. There is thus a particularly delicate relationship between an individual foraging bee and the flowers it visits; and the size of a bee determines not only its value to the nest as a forager, but also its PE value to the plants that it chooses to visit.

ACKNOWLEDGEMENTS

We are grateful for comments on the manuscript from Dr G Ballantyne, H Cunnold, and Prof G Ruxton (who also gave valued advice on statistical analyses), and from two anonymous referees. Funding for KF was provided by the School of Biology, University of St Andrews.

REFERENCES

Alford DV (1975) Bumblebees. Davis-Poynter, London.

Baron GL, Raine NE & Brown MJF (2014) Impact of chronic exposure to a pyrethroid pesticide on bumblebees and interactions with a trypanosome parasite. Journal of Applied Ecology 51:460-469.

Benton T (2006) Bumblebees. Harper Collins, London.

Bishop JA & Armbruster WS (1999) Thermoregulatory abilities of Alaskan bees: effects of size, phylogeny and ecology. Functional Ecology 13:711–724.

Butz Huryn VM (1997) Ecological impacts of introduced honeybees. Quarterly Review of Biology 72:275-297.

Cane JH & Sipes S (2006) Characterizing floral specialization by bees: analytical methods and a revised lexicon for oligolecty. In: Waser NM & Ollerton J (eds) Plant pollinator interactions: from specialization to generalization, Chicago University Press, Chicago, pp 99-112.

Castellanos MD, Wilson P, Keller SJ, Wolfe AD & Thomson JD (2006) Anther evolution: pollen presentation strategies when pollinators differ. American Naturalist 167:288-296.

Couvillon MJ & Dornhaus A (2009) Location, location, location: larvae position inside the nest is correlated with adult body size in worker bumble-bees (Bombus impatiens). Proceedings of the Royal Society B 276:2411-2418.

Couvillon MJ & Dornhaus A (2010). Small worker bumblebees (Bombus impatiens) are hardier against starvation than their larger sisters. Insectes Sociaux 57:193-197.

Darwin C (1861) Fertilization of Vincas. Gardeners' Chronicle 552

Dawson EH, Avargues-Weber A, Chittka L & Leadbeater E (2013) Learning by observation emerges from simple associations in an insect model. Current Biology 23:727-730.

del Castillo RC & Fairbairn D (2011) Macroevolutionary patterns of bumblebee body size: detecting the interplay between natural and sexual selection. Ecology & Evolution 2:46-57.

Durisko Z, Shipp L & Dukas R (2011) Effects of experience on short and long-term foraging performance in Bumblebees. Ethology 117:49-55.

Fjell I (1983) Anatomy of the xeromorphic leaves of Allamanda neriifolia, Thevetia peruviana and Vinca minor (Apocynaceae). Nordic Journal of Botany 3:383-392.

Fontaine C, Collin CL & Dajoz I (2008) Generalist foraging of pollinators: diet expansion at high density. Journal of Ecology 96:1002-1010.

Fryxell P (1957) Mode of reproduction of higher plants. Botanical Review 23:135-233.

Gonzalez A, Rowe CL, Weeks PJ, Whittle D, Gilbert FS & Barnard CJ (1995) Flower choice by honeybees (Apis mellifera L) – sex phase of flowers and preferences among nectar and pollen foragers. Oecologia 101:258-264.

Goulson D (2010) Bumblebees: behaviour, ecology, and conservation (2nd ed.) Oxford University Press, Oxford.

Goulson D, Peat J, Stout JC, Tucker J, Darvill B, Derwent LC & Hughes WOH (2002) Can alloethism in workers of the bumblebee, Bombus terrestris, be explained in terms of foraging efficiency? Animal Behaviour 64:123-130.

Goulson D & Sparrow KR (2009) Evidence for competition between honeybees and bumblebees; effects on bumblebee worker size. Journal of Insect Conservation 13:177-181.

Greenleaf S, Williams N, Winfree R & Kremen C (2007) Bee foraging ranges and their relationship to body size. Oecologia 153:589–596.

Harder L (1982) Measurement and estimation of functional proboscis length in bumblebees (Hymenoptera: Apidae). Canadian Journal of Zoology 60:1073-1079.

Heinrich B (1979) 'Majoring' and 'minoring' by foraging bumblebees, Bombus vagans; an experimental analysis. Ecology 60:245-255.

Heinrich B & Heinrich MJE (1983) Size and caste in temperature regulation by bumblebees. Physiological Zoology 56:552–562.

Horwood A (1919) British Wild Flowers – In Their Natural Haunts. Gresham Publ Co.

Inoue MI & Yokoyama J (2006) Morphological variation in relation to flower use in bumblebees. Entomological Science 9:147–159.

Jander R (1976) Grooming and pollen manipulation in bees (Apoidea): the nature and evolution of movements involving the foreleg. Physiological Entomology 1:179-194.

Kandori I (2002) Diverse visitors with various pollinator importance and temporal change in the important pollinators of Geranium thunbergii (Geraniaceae). Ecological Research 17:283-294.

Kapustjanskij A, Streinzer M, Paulus HF & Spaethe J (2007) Bigger is better: implications of body size for flight ability under different light conditions and the evolution of alloethism in bumblebees. Functional Ecology 21:1130-1136.

King C, Ballantyne G & Willmer P (2013) Why flower visitation is a poor proxy for pollination: measuring single-visit pollen deposition, with implications for pollination networks and conservation. Methods in Ecology & Evolution 4:811-818.

Klinkhamer PGL, de Jong TJ & Wesselingh RA (1991) Implications of differences between hermaphrodite and female flowers for attractiveness to pollinators and seed prodcution. Netherlands Journal of Zoology 41:130-143.

Laverty TM (1994) Bumblebee learning and flower morphology. Animal Behaviour 47:531-545.

Leadbeater E & Chittka L (2009) Bumble-bees learn the value of social cues through experience Biology Letters 5:310-312.

Macuda T, Gegear RJ, Laverty TM & Timney B (2001) Behavioural assessment of visual acuity in bumblebees (*Bombus impatiens*). Journal of Experimental Biology 204:559–564.

Mares S, Ash L & Gronenberg W (2005) Brain allometry in bumblebee and honey bee workers. Brain Behaviour & Evolution 66:50-61.

Morse D (1977) Estimating proboscis length from wing length in bumblebees (*Bombus* spp.). Annals of the Entomological Society of America 70:311-315.

Morse DH (1978) Size-related foraging differences of bumble-bee workers. Ecological Entomology 3:189-192.

Ne'eman G, Jürgens A, Newstrom-Lloyd L, Potts S, & Dafni A (2010) A framework for comparing pollinator performance: effectiveness and efficiency. Biological Reviews 85:435-451.

Peat J, Darvill B, Ellis J & Goulson D (2005) Effects of climate on intra- and interspecific size variation in bumble-bees. Functional Ecology 19:145-151.

Peat J & Goulson D (2005) Effects of experience and weather on foraging rate and pollen versus nectar collection in the bumblebee, *Bombus terrestris*. Behavioral Ecology & Sociobiology 58:152-156.

Peat J, Tucker J & Goulson D (2005) Does intraspecific size variation in bumblebees allow colonies to efficiently exploit different flowers? Ecological Entomology 30:176–181.

Persson AS & Smith HG (2011) Bumblebee colonies produce larger foragers in complex landscapes. Basic & Applied Ecology 12:695-702.

Plowright RC & Jay SC (1968) Caste differentiation in bumblebees (*Bombus latr.*: Hym.) I. The determination of female size. Insectes Sociaux 2:171–192.

Pouvreau A (1989) Contribution à l'étude du polyéthisme chez les bourdons, *Bombus Latr.* (Hymenoptera, Apidae). Apidologie 20:229-244.

Rademaker MCJ, de Jong TJ & Klinkhamer PGL (1997) Pollen dynamics of bumble-bee visitation on *Echium vulgare*. Functional Ecology 11:554-563.

Rademaker MCJ, de Jong TJ & van der Meijden E (1999) Selfing rates in natural populations of *Echium vulgare*: a combined empirical and model approach. Functional Ecology 13:828-837.

Raine N & Chittka L (2007). Pollen foraging: learning a complex motor skill by bumblebees (*Bombus terrestris*). Naturwissenschaften 94:459-464.

Roulston TH & Cane JH (2000) The effect of diet breadth and nesting ecology on body size variation in bees (Apiformes). Journal of the Kansas Entomological Society 73:129–142.

Sahli HF & Conner JK (2007) Visitation, effectiveness, and efficiency of 15 genera of visitors to wild radish, *Raphanus raphanismum* (Brassicaceae). American Journal of Botany 94:203-209.

Seeley TD (1982) Significance of the age polyethism schedule in honeybee colonies. Behavioral Ecology & Sociobiology 11:287-293.

Shelly TE & Villalobos E (2000) Buzzing bees (Hymenoptera: Apidae, Halictidae) on *Solanum* (Solanaceae): floral choice and handling time track pollen availability. Florida Entomologist 83:180-187.

Smeets P & Duchateau M (2003) Longevity of *Bombus terrestris* workers (Hymenoptera: Apidae) in relation to pollen availability, in the absence of foraging. Apidologie 34:333-337.

Spaethe J, Brockmann A, Halbig C & Tautz J (2007) Size determines antennal sensitivity and behavioural threshold to odors in bumblebee workers. Naturwissenschaften 94:733-739.

Spaethe J & Weidenmüller A (2002) Size variation and foraging rate in bumblebees (*Bombus terrestris*) Insectes Sociaux 49:142-146.

Stout JC (2000) Does size matter? Bumblebee behaviour and the pollination of *Cytisus scoparius* L. (Fabaceae). Apidologie 31:129-139.

Sutcliffe GH & Plowright RC (1988) The effects of food supply on adult size in the bumble bee *Bombus terricola* Kirby (Hymenoptera: Apidae). The Canadian Entomologist 120:1051–1058.

Thomson D (2004) Competitive interactions between the invasive European honey bee and native bumble bees. Ecology 85: 458-470.

Thomson J & Goodell K (2001) Pollen removal and deposition by honeybee and bumblebee visitors to apple and almond flowers. Journal of Applied Ecology 38:1032-1044.

Thomson JD & PlowrightRC (1980) Pollen carryover, nectar rewards and pollinator behaviour with special reference to *Diervilla lonicera*. Oecologia 46:68-74.

Tur C, Vigalondo B, Trojelsgaard K, Olesen J & Traveset A (2014) Downscaling pollen-transport networks to the level of individuals. Journal of Animal Ecology 83:306-317.

Waddington KD, Herbst LH & Roubik DW (1986) Relationship between recruitment systems of stingless bees and within-nest worker size variation. Journal of the Kansas Entomological Society 59:95–102.

Willmer PG (1983) Thermal constraints on activity patterns in nectar-feeding insects. Ecological Entomology 8:455-469.

Willmer PG (2011) Pollination and floral ecology. Princeton University Press, Princeton.

Willmer PG, Bataw AAM & Hughes JP (1994) The superiority of bumblebees to honeybees as pollinators: insect visits to raspberry flowers. Ecological Entomology 19:271-284.

Willmer PG & Stone GN (2005) Behavioural, ecological and physiological determinants of the activity patterns of bees. Advances in the Study of Behavior 34:347-466.

Worden BD, Skemp AK, & Papaj DR (2005) Learning in two contexts: the effects of interference and body size in bumblebees. Journal of Experimental Biology 208:2045-2053.

6

Dynamics of insect pollinators as influenced by cocoa production systems in Ghana

Eric A. Frimpong[1]*, Barbara Gemmill-Herren[2], Ian Gordon[3] and Peter K. Kwapong[1]

[1]Department of Entomology and Wildlife, School of Biological Sciences, University of Cape Coast, Cape Coast, Ghana
[2]AGPS – FAO, Viale Terme di Caracalla, Roma, Italy
[3]International Centre of Insect Physiology and Ecology (ICIPE), P.O. Box 30772-00100 Nairobi, Kenya

Abstract—Cocoa is strictly entomophilous but studies on the influence of the ecosystem on insect pollinators in cocoa production systems are limited. The abundance of cocoa pollinators and pod-set of cocoa as influenced by a gradient of farm distances from natural forest and proportion of plantain/banana clusters in or adjacent to cocoa farms were therefore investigated. Cocoa pollinators trapped were predominantly ceratopogonid midges hence, analyses were based on their population. Variation in farm distance to forest did neither influence ceratopogonid midge abundance nor cocoa pod-set. However, we found a positive relationship between pollinator abundance and pod set and the proportion of plantain/banana intercropped with cocoa. The results suggest appropriate cocoa intercrop can enhance cocoa pollination, and the current farming system in Ghana can conveniently accommodate such interventions without significant changes in farm practices.

Keywords: *Pollination, cocoa pod-set, ceratopogonid midges, plantain/banana, forest.*

Introduction

The estimated value of food crops directly consumed by humans attributed to insect pollination services in 2005 was US$ 153 billion, representing about 9.5% of total world production of human food (FAO, 2008). Insect exclusion experiments have shown that cocoa is strictly entomophilous and obligatorily requires insect pollinators (Cilas 1988; Ibrahim 1988; Posnette 1950). Klein et al (2007) have categorized cocoa among the 13 leading crops whose production would be reduced by over 90% in the absence of animal pollinators. Moreover, pollination in cocoa has been evaluated to be a higher order limiting factor in cocoa yield than agronomic resources (Groeneveld et al. 2010). Reports of decline in pollinator populations in agro-ecosystems and consequential decrease in food crop production (Ahmad et al. 2006; FAO 2008) suggest the languid nature of studies on natural pollination of cocoa should be intensified.

In recognition of the fact that pollination is a vital crop production factor, integrated crop production strategies are incorporating pollinator-friendly and conservation modules to enhance production. Landscape approaches have hitherto been the most frequently emphasized interventions, particularly through the conservation of native habitats (Aidoo 2008; Gemmill-Herren & Ochieng 2008; Klein et al. 2003a; Kremen et al. 2007). Studies focusing on bee pollinated crops such as melon (Kremen et al. 2002), grapefruit (Chacoff & Aizen 2006), eggplant (Gemmill-Herren & Ochieng 2008) and coffee (Klein et al. 2003a,b) show pollination services are influenced by gradients of distances between agricultural landscapes and natural forests.

It is not sure that this phenomenon holds at all in cocoa, which is pollinated by ceratopogonid flies (Posnette 1950; Young 1982a). However, South and Central American cocoa plantations have been postulated to be under-pollinated due to a shift from more diverse agroforest systems to simple cocoa monocultures (Young 1986). Young (1982a; 1986) subsequently postulated that critical associations exist between cocoa pollinators and natural forest, because cocoa in its native wild in the Amazon occurs as understory tropical tree distributed in aggregates along small streams.

As noted by Klein et al. (2008), studies on cocoa pollinators have centred on breeding substrates rather than the role of landscape matrices. Artificial introduction of slices of banana stems have been found to be a good breeding substrate for midges, which increases their population and pod-set in cocoa farms (Elizondo & Enriquez 1988; Young 1982b). Its practical application is however yet to be developed. Most newly established cocoa farms in Ghana are intercropped with plantain or banana as temporal shade cover, as staple food, and for income prior to and at initial fruiting stages (Acquaah, 1999). Evaluating the impact of plantain or banana stands on cocoa pollinators may help develop its mass application. This study therefore assessed two landscape features, the relative contribution of natural forest and proportion of cocoa and plantain/banana intercrop to cocoa pollination, which are familiar components of cocoa cropping practices in Ghana.

Materials and Methods

Study Areas and Farm Management

The study was carried out in 18 small scale (1.6 - 4.0 ha) farmer managed cocoa farms in three cocoa growing areas in

*Corresponding author
Email: nanakofy@yahoo.com

TABLE I: Characteristics of the experimental cocoa farms

Site	Focal forest reserve	Farm code	Farm distance from forest (km)	Distance category	Latitude	Longitude	Elevation (m)	Variety	Plantain or banana cluster/ha
Kubease-Wuraponso	Bobiri	A1	0	Adjacent	N06°40.899'	W001°21.217'	245	Hybrid, Amazonia	8.0
		A2	0	Adjacent	N06°41.550'	W001°21.859'	237	Amazonia, Amelonado	2.0
		B1	0.81	0.8 - 1.0	N06°40.621'	W001°21.240'	241	Amazonia	9.0
		B2	0.85	0.8 - 1.0	N06°40.452'	W001°20.777'	243	Amazonia, Amelonado	3.4
		C1	1.80	1.5 - 2.0	N06°40.163'	W001°20.528'	248	Hybrid, Amazonia	3.2
		C2	1.60	1.5 - 2.0	N06°39.947'	W001°20.321	231	Amazonia	0.8
Abrafo-Ebekawopa	Kakum	A3	0	Adjacent	N05°19.432'	W001°24.107'	161	Hybrid, Amazonia	2.8
		A4	0	Adjacent	N05°19.212'	W001°24.743'	174	Amazonia	1.2
		B3	1.00	0.8 - 1.0	N05°19.873'	W001°22.753'	128	Hybrid, Amazonia	≥10
		B4	0.93	0.8 - 1.0	N05°19.744'	W001°22.634'	127	Hybrid, Amazonia	2.0
		C3	1.90	1.5 - 2.0	N05°19.410'	W001°24.211'	159	Amazonia	3.6
		C4	1.95	1.5 - 2.0	N05°19.516'	W001°24.107'	171	Amazonia, Hybrid	9.0
Edwenease	Pra-Suhyen	A5	0	Adjacent	N05°14.486'	W001°29.619'	198	Hybrid, Amazonia	1.0
		A6	0	Adjacent	N05°14.575	W001°29.68'	124	Hybrid	2.0
		B5	0.97	0.8 - 1.0	N05°14.517	W001°29.140'	237	Hybrid, Amazonia	2.4
		B6	0.90	0.8 - 1.0	N05°14.348	W001°29.376'	231	Amazonia, Amelonado	0.9
		C5	1.55	1.5 - 2.0	N05°14.379	W001°28.744'	266	Amazonia	0.0
		C6	2.00	1.5 - 2.0	N05°14.619	W001°28.714'	265	Hybrid, Amazonia	9.4

Ghana. The areas are within the semi-deciduous rainforest belt with dual rainfall in April-July and September-November. These areas had scattered cocoa farms at varying distances from natural forest reserves. The areas were Kubease (Ashanti Region), Abrafo-Ebekawopa (Central Region) and Edwenease (Western Region) each of which has Bobiri, Kakum and Pra-Suhyen as focal forest reserves respectively (Tab. I). Farms were selected such that they fell within three specified distances of 0 km (adjacent), 0.8 - 1.0 km, 1.5 - 2.0 km) from the forest reserves. Farms were subsequently grouped under those three distances.

Varieties grown were Upper Amazon and hybrids. Farms were ten to twenty-five years old with varying plantain/banana intercrop distribution. Standing plantain/banana was included due to the observation by Young (1982a) that artificially provided banana stems favours breeding cocoa pollinating midges and for the fact that these crops are planted as temporary shade for cocoa in Ghana. Insecticides were sprayed monthly, from September to

December 2007, to conform to the recommended cocoa management practices in Ghana (Opoku et al. 2008). Weeds were manually cleared twice while parasitic mistletoe *Tapinanthus bangwensis* (Engl. and Krause) was pruned off the cocoa trees once within the study period.

Estimation of Pollinator Abundance

A pair of farms from each distance (adjacent, 0.8 - 1.0 km, 1.5 - 2.0 km) from a focal forest was selected from each site and every farm was divided into four quadrants (plots) of mean size (± SD) of 0.5 ± 0.1 ha. Three pollinator sampling methods described below were used concurrently at monthly intervals from April 2007 through April 2008 (Frimpong et al. 2009).

(1) Focal tree observation and sampling with motorized aspirator: Stratified tree sampling was used; 4 cocoa trees (1 per plot) with 10 – 20 open flowers were selected from each farm. It must be noted that flowers are scarcely available from August to November hence lesser number (average of 11.3 flowers) were sampled. A mean of 19.5 flowers were, however, sampled during normal flowering. Thus, trees were selected monthly based on availability of open-flowers on the lower section of the trunk. Open-flowers within 0.3 m above the soil to 1.3 m section of the trunk were observed for 10 minutes and all visiting insects were collected using a motorized suction pump. Samplings were conducted between 07:00 h and 11:00 h, and collected insects (except Lepidopterans) were preserved in 70% alcohol.

(2) UV-bright Painted Pan Traps (UVPPT): Another set of cocoa trees, 1 per plot, were randomly selected from each farm and marked. A set of UVPPT, comprising yellow, blue and white were filled to three-quarters full with soapy water and hung in the canopy of each experimental tree. Traps were removed after 48 hours and trapped insects were sieved off with muslin cloth, collected using fine camel hair brush and preserved.

(3) McPhail trap: A third set of 4 cocoa trees (1 per plot) were again randomly selected from each farm. Steam distilled cocoa floral oil was inoculated into cotton wool suspended in McPhail traps (Young et al. 1988). The trough of the traps were filled with soapy water and hung in the canopies of the third set of trees. Traps remained in the canopy for 48 hours and trapped insects were collected and preserved.

The three complementary methods were used in order to increase sampling efficiency because midge populations are generally low, especially during the dry season. Moreover, efficiency and ease of application of each method varies with respect to the vertical plane of the cocoa tree. Whilst the motorized aspirator easily and efficiently samples sections below the canopy, UVPPT and McPhail traps are more efficient at the canopy level (Frimpong et al. 2009).

Counts of ceratopogonid midges were made after they were sorted out in the laboratory using dipteran taxonomic identification key (Scudder & Cannings 2006). Some samples were then barcoded by Barcoding of Life Datasystems (BOLD), for further identification but specimens could only be identified to families (Appendix I; Anon 2008).

Estimation of Cocoa Pod-set

A fourth set of 4 cocoa trees per farm were randomly selected and 0.3 m - 1.3 m section of the trunk above the soil marked (Sarfo et al. 2003). All open flowers, cherelles (young pods) and pods were excised from the marked section on the first study month (April). Flower buds, open-flowers, cherelles (both viable and wilted) were counted at 30 day intervals (Appendix II). This interval was based on the 28 days that a flower bud takes to fully develop and open (Swanson et al. 2005) and the approximately 2 days survival span of open-flowers (McKelvie, 1962). To ensure that all new cherelles and ripe pods which might have been incidentally removed by farmers prior to a sampling date were counted, stalks of newly formed cherelles were carefully marked with permanent marker, and matured unripe pods were remarked during each sampling. Wilted cherelles and ripe pods were excised from the trees on each sampling occasion. This was to ensure that cherelles which wilted within the month were also counted. Monthly percent pod-set P_s of the cocoa trees was calculated as:

$$P_s = \frac{[(C_u + C_w + P_u + P_r) - (C_m + P_u)]100}{F_b + F_o}$$

Where C_u, unmarked cherelles for the month; C_w, wilted cherelles; P_u, unripe pods; P_r, ripe pods; C_m, previous months' cherelles; P_u, unripe pods; F_b. 95% of flower buds [according to McKelvie (1962), estimated 95% of flower buds become open flowers]; F_o, open flowers of the previous month.

Data Analysis

The numbers of midges and percent pod-set of cocoa were normalized through square root and arcsine transformations respectively, after testing for normality and homogeneity by plotting mean against variance (Gomez & Gomez 1984). A Multiple regression was run to determine the relationship between farm distance to forest, availability of plantain/banana clusters, and abundance of pollinators and cocoa pod-set, using Minitab release 13.2. All data were back-transformed to original scales before interpreting.

Data for farm distance to forest were re-categorised to adjacent, 0.8 - 1.0 km and 2.0 - 2.5 km and availability of plantain/banana to abundant (> 8 clusters) and scanty/absent (< 4 clusters). This allowed the dynamics of monthly pollinator populations and cocoa pod-sets under these landscape parameters to be analyzed.

RESULTS

Abundance of Cocoa Pollinators

Midges belonging to Ceratopogonidae and Cecidomyiidae families were the predominant cocoa flower visitors recorded for all the trapping methods, with the former being overly abundant (see details in Frimpong et al. 2009). The number of *Liotrigona parvula* Darchen (Hymenoptera: Apidae: Meliponini), the only recorded bee with the potential to pollinate cocoa (Frimpong et al. 2009) occurred in very low numbers — 38 individuals over the whole sampling period. Detailed analysis and discussion of the results presented here therefore focused on the ceratopogonids, conventionally acknowledged prime pollinators of cocoa (Kaufmann 1975; Posnette 1950; Young 1982a). The validity of this is justified by the significantly positive correlation between the number of ceratopogonids midges and cocoa pod-set (Fig. 1a).

The abundance of ceratopogonid midges did not correspond to variation in distance of farm from natural forest (Fig. 1b). Availability of plantain/banana, however, had a marked influence on the abundance of ceratopogonids. We obtained a positive association between ceratopogonid midges abundance and the number of plantain/banana clusters intercropped with cocoa (or within 50 m radius from farm; Fig. 1c). Thus, farms which had higher proportions of plantain/banana intercropped with cocoa or close by, had more abundant ceratopogonid midges compared to farms with no or scanty remnants of plantain/banana.

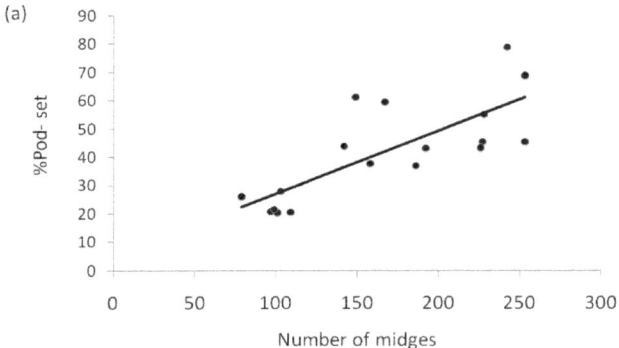

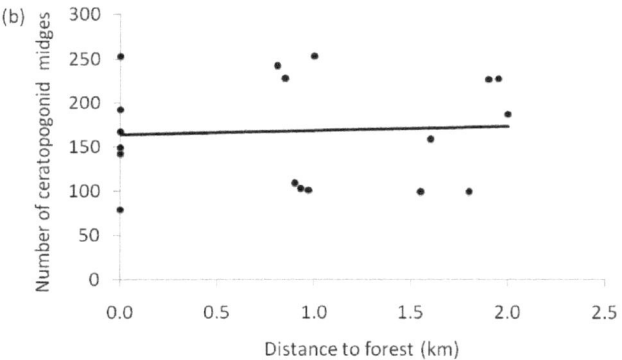

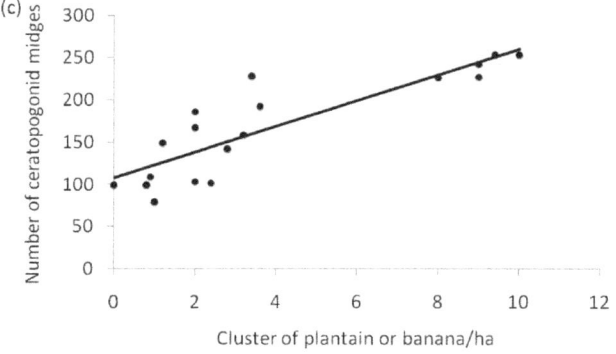

FIG. 1. Relationship between number of cocoa pollinating ceratopogond midges: (a) Pod-set of cocoa; $y = 4.839 + 0.221 x$, $r2 = 0.59$, n =18, f = 22.77, p < 0.001. (b) Distance of cocoa farm from forest; $y = 154.88 + 0.04 x2$, $r2 = 0.02$, n = 18, f = 0.02, p = 0.884. (c) Cluster of plantain/banana in or near cocoa farm; $y = 107.5 + 15.24 x$, $r2 = 0.75$, n = 18, f = 45.11, p < 0.001.

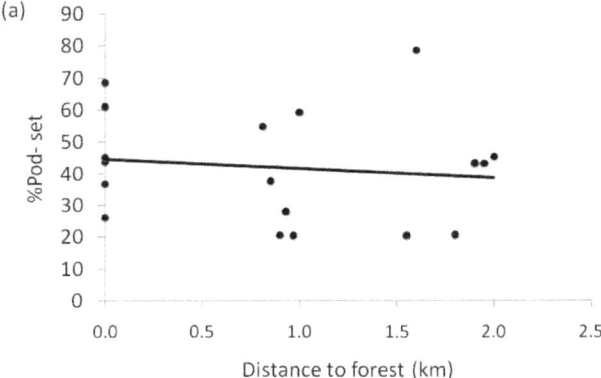

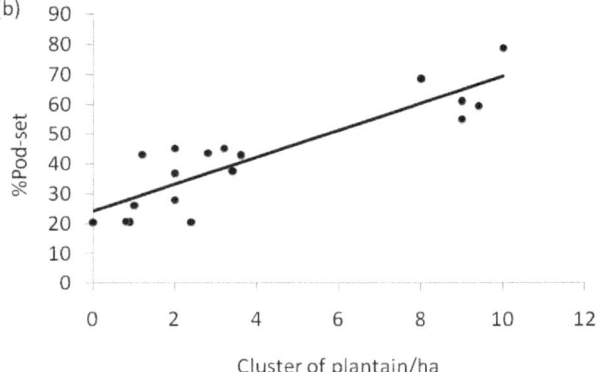

FIG. 2. Relationship between cocoa pod-set: (a) Distance of farm from forest: $y = 44.52 - 0.03 x2$, $r2 = 0.02$, n = 18, f = 0.38, p = 0.544. (b) Cluster of plantain/banana in or near cocoa farm: $y = 24.06 + 4.525 x$, $r2 = 0.78$, n = 18, f = 58.00, p < 0.001.

Cocoa Pod-set

The pod-sets were also independent of the proximity of the cocoa farms to natural forest (Fig. 2a). Nevertheless, we found a positive relationship between availability of plantain/banana clusters and pod-set (Fig. 2b).

Dynamics of ceratopogonid midge populations and pod-set

There were marked monthly variations in ceratopogonid populations and these corresponded with cocoa pod-sets. The dynamics of these two variables were similar under both varying farm distances from forest and availability of plantain/banana (Figs. 3a-b and 4a-b). We recorded high numbers of ceratopogonid midges and cocoa pod-set from June to November before dropping sharply to a minimum in February and March, under the two farm characteristics investigated. Abundance of ceratopogonid midges and pod-sets were comparable under the three distances relative to the focal forests (Figs. 3a and 4a). Farms with high proportions of plantain/banana intercropped with cocoa, however, exhibited significantly bigger ceratopogonid populations and higher cocoa pod-set throughout the season (Figs. 3b and 4b).

DISCUSSION

Clusters of wild cocoa found in the Amazon forest are postulated to provide the right proportion of cocoa

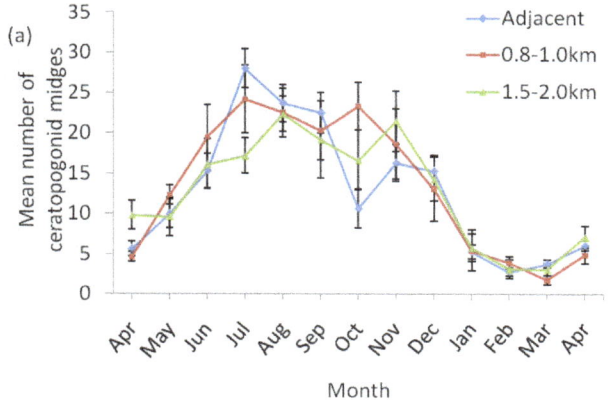

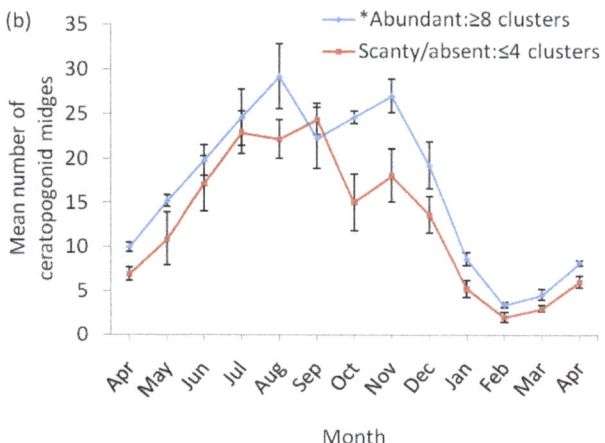

FIG. 3. Monthly population dynamics of ceratopogonid midges under: (a) Varying farm distance from natural forest. (b) Availability of plantain/banana [*see figs. 1c & 2, basis for re-categorizing to abundance and scanty/absent].

pollinating midges compared to large plantations with numerous flowers which have to be pollinated by relatively few midges. The insects thus get satiated by sheer abundance of resources in large cocoa monocultures (Young 1982a). This suggests that cocoa farms at close proximity to natural forests potentially have supplementary midge populations from the forest to enhance pollination, hence increased pod-set. Results obtained in this study however imply that natural forest adjacent to cocoa farms did not significantly increase the population of ceratopogonid midges and therefore cocoa pod-set. All the experimental farms, however, had secondary or regenerating forest patches close by (common to most cocoa farms in Ghana) which possibly offered resources and conditions similar to the focal primary forests. Moreover, 16 out of the 18 farms were established along small streams similar to cocoa stands in the native wild forests hence conditions would possibly be closely related, particularly in the wet season. Streams, however, dry out during the dry season and conditions might deviate during this period from those of the native wild, where streams are mostly perennial (Young 1982a). Although conditions suitable for efficient pollination of cocoa under agro system conditions are widely discussed (Brew 1988; Cilas 1988; Falque et al. 1995; Groeneveld et al. 2010; Ibrahim 1988; Kaufmann 1975;

Young 1982a,b; 1983; 1986) empirical data on the optimal pollination of wild cocoa in its native habitats are scarcely available. Cocoa is often designated as under-pollinated due to the small fraction of flowers pollinated (Paulin et al. 1985; Cilas 1988) although the proportion of pollinated flowers of the wild tree has not yet been established. It would be desirable to study pod-set rate of wild cocoa in the Amazon forest, as this will enlighten whether natural pollination deficits of the crop exist.

The results show availability of plantain/banana significantly influenced ceratopogonid midge population and pod-set. Remains of rotting plantain/banana stumps and stems after harvesting provide ideal breeding sites for midges (Elizondo & Enriquez 1988; Young 1982b) which explains high pollinator abundance with corresponding pod-set in farms where such substrates were available in substantial quantities. Most mature cocoa farms, however, does not benefit from cocoa/plantain or banana intercrop because plantain and banana planted as temporary shade crop at the early establishment of cocoa become stunted or die when the cocoa canopy closes. Development of a more persistent cocoa/plantain or cocoa/banana intercrop system will help augment population of pollinating ceratopogonids and therefore increase pod-set of cocoa. Nevertheless, some farmers fill wide gaps within mature cocoa farms with plantain as food crop and thus inadvertently increase

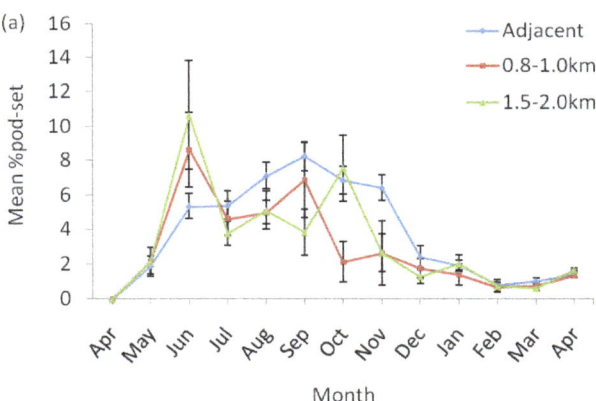

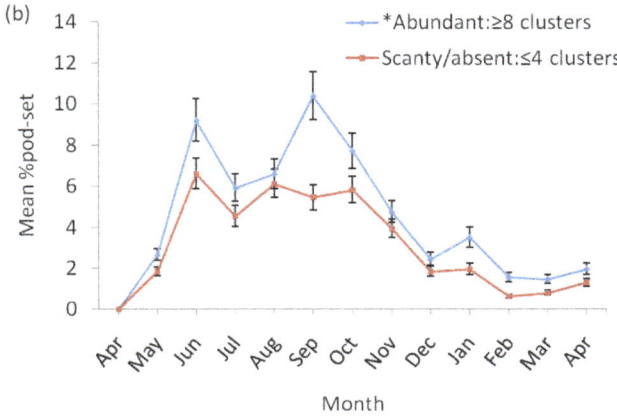

FIG. 4. Monthly dynamics of cocoa pod-set under: (a) Varying farm distance from natural forest. (b) Availability of plantain/banana [*see figs. 1c & 2].

pollination of their crop. Additionally, the plantain/banana could be planted at boundaries of already established cocoa farms.

Unlike the farm-forest proximity experiment where no clear differences in pollinator population and pod-set were recorded, the differences between farms with abundant and scanty plantain/banana were significantly higher in both the rainy (June through November) and dry periods (December-March). This corroborates other findings that rotten banana stems sustain adequate moisture for the midges to breed in during the dry season (Young 1982b). Standing plantain and banana in cocoa farms could thus be manipulated to augment midge population in the dry season. This is particularly important because cocoa tends to produce flowers profusely around this period and at the on-set of rains, at which time midge population is minimal (Frimpong et al. 2009; McKelvie, 1962).

Conclusion

Results from the landscape features studied suggest farm practices such as small scale farming and intercropping cocoa with plantain/banana favours pollination services in the cocoa agro-ecosystem. We identified a relationship between standing plantain/banana and cocoa pollination, and therefore a proper spatial cocoa-plantain/banana intercrop outlay will help boost pollination of the crop. Nearness of farm to natural forest did not offer any pollination advantage and that secondary forest patches surrounding cocoa farms possibly offer pollinator resources similar to that of natural forest. We suggest that assessment of natural pollination of wild cocoa in its native forest habitat and also pollination under large cocoa plantations with varying vegetation interface should be conducted.

Acknowledgements

The authors are grateful to the Dutch Programme for Cooperation with International Institutions (Netherlands SII) for funding this research through ICIPE, and Entomology and Wildlife Department, University of Cape Coast for hosting the research. We also thank Dr. Laurence Parker (York University, Toronto) and Dr. Daniel Masiga (ICIPE, Nairobi) for the barcoding, and farmers whose farms were used. This research has been part of FAO Project on "conservation and management of pollinators for sustainable agriculture, through an ecosystem approach", supported by the Government of Norway, United Nations Environment Programme UNEP and Global Environment Facility (GEF).

Appendices

Additional supporting information may be found in the online version of this article:

Appendix I-Taxon data
Appendix II. Monthly cocoa flower production

References

Acquaah B (1999) Cocoa development in West Africa: Early development with special reference on Ghana. Ghana Universities Press, Accra. 187 pp.

Ahmad F, Banne S, Castro M, Chavarria G, Clarke J, Collette L, Eardley C, Fonseca V, Freitas BM, French C, Gemmill-Herren B, Griswold T, Gross C, Kwapong P, Lundall-Magnuson E, Medellin R, Partap U, Potts SG, Roth D, Ruggiero M, Urban R, Willemse G (2006) Pollinators and Pollination: A resource book for policy and practice. Eardley C, Roth D Clarke J. Buchmann S, Gemmill-Herren B. (eds.), African Pollinator Initiative, Pretoria.

Aidoo SK (2008) Pollination and cashew (Anarcadium occidentale L) production in Ghana. PhD Thesis, University of Cape Coast. 162p.

Anon (2008) http://www.boldsystems.org/views/taxbrowser. php?taxid=567. (Last accessed: June, 2011).

Brew AH (1988) Cocoa pod husk as breeding substrate for forcipomyia midges and related species which pollinate cocoa in Ghana. Cocoa Growers Bulleting 40: 40-42.

Chacoff NP, Aizen MA (2006) Edge effects on flower-visiting insects in grapefruit plantations bordering premontane subtropical rainforest. Journal of Applied Ecology 43: 18-27.

Cilas C (1988) Study of natural cacao pollination in Togo and its implication for production. Proceedings of the 10th International Cocoa Research Conference 1987, Santo Domingo, pp283-286.

Elizondo JE, Enriquez GA (1988) Evaluation of 12 different types of musaceae as breeding sites for the cacao pollinating insects (Forcipomyia spp.) in shade and full sun at La Lola, Costa Rica. Proceedings of the 10th International Cocoa Research Conference 1987, Santo Domingo, pp297-302.

Falque M, Vincent A, Vaissiere BE, Eskes AB (1995) Effect of pollination intensity on fruit and seed set in cacao (Theobroma cacao L.). Sexual Plant Reproduction 8 (6): 354 – 360.

FAO (2008) Rapid Assessment of Pollinators' Status. Food and Agriculture Organization, Rome, Italy.

Frimpong EA, Gordon I, Kwapong PK, Gemmill-Herren B (2009) Dynamics of cocoa pollination: tools and applications for surveying and monitoring cocoa pollinators. International Journal of Tropical Insect Science 29: 62-69.

Gemmill-Herren B, Ochieng AO (2008) Role of native bees and natural habitats in eggplant (Solanum melongena) pollination in Kenya. Agriculture Ecosystem and Environment. 127: 31-36.

Gomez KA, Gomez AA (1984) Statistical procedures for agricultural research. 2nd edition. John Wiley and Sons Inc. New York.

Groeneveld JH, Tscharntke T, Moser G, Clough Y (2010) Experimental evidence for stronger cacao yield limitation by pollination than by plant resources. Perspectives in Plant Ecology, Evolution and Systematics 12:183-191.

Ibrahim GA (1988) Effects of insect pollinators on fruit set of cocoa flowers. Proceedings of the International Cocoa Research Conference 1987, Santo Domingo: 303-306.

Kaufmann T (1975) Studies on the ecology and biology of a cocoa pollinator, Forcipomyia squamipennis I. and M. (Diptera, Ceratopogonidae), in Ghana. Bulleting of Entomological Research 65: 263-268

Klein AM, Stefan-Dewenter I, Tscharntke T, (2003a) Bee pollination and fruit set of Coffea arabica and C. canephora (Rubiaceae). American Journal of Botany 90: 153-157.

Klein AM, Stefan-Dewenter I, Tscharntke T (2003b) Pollination of Coffea canephora in relation to local and regional agroforestry management. Journal of Applied Ecology 40: 837 – 845.

Klein AM, Vaissie're BE, James H. Cane JH, Steffan-Dewenter I, Cunningham SA, Kremen C, Tscharntke T (2007) Importance of pollinators in changing landscapes for world crops. Proceedings of the Royal Society B: Biological Sciences 274: 303-313.

Klein AM, Cunningham SA, Bos M, Steffan-Dewenter I (2008). Advances in pollination ecology from tropical plantation crops. Ecological Society of America 89: 935-943.

Kremen C, Williams NM, Thorp RW (2002) Crop pollination from native bees at risk from agricultural intensification. Proceedings of the National Academy of Science 99: 16812-16816.

Kremen C, Williams NM, Aizen MA, Gemmill-Harren B, LeBuhn G, Minckley R, Packer L, Potts SG, Roulston T, Steffan-Dewenter I, Vazquez DP, Winfree R, Adams L, Crone EE, Greenlead SS, Keitt TH, Klein AM, Regetz J, Ricketts TH (2007) Pollination and other ecosystem services produced by mobile organisms: a conceptual framework for the effects of land-use change. Ecology Letters 10: 299-314.

McKelvie AD (1962) Cocoa physiology. In: Willis JB (ed) Agriculture and land use in Ghana. Oxford University Press, London, pp256-260.

Opoku IY, Frimpong-Ofori K, Sarfo JE (2008) The role of the national cocoa diseases and pests control (CODAPEC) and the hi-tech programmes in sustainable cocoa economy. In: Owusu GK (ed) Plenary presentations, 25th Biennial Conference of Ghana Science Association 2007, Tafo/Bunso, pp66-76.

Paulin D, Decazy B, Coulibaly N (1985) Seasonal variations in pollination and pod production in a cacao smallholding in Ivory Coast. Proceedings of the 9th International Cocoa Research Conference, 1984, Lome, pp549-556.

Posnette AF (1950) The pollination of cacao in the Gold Coast.

Journal of Horticultural Science 25: 155 – 163.

Sarfo JE, Padi B, Oppong FM, Opoku IY, Akrofi AY (2003) Effects of two herbicides and four fungicides on insect pollination of cocoa. Proceedings of the 14th International Cocoa Research Conference 2, 1983, Accra, pp1387-1392.

Scudder GGE, Cannings RA, (2006) Diptera families of British Columbia. http://ditera.info/download (Last accessed on March 2007).

Young AM (1982a) Population biology of tropical insects. Plenum Press, New York.

Young AM (1982b). Effects of shade cover and availability of midge breeding sites on pollinating midge populations and fruit set in two cocoa farms. Journal of Applied Ecology 19: 47-63.

Young AM (1983) Seasonal differences in abundance and distribution of cocoa pollinating midges in relation to flowering and fruit set of between shaded and sunny habitat of the La Lola cocoa farms in Costa Rica. Journal of Applied Ecology 20: 801-831.

Young AM (1986) Habitat differences in cocoa tree flowering, fruit-set and pollinator availability in Costa Rica. Journal of Tropical Ecology 2: 163-186.

Young AM, Erickson BJ, Erickson EH (1988) Steam distilled floral oils of *Theobroma* sp. (Sterculiaceae) as attractants to flying insects during dry and wet seasons in a Costa Rican cocoa plantation. Proceedings of the 10th International Cocoa Research Conference 1987, Santo Domingo, pp289-296.

7

Pollinator dependency, pollen limitation and pollinator visitation rates to six vegetable crops in Southern India

Saranya Arwen Carr and Priya Davidar*

Department of Ecology and Environmental Sciences, Pondicherry University, Kalapet, Pondicherry 605014, India

Abstract—We investigated levels of pollinator dependency and pollinator visitation rates to flowers of six vegetable crops: brinjal (aubergine), tomato, chilli pepper (Solanaceae), okra (Malvaceae), bitter and snake gourds (Cucurbitaceae) in six small family farms in the Coimbatore region of southern India. We tested the null hypothesis that fruit set in these crops would be independent of pollinators. We assessed fruit set through self and cross pollination by pollen augmentation, by pollinator exclusion and open pollination. We evaluated pollen limitation by comparing percentage fruit set by hand outcrossed pollen with open pollination; pollinator dependency by differences in percentage fruit set by open pollination and autogamous pollination; and visitation rates to flowers by pollinating insects. Tomato, chilli and okra produced self-compatible hermaphrodite flowers, with higher levels of autogamous fruit set (32-76%) and significantly lower levels of pollinator dependency (0-37%), whereas andromonoecious brinjal and monoecious gourds had significantly lower levels of fruit set through autogamy, and higher levels of pollinator dependency. Pollen limitation was not evident in any crop. Diverse pollinating insects visited the flowers, and the frequency of visits by different pollinator taxa differed with crop type. Native vegetation and uncultivated land may enhance pollinator diversity in small farms.

Keywords: Agro-biodiversity, Coimbatore, India, pollinator dependency, pollination services, vegetable crops

INTRODUCTION

Pollinators are critical for the reproduction of many plants, and about 94% of plant species in tropical communities and a third of global food crops are likely to rely on animal pollination (McGregor 1976; Klein et al. 2007; Ollerton et al. 2011). Aizen et al. (2009) suggested that in the absence of pollinators, reduction in crop production would range from 3-8% (Allen-Wardell et al. 1998; Kearns et al. 1998; Klein et al. 2007; Garibaldi et al. 2013). Tropical agriculture could be particularly susceptible to pollinator declines since the cultivation of pollinator dependent crops, and the use of pesticides has increased (Roubik 1995; Aizen et al. 2009). As proximity to forests has been shown to increase pollinator activity and enhance crop production (De Marco & Coelho 2004; Ricketts et al. 2004, 2008; Blanche et al. 2006), tropical deforestation could further imperil pollination services (Bradshaw et al. 2009). In the Indian subcontinent plantation crops such as cardamom and coffee and many of the vegetable and fruit crops are dependent on wild bees for pollination (Partap 1999; Chandel et al. 2004; Sinu & Shivanna 2007a, b; Davidar 2009; Krishnan et al. 2012; Davidar et al. 2015). The Indian green revolution which started in the 1960's with the introduction of high yielding varieties and the intensive use of organophosphates, carbamates, synthetic pyrethroids

and organochlorine pesticides (Kumari et al. 2002; Bhanti & Taneja 2007; Roy et al. 2007), could have adversely affected pollinating insects thereby reducing pollination services (Basu et al. 2011).

The aim of this study is to document and assess levels of pollinator dependency, pollen limitation and identify major pollinating insects visiting flowers of six vegetable crops in the Coimbatore district of Tamil Nadu (Tab. 1, Fig. 1), a major industrial and agricultural region in southern India. The six crops were widely used vegetables such as aubergine known in India as brinjal (*Solanum melongena* L., 1753), tomato (*Solanum lycopersicum* L., 1753), and chilli pepper (*Capsicum annuum* L., 1753) of the family Solanaceae; tropical gourds such as the bitter gourd (*Momordica charantia* L., 1753) and snake gourd (*Trichosanthes cucumerina* var. *anguina* (L.) Haines, 1922) belonging to Cucurbitaceae; and okra (*Abelmoschus esculentus* (L.) Moench, 1794), a member of the Malvaceae.

We used hand pollination experiments to assess levels of pollinator dependency and observed flowers to record visitation rates by pollinating insects. We tested the null hypothesis that fruit set in these six crops, regardless of sexual systems would be independent of pollinators.

MATERIALS AND METHODS

Study area

The Coimbatore district of the southern Indian state of Tamil Nadu has a dry and hot climate with annual ambient

---

*Corresponding author; email: pdavidar@gmail.com

TABLE 1. Description of the farms where the study was conducted with geographical coordinates, area of farm, type of agriculture, irrigation mode, matrix vegetation type and crops studied.

Site name	Latitude °N	Longitude °E	Farm size (km²)	Conventional/ Organic	Type of irrigation	Matrix type	Crops studied
Edayarpalayam	11.22	76.55	0.5	Conventional	Bore well, Municipality Piped water	Shrubbery	Bitter gourd, snake gourd
Karumathampatti	11.07	77.11	0.5	Conventional	Bore well	Pastures and shrubbery	Brinjal, bitter gourd, snake gourd
Madukarai	10.54	76.55	0.04	Organic	Bore well	Grassland, coconut farms	Tomato
Nehru Nagar	11.33	77.24	0.004	conventional	Municipality Piped water	Urban housing unit	Bitter gourd, snake gourd
Pappampatti	10.56	77.65	0.4	Conventional	Bore well, rain fed	Pastures, open woodland	Chili, brinjal, tomato, bitter gourd
Sirumugai	11.22	77.05	0.02	Organic/ conventional	River	Dry forest	Chili, okra

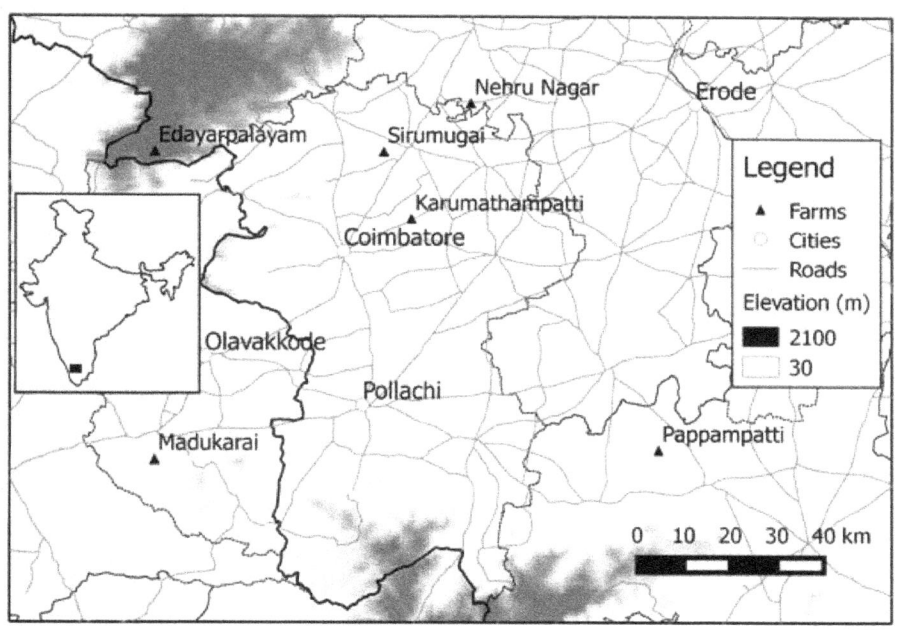

FIGURE 1. Map of the study site in Coimbatore indicating location of the farms.

temperature ranging from 18° to 35° C. Average annual rainfall is around 700 mm from the SW and NE monsoons. The farmlands have a rich fertile soil composed of mostly alfisols and vertisols where grain, pulses and vegetable crops are grown (Surendran & Murugappan, 2008).

Six small scale family farms were selected for the study in the following localities: Edayarpalayam, Karumathampatti, Nehru Nagar, Papampatti, Sirumugai, and Madukarai (Tab. 1; Fig. 1). Each farm has a different mix of crops and often the variety grown and timing of crop production differed, which made it difficult to standardise a study design. All the farms were irrigated and used pesticides except for two organic farms in Sirumugai and Madukkarai. Five of the farms had shrubbery, forests, grasslands and pastures in the surrounding matrix, whereas one farm in Nehru Nagar was located in an urban area (Tab. 1). Small scale agriculture is widely practised across India and large scale intensive agriculture is less common due to regulations on land holdings.

The study was conducted from December 2011 to June 2012, from November to December 2012, and from January to March 2013 during the main growing season for vegetables. The vegetable varieties cultivated in the farms (Tab. 2) were developed by the Tamil Nadu Agricultural University and the Sugarcane Institute at Coimbatore. We classified each crop according to its sexual system.

TABLE 2. Results of the pollination experiments indicating the fruit set [in % with number of investigated flowers in brackets] by hand augmented cross (CP) and self pollination (SP), open pollination (OP) and autogamy (AUT). Levels of pollinator dependency (PD), index of self incompatibility (ISI) and pollinator limitation (PL) in six crops. The decimal values have been rounded to the nearest whole number. Abbreviations of site names: Edyarpalayam (E), Karumathampatti (K), Madukarai (M), Nehru Nagar (N), Papampatti (P), Sirumugai (S).

Name of crop	Farm	Month-Year	Variety	CP	SP	OP	AUT	% PD (%OP-%AUT)	ISI	% PL
Brinjal (*Solanum melongena*)	E	Nov-12	Shiva	30 (23)	25 (24)	39 (31)	9 (43)	29	I	0.8
	N	Dec-12	Aruki25	9 (11)	0 (9)	27 (15)	6 (17)	21	0	0.3
	P	Dec-12	Shiva	20 (15)	0 (11)	33 (21)	7 (14)	26	0	0.6
	P	Feb-13	Shiva	50 (34)	50 (22)	48 (50)	6 (17)	42	I	1.0
	K	Mar-13	Kathri25	38 (29)	23 (31)	42 (43)	10 (51)	32	I	0.9
	Total			35 (112)	25 (97)	41 (160)	8 (142)	30		
Tomato – (*Solanum lycopersicum*)	P	Jan-12	Lakshmi 5005	68 (22)	60 (20)	58 (24)	76 (34)	0	I	1.2
	M*	Feb-12	NS 25	67 (24)	88 (24)	27 (26)	32 (57)	0	I	2.5
	Total			67 (46)	75 (44)	42 (50)	48 (91)	0		
Chili – (*Capsicum annuum*)	P	Nov-11	S7	86 (14)	-	86 (14)	-	-	I	1.0
	S*	Mar-12	Sannam	50 (20)	50 (28)	60 (15)	36 (28)	24	I	0.8
	Total			65 (34)	50 (28)	72 (29)	36 (28)	36		
Bitter gourd – (*Momordica charantia*)	N	Nov-12	Neelam105	13 (16)	23 (13)	11 (9)	0 (11)	11	I	1.1
	P	Nov-12	Neelam105	0 (7)	11 (9)	8 (12)	0 (9)	8	0	0.6
	N	Jan-13	Neelam105	44 (25)	57 (21)	76 (17)	0 (20)	76	I	0.8
	P	Jan-13	Neelam105	26 (34)	29 (38)	33 (43)	0 (51)	33	I	0.8
	K	Mar-13	Raja	34 (56)	29 (55)	43 (44)	0 (54)	43	I	1.1
	Total			30 (138)	32 (136)	38 (125)	0 (145)	38		
Snake gourd – (*Trichosanthes cucumerina var. anguina*)	N	Nov-12	Lakshmi7	30 (27)	38 (24)	41 (32)	11 (19)	30	I	0.7
	P	Nov-12	S25	17 (12)	14 (14)	11 (9)	0 (10)	11	I	1.5
	P	Jan-13	S25	41 (41)	38 (37)	50 (46)	0 (32)	50	I	0.8
	K	Mar-13	Bhuvan	23 (40)	30 (37)	36 (55)	3 (33)	33	I	0.6
	Total			30 (120)	32 (112)	40 (142)	3 (94)	37		
Okra – (*Abelmoschus esculentus*)	S	Mar-12	Shakthi	38 (40)	32 (34)	55 (40)	35 (20)	20	I	0.68

* organic

Pollination treatments

We evaluated the breeding systems of the plants, levels of autogamy and pollinator dependency using hand pollination experiments. Plants were randomly selected in a field, and 1-3 flowers were selected for hand pollinations on target plants. Usually just one flower was used for the pollination treatment. However, in some cases based on the size of the plant, 2-3 flowers were used for different treatments, one of which was to document fruit set through open pollination (OP). The mode of collection of pollen varied for each crop type. For tomatoes, brinjal and okra, the stamens are fused into a cylinder around the pistil, and therefore a razor blade was used to slit the cylinder on one side and around the base, and remove it using tweezers. For chilli, the stamens were separated and removed using curved scissors and tweezers. The pollen was collected using a fine brush or forceps for deposition onto the target stigma. Cotton mosquito mesh cloth bags of 15 cm in length and 7.5 cm in width were used to exclude pollinators after treatment of the flowers and tagged. Fruit set was monitored until maturity.

The following pollen augmentation treatments were used to assess the extent of selfing and outcrossing. For the self pollination treatment (SP), pollen from the dehisced anthers of a freshly opened flower was transferred onto the receptive stigma of the same flower in the case of hermaphrodites and andro-monoecious crops, or onto the stigma of another isolated flower on the same plant in monoecious crops, and the flowers were re-bagged. For augmented hand cross pollination treatment (CP), the flower buds were teased opened the previous evening and the anthers excised to prevent self pollination. The buds were enclosed and tagged. The next morning fresh pollen from flowers of a different plant was collected and deposited on the stigma. To test for parthenocarpy or autogamous self pollination (AUT), a tagged flower bud was enclosed in a mosquito mesh bag. For open pollination, fresh flowers were tagged and fruit set noted.

Breeding system, pollinator dependency and pollen limitation

The breeding system of the crop varieties was assessed using the index of self-incompatibility (ISI) which is the ratio between % fruit set from augmented hand self pollination over augmented hand cross pollination (Zapata & Arroyo 1978). Crops with ratios < 0.25 were considered self incompatible and those > 0.25 as self compatible (Bawa 1974).

Pollinator dependency (PD) was estimated by subtracting the % fruit set by autogamous pollination from % fruit set by open pollination (PD = OP - AUT, Tur et al. 2013). PD ranges from 0 for plants that are not dependent on pollinators to 100 for plants that completely rely on pollinators for fruit production. Negative values were represented by 0 = no pollinator dependency.

Pollen limitation was assessed as the ratio of % fruit set from augmented hand cross pollination by open pollination (Larson and Barrett 2000; Rathcke 2000). The scale was from 0 to 100 with 0 indicating no pollen limitation and 100 indicating complete pollen limitation. Negative values resulted where in some cases % fruit set from OP was higher than that of % from CP and was represented as 0 which indicated no pollen limitation.

Pollinator Visitation Rates

Usually the study crop was grown in only one field. Flowering plants were selected for observation in different areas of the field to be representative of site conditions. One or two freshly opened flowers in a plant were observed continuously for 5-minute blocks separated by intervals of 5-minutes, for a total of 30 minutes and all flower visitors were recorded. Then the observations were continued on a different plant. Observations were carried out from dawn (0700 hours) till dusk (1800 hours), except for okra flowers that closed at around 1500 hours. All insects visiting the flowers and touching the reproductive parts were noted; however, it was not possible to definitively ascertain whether pollen was transferred. The visitation data was summed for

that field in a farm totalling minutes of observation and converting to hours of observation. The visitation rate (mean ± standard deviation) was estimated by dividing the number of visits by each pollinator group by the total hours of observation in that field.

The pollinators were grouped into the following categories: *Apis* honeybees, Meliponine bees such as *Heterotrigona*, solitary bees of the genus *Xylocopa* and *Amegilla*, Lepidoptera which included butterflies and moths, and wasps of different families. The overall average visitation rate for a crop was also calculated using the data from individual farms.

Data analysis

We used non parametric statistical analysis to test for differences between fruit set from the pollination treatments between crops, and visitation rates of different pollinator taxa. All analyses were conducted using SYSTAT (SPSS 2000).

RESULTS

The flowers of all the crops opened in the morning hours and closed in the late afternoon, except in okra which closed earlier in mid afternoon. Tomato, chilli and okra were hermaphroditic with both male and female sexual parts in the same flower, the brinjals were andro-monoecious with male and hermaphrodite flowers on the same plant and the gourds were monoecious with male and female flowers on the same plant. Pollen was the major reward for all the 6-crops and only okra produced nectar.

Fruit set, breeding systems and pollinator limitation

All crops were self-compatible except for two cases in brinjal and one in bitter gourd (Tab. 2). In many cases the fruit set through CP was lower than that resulting from OP, which could be due to damage to floral parts while emasculation, or poor quality pollen. Pollinator dependency was 26 ± 21 on average, except for tomato where it was 0. PD ranged from 8-76% across all crop varieties except tomato (Tab. 2).

Fruit set through OP ranged from 40-72% and did not differ with crop type (Kruskal-Wallis One Way ANOVA: F = 0.52, $N = 18$, $P = 0.72$ excluding chilli due to inadequate data), although levels of autogamy significantly differed (Kruskal-Wallis One Way ANOVA: F = 13.99, $N = 18$, $P = 0.0001$). Monoecious/andromonoecious crops had significantly lower autogamous fruit set than hermaphrodite crops (MWU test: Z = 6, $N = 19$, $P = 0.009$), higher levels of pollinator dependency (MWU test: Z = 70, $N = 18$, $P = 0.001$), and marginally higher levels of fruit set through hand augmented self pollination (MWU test: Z = 56.5, $N = 19$, $P = 0.05$) indicating potential for geitonogamous pollination (Tab. 2). Pollen limitation was not evident in most crops except for brinjal in one site (Tab. 2).

Pollinators

The study identified five major flower visiting taxa: Social bees of the family Apidae: *Apis cerana* Fabricius 1793,

TABLE 3. Visitation rates (total visits/hour) of different pollinator taxa to each crop in farms in the Coimbatore region. The decimal values have been rounded to the nearest whole number. Abbreviations of site names: Edyarpalayam (E), Karumathampatti (K), Madukarai (M), Nehru Nagar (N), Papampatti (P), Sirumugai (S).

Name of crop	Farm	Month-Year	Sample sizes			Visits per hour (mean ± SD)				
			No of hours	No of plants	No of flowers	Apis spp	Stingless bees	Solitary bees	Butterflies Moths	Wasps
Brinjal (Solanum melongena)	E	Nov-12	13	23	57	0.5 ± 0.7	0	3	1.5 ± 0.7	0
	N	Dec-12	20	74	198	2.7 (0.6)	1 ± 1	3 ± 2	2.7 ± 0.6	0
	P	Feb-13	21	364	916	1 ± 1	1.3 ± 1.2	8 ± 1	6 ± 2.6	1
	K	Mar-13	15	54	135	3 ± 4	0.5 ± 0.7	2.5 ± 0.7	9 ± 7	2.5 ± 0.7
	Total		69	515	1306	1.8	0.7	4.1	4.8	0.9
Tomato (Solanum lycopersicum)	P	Jan-12	78	212	532	2.2 ± 0.9	1.3 ± 1.2	2 ± 1.1	6.5 ± 2.2	1.5 ± 1.8
	M*	Mar-Jun-12	53	106	537	9.3 ± 7.4	0.9 ± 1.5	1.1 ± 1.2	5.3 ± 2.4	0.2 ± 0.4
	Total		131	318	1069	5.8	1.1	1.6	5.9	0.9
Chili (Capsicum annuum)	P	Dec-11	53	206	516	8.4 ± 3.4	0	0	5.9 ± 2	0.6 ± 0.8
	S*	Feb-12	29	91	217	15 ± 4	0	0	4.8 ± 2.3	0
	Total		82	297	733	11.7	0.00	0.00	5.4	0.3
Bitter gourd (Momordica charantia)	N	Nov-12	19	45	199	5.3 ± 1.5	4 ± 1	0	9 ± 2	0.3 ± 0.6
	P	Dec-12	12	506	1256	7 ± 2.7	1 ± 1	3.3 ± 1.2	12.3 ± 3.5	0.3 ± 0.6
	K	Mar-13	18	64	168	10.3 ± 1.5	1.3 ± 0.6	0	12 ± 1	0.3 ± 0.6
	Total		49	615	1623	7.5	2.1	1.1	11.1	0.3
Snake gourd (Trichosanthes cucumerina var. anguina)	P	Jan-13	16	76	266	12.5 ± 0.7	2.5 ± 0.7	5.5 ± 0.7	15.5 ± 3.5	3.5 ± 0.7
	K	Mar-13	29	58	297	7 ± 2.6	2.3 ± 1.3	3 ± 0.8	20.3 ± 3.9	0
	N	Nov-12	21	42	213	1 ± 0.1	0.5 ± 0.2	0	1.8 ± 0.2	0.2 ± 0.2
	Total		66	176	776	6.8	1.8	2.8	12.5	1.2
Okra (Abelmoschus esculentus)	S	Aug-Oct-12	47	193	496	10.1 ± 3.5	0.1 ± 1.4	3.4 ± 1.8	± 1.5	0

*organic

A. florea Fabricius 1787, and A. dorsata Fabricius 1793; the stingless bee Heterotrigona iridipennis Smith 1854 (Meliponinae); solitary bees, Xylocopa Latreille, 1802, Ceratina Latreille, 1802 and Amegilla Friese, 1897 (Anthophorini); Lepidoptera which included butterflies and moths; and wasps of the families Sphecidae, Braconidae, Chalcididae and Vespidae (Ollerton et al. 2014) (Tab. 3). The sweat bees of the family Halictidae represented by Nomia Latreille, 1804, and Syrphid flies (Syrphidae) were minor visitors (Carr 2012).

Average rates of visitation to crops differed significantly between pollinator taxa (Friedman's two way nonparametric ANOVA: F = 25.53, df = 3, P < 0.0001). Apis honeybees (mean visits hr-1 6.4 ± 4.6) and butterflies (mean visits hr-1 7.8 ± 5.3) were major visitors and visited the flowers of all six crops. Apis honeybees and butterflies had higher visitation rates to tomato and chilli; butterflies and Apis honeybees to gourds; solitary bees and butterflies to brinjal, and Apis honeybees to okra (Tab. 3).

DISCUSSION

Our study shows that levels of pollinator dependency ranged from 0% in tomato to 76% in bitter gourd, and pollinators are required for fruit set in five of the six crops. Monoecious and andromonoecious crops were more reliant on pollinators than hermaphrodite crops. There was considerable variation in levels of pollinator dependency which could be due to effects of site, varieties of crops grown and inputs, which we could not test because of limitations of study design.

Our results did not support many cases in literature: For example brinjal and chilli, listed as having low levels of reliance on pollinators for crop production (Klein et al. 2007) showed significant levels of pollinator dependency in our study (32% and 37%); okra, described as having moderate levels of pollinator dependence (Klein et al. 2007) was lower in our assessment (20%). The tropical monoecious gourds which were not assessed earlier, showed significant levels of pollinator dependency because pollinators are needed to move pollen within and between plants. Tomato that has been demonstrated to have enhanced fruit set with wild pollinators (Greenleaf & Kremen 2006; Hogendoorn et al. 2006) showed no pollinator dependency in our study. Therefore, more data from different sites are required to get a realistic picture of pollinator dependence of tropical crops.

However, the good news is that there was minimal pollen limitation and diverse pollinator assemblages visiting flowers in this study. This could be due to small scale farms having adequate shrubbery and weedy vegetation that could provide foraging and nesting habitats for pollinators despite the extensive use of pesticides. Uncultivated areas and shrubbery around farms are an important refuge for insects and buffer the effects of insecticides (Lee et al. 2001).

The visitation rates of pollinator taxa differed with crop type although we could not evaluate their effectiveness. Brinjal like many of the Solanaceae is buzz pollinated, and buzzing bees such as *Xylocopa* and *Amegilla*, are probably the major pollinators (Davidar et al. 2015). Our study also identifies butterflies as a possible major pollinator group. Bees, particularly honeybees have rightly been given the key role of pollinating crops worldwide (Potts et al. 2010); however, butterflies visited flowers as frequently and were common visitors to five of the six crops. Butterflies could be important as pollinators because Carr (2012) recorded more species of butterflies and at greater size ranges than the other pollinator taxa. Wing span ranged from about 20-30 mm in the lycaenids to 90-100 mm in the papilionids (Kehimkar 2008). This diverse butterfly assemblage requires host plants for reproduction which are probably available in unmanaged hedges and fallow land with shrubbery and native plants that provide food, shelter and breeding habitats for pollinators. Indian small family farms have been recognized to promote biodiversity by incorporating forest and fallow lands with non domesticated plant species that are used by people for wood, fodder and medicine (Robbins 2001). The overall importance of the landscape for supporting diverse pollinator fauna should not be underestimated.

Conclusion

Our study demonstrates that overall levels of pollinator dependency among the six crops were significant in small family farms across the Coimbatore region in southern India. We did not find any evidence of pollen limitation, and the diverse pollinator assemblage probably enhances pollination services to crops.

ACKNOWLEDGEMENTS

This study was partially funded by a grant from the Lillian Goldman Charitable Trust to Dr. Allison Snow, Ohio State University. We thank Mr. Raghupathy, Mr. Naveen Bala, Mr. Vishwanath, Mrs. Pushpa, Mr. Ravi, Mr. Anand George, and Mr. Vasanthakumar for permitting us to work in their farms; Dr. J.-Ph. Puyravaud for the map and anonymous referees for helpful comments on earlier versions of the manuscript.

REFERENCES

Aizen MA, Garibaldi LA, Cunningham SA, Klein AM (2009) Long-term global trends in crop yield and production reveal no current pollination shortage but increasing pollinator dependency. Current Biology 18: 1572-1575.

Allen-Wardell G, Bernhardt P, Bitner R, Burquez A, Buchmann S, Cane J, Cox PA, Dalton V, Feinsinger P, Ingram M, Inouye D, Jones CE, Kennedy K, Kevan P, Koopowitz H, Medellin-Morales S, Nabhan GP, Pavlik B, Tepedino V, Torchio P, Walker S (1998) The potential consequences of pollinator declines on the conservation of biodiversity and stability of food crop yields. Conservation Biology 12: 8–17.

Basu P, Bhattacharya R, Iannetta PPM (2011) A decline in pollinator dependent vegetable crop productivity in India indicates pollination limitation and consequent agro-economic crises. Nature Precedings URL: http://www.farmlandbirds.net/sites/default/files/Basu%20et%20al%202011.pdf (accessed May, 2013)

Bawa KS (1974) Breeding systems of tree species of a lowland tropical community. Evolution 28: 85-92.

Bhanti M, Taneja A (2007) Contamination of vegetables of different seasons with organo phosphorus pesticides and related health risk assessment in North India. Chemosphere 69: 63-68.

Blanche KR, Ludwig JA, Cunningham SA (2006) Proximity to rainforest enhances pollination and fruit set in orchards. *Journal of Applied Ecology* 43: 1182-1187.

Bradshaw CJ, Sodhi NS, Brook BW (2008) Tropical turmoil: a biodiversity tragedy in progress. Frontiers in Ecology and the Environment 7: 79-87.

Carr SA (2012) Pollination of selected crop plants in Coimbatore district of Tamil Nadu. Unpublished M. Sc. dissertation, Pondicherry University, India.

Chandel RS, Thakur RK, Bhardwaj NR, Pathania N (2004) Onion seed crop pollination: a missing dimension in mountain horticulture. Acta Horticulturae 631:79-86.

Davidar P (2009) Pollination services to NTFP and cultivated plants in the Nilgiri Biosphere Reserve., In: Dutt R, Seeley J, Roy, P (Eds.) Proceedings of the Biodiversity and Livelihoods Conference, pp 106-111.

Davidar P, Snow AA, Rajkumar M, Pasquet R, Daunay MC, Mutegi E (2015) The potential for crop-to-wild hybridization in eggplant (*Solanum melongena*, Solanaceae) in southern India. American Journal of Botany 102: 140-148.

De Marco Jr P, Coelho FM (2004) Services performed by the ecosystem: forest remnants influence agricultural cultures' pollination and production. Biodiversity & Conservation, 13: 1245-1255.

Garibaldi LA, Steffan-Dewenter I, Winfree R , Aizen MA, Bommarco R, Cunningham SA, Kremen C et al. (2013) Wild pollinators enhance fruit set of crops regardless of honey bee abundance. Science 339: 1608-1611.

Greenleaf SS, Kremen C (2006) Wild bee species increase tomato production and respond differently to surrounding land use in northern California. Biological Conservation 133: 81–87.

Hogendoorn K, Gross CL, Sedgley M, Keller MA (2006) Increased tomato yield through pollination by native Australian *Amegilla chlorocyanea* (Hymenoptera: Anthophoridae). Journal of Economic Entomology 99: 828-833.

Kearns CA, Inouye DW, Waser NM (1998) Endangered mutualisms: the conservation of plant-pollinator interactions. Annual Review of Ecology and Systematics 29: 83-112.

Kehimkar I (2008) The book of Indian butterflies. BNHS and Oxford University Press, New Delhi.

Klein AM, Vaissiere BE, Cane JH, Steffan-Dewenter I, Cunningham SA, Kremen C, Tscharntke T (2007) Importance of pollinators in changing landscapes for world crops. Proceedings of the Royal Society of London. Series B: Biological Sciences 274: 303–313.

Krishnan S, Kushalappa CG, Shaanker RU, Ghazoul J (2012) Status of pollinators and their efficiency in coffee fruit set in a fragmented landscape mosaic in South India. Basic and Applied Ecology 13: 277-285.

Kumari B, Madan VK, Kumar R, Kathpal TS (2002) Monitoring of seasonal vegetables for pesticide residues. Environmental Monitoring and Assessment 74: 263-270.

Lee JC, Menalled FD, Landis DA (2001). Refuge habitats modify impact of insecticide disturbance on carabid beetle communities. Journal of Applied Ecology 38: 472–483.

McGregor SE (1976) Insect pollination of cultivated crop plants. Agriculture Handbook 496, SDA-ARS, Washington, D.C.

Ollerton J, Winfree R, Tarrant S (2011) How many flowering plants are pollinated by animals? Oikos 120: 321-326.

Ollerton J, Erenler H, Edwards M, Crockett R (2014) Extinctions of aculeate pollinators in Britain and the role of large-scale agricultural changes. Science 346: 1360-1362.

Partap U (1999) Pollination management of mountain crops through beekeeping: trainer resource book. International Centre for Integrated Mountain Development, Kathmandu, Nepal.

Potts SG, Biesmeijer JC, Kremen C, Neumann P, Schweiger O, Kunin WE (2010) Global pollinator declines: trends, impacts and drivers. Trends in Ecology & Evolution 25: 345-353.

Ricketts TH, Daily GC, Ehrlich PR, Michener CD (2004) Economic value of tropical forest to coffee production. Proceedings of the National Academy of Sciences 101: 12579–12582.

Ricketts TH, Regetz J, Steffan Dewenter I, Cunningham SA, Kremen C, Bogdanski A, Gemmill-Herren B, Greenleaf SC, Klein AM, Mayfield MM, Morandin LA, Ochieng A, Viana BF (2008) Landscape effects on crop pollination services: are there general patterns? Ecology Letters 11: 499-515.

Robbins P (2001) Tracking invasive land covers in India, or why our landscapes have never been modern. Annals of the Association of American Geographers 91: 637-659.

Roubik DW (ed) (1995). Pollination of cultivated plants in the tropics (Vol. 118). Food & Agriculture Organisation, Rome, Italy.

Roy SS, Mahmood R, Niyogi D, Lei M, Foster SA, Hubbard KG, Douglas E, Pielke R (2007) Impacts of the agricultural Green Revolution–induced land use changes on air temperatures in India. Journal of Geophysical Research 112 D21, DOI: 10.1029/2007JD008834.

Sinu PA, Shivanna KR (2007a) Pollination biology of large cardamom (*Amomum subulatum*). Current Science 93: 548-552.

Sinu PA, Shivanna KR (2007b) Pollination ecology of cardamom (Elettaria cardamomum) in the Western Ghats, India. Journal of Tropical Ecology 23: 493-496.

SPSS Inc. (2000) Systat, Version 10. SPSS Inc., Chicago, IL.

Surendran U, Murugappan V (2008) A micro- and meso-level modeling study for assessing sustainability in semi-arid tropical agro ecosystem using NUTMON-tool box. Journal of Sustainable Agriculture 29: 151-179.

Tur C, Castro-Urgal R, Traveset A (2013) Linking plant specialization to dependence in interactions for seed set in pollination networks. PLoS ONE 8: e78294.

Zapata TR, Arroyo MTK (1978) Plant reproductive ecology of a secondary deciduous tropical forest in Venezuela. Biotropica 10: 221-230.

8

BIOTIC AND ABIOTIC FACTORS CONTRIBUTE TO CRANBERRY POLLINATION

Hannah R. Gaines-Day* and Claudio Gratton

University of Wisconsin-Madison, Department of Entomology, 237 Russell Labs, 1630 Linden Dr., Madison, WI 53706, USA

Abstract—As bee populations continue to decline, farmers face possible crop failures due to insufficient pollination. Crops, however, vary in the degree to which they depend on pollinators, suggesting that some crops may not be as sensitive to variation in pollinator availability and/or abundance as others. The objective of this study was to determine the contribution of biotic and abiotic factors to cranberry pollination. We performed field and greenhouse experiments to compare the effect of biotic (i.e., bee or hand pollination) and abiotic (i.e., wind, agitation) factors on yield. We found that even in the absence of bees, cranberry is able to produce a significant yield. In the field, plants in the abiotic treatments produced higher yields (wind 230 bbl/ac [barrels per acre], agitation 200 bbl/ac) than the closed control treatment (108 bbl/ac), although these yields were not as high as the open, biotic treatment (367 bbl/ac). This corresponds to a contribution of 41% by bees, 30% by non-bee insects, and 29% by mechanical agitation. In the greenhouse, the agitation treatment had, on average, higher berry weight per upright (0.6 g/upright) than the undisturbed control treatment (0.04 g/upright), but again, not as high as the biotic treatment (3.0 g/upright). This confirmed that cranberry does not autogamously self-pollinate indicating that all yields are due to biotic or abiotic vectors moving pollen between flowers. Although bees clearly contribute to cranberry pollination, previous studies have understated the contribution of alternative mechanisms by which cranberry pollen can move between flowers.

Keywords: Vaccinium macrocarpon, *pollinator decline, physical factors, agitation, wind pollination*

INTRODUCTION

Insect pollination is an important ecosystem service required by or benefitting two-thirds of global crops (Klein et al. 2007). These crops, however, vary in the degree to which they depend on pollinators, suggesting that not all crops are equally susceptible to variability in the abundance of pollinators. In the absence of pollinators, some plants produce no fruit (e.g., watermelon, Delaplane & Mayer 2000), while others produce misshapen (e.g., strawberries, Free 1968; Jaycox 1970) or small fruit (e.g., cherry tomato, Greenleaf & Kremen 2006). Determining the role of insect and non-insect factors in the pollination of specific crops will provide a better understanding of how the decline of pollinators may affect crop production.

Cranberry (*Vaccinium macrocarpon*) is one crop that is considered pollinator-dependent (reviewed by McGregor 1976; Eck 1986, 1990; Free 1993; Delaplane & Mayer 2000). Although the flowers are self-compatible (Reader 1977; Dana et al. 1989; Sarracino & Vorsa 1991), pollen is released before the stigma is receptive making self-pollination unlikely (Rigby & Dana 1972). Bees are effective pollinators of cranberry (Mohr & Kevan 1987), depositing enough pollen to produce a full-sized fruit in one or two visits (Cane & Schiffhauer 2003). As a result, individual cranberry growers spend thousands of dollars each year on honey bee rentals to ensure sufficient pollination (USDA NASS WASS 2006).

However, despite over one hundred years of research on cranberry pollination and the widespread use of honey bees, the degree to which cranberry depends on pollinators remains unclear. Previous studies have estimated that 30% to 100% of the cranberry crop would be lost in the absence of bees (Southwick & Southwick 1992; Williams 1994; reviewed by Delaplane & Mayer 2000). At the lower end of this range, pollinator decline would have a minimal effect on cranberry yield, while at the upper end, complete dependence on pollinators would suggest crop failure in their absence. The high variability in these crop-loss estimates may suggest that other, non-bee, factors such as wind or mechanical agitation also contribute to cranberry pollination. If non-bee factors provide significant pollination, cranberry growers may not need to stock as many hives of honey bees as they currently use.

Few studies, however, have considered the role of non-bee factors in cranberry pollination. Of those that do, the results are conflicting. For example, despite the claim that wind is of minor importance to cranberry pollination (e.g., Filmer 1949), Papke et al. (1980) demonstrated that there is enough cranberry pollen being carried in the wind to provide a significant contribution to pollination. Moreover, there is some evidence that manually agitated cranberry plants could set fruit in the absence of bees (Roberts & Struckmeyer 1942). In contrast, several studies found contrary evidence and concluded that wind or the manual agitation of plants does not contribute to cranberry pollination (Filmer 1949;

*Corresponding author; email: hgaines@gmail.com

MacKenzie 1994). The variability among these studies may be partly due to methodological differences in the agitation treatments, differences between cultivars, or variation between growing regions where the studies were conducted. In sum, the paucity of research investigating non-bee mechanisms of pollination, and the inconsistencies in findings between the studies that do, suggest that we still lack a clear understanding of the factors contributing to cranberry pollination.

The goal of this study was to understand the relative importance of biotic and abiotic factors to cranberry pollination. We established field and greenhouse experiments in which we manipulated biotic (i.e., bees) and abiotic (i.e., wind, manual agitation) factors that could contribute to cranberry pollination. Pollination success was measured using yield, berry weight, berries per upright, and seeds per berry since these are measurements relevant to agricultural production.

MATERIALS AND METHODS

Cranberry (Ericaceae: *Vaccinium macrocarpon* Aiton) is a perennial fruit crop native to North America. It is grown commercially in artificially created marshes with sandy, acidic soil. The main production region, where the field component of this study was conducted, is Wood County in central Wisconsin, USA (44.30 °N, 90.11 °W). This region is characterized by sandy soil and flat open terrain (Dott & Attig 2004). The area is heavily agricultural and produces most of the states' cranberry and potato crops. The mean annual temperature is 7°C (mean low -9°C, mean high 20°C) with average summer temperatures of 20°C, and mean summer precipitation of 290 mm (Wisconsin State Climatology Office), which is augmented with irrigation in production areas. Cranberry grows as a vine along the ground and sends up vegetative and flowering shoots ("uprights"). Each flowering upright produces up to 8 flowers that bloom sequentially from the bottom of the upright upwards over the course of several weeks in late June and early July (Eck 1990). Honey bees are commonly brought to commercial marshes for the duration of bloom (Delaplane & Mayer 2000).

Field experiment

To assess the influence of biotic and abiotic factors on cranberry yield in a field setting, we established a cage study in a commercial cranberry marsh in Wood County, WI (USA). Four treatments were established in a single bed of the "Stevens" cultivar: (1) "open", which allowed both insect visitation and movement of plants by wind, (2) "wind", which blocked insect visitation with a fine nylon mesh (bridal veil) but allowed wind to agitate the plants or move pollen, (3) "closed", which prevented insect visitation and wind using floating row cover (Agribon+ AG-15 Insect Barrier, Johnny's Selected Seeds, Fairfield, Maine) and a corrugated plastic wind block surrounding each cage. And (4), an "agitation" treatment was established which used the same cage design as the "closed" treatment but received manual agitation twice per week during bloom. Each agitation consisted of lifting the row cover material and brushing a PVC tube (30 cm × 2 cm diameter) 20 times across the top of the cranberry uprights. Each treatment was replicated 10 times. Cage frames used in treatments 2 through 4 measured ~45 × 45 cm by ~40 cm tall (for further information on construction see Appendix I).

In order to account for possible differences in local growing conditions (e.g., upright density, soil moisture) as a function of location within the cranberry bed, cages were arranged in a grid and treatments were assigned using a modified Latin Square design with each treatment occurring once per column and twice per row. Cages were set up before cranberry bloom (May 25, 2012) and removed after bloom was complete (July 9, 2012). All berries from within a 30 cm × 30 cm (0.09 m2) plot in the centre of each cage were harvested (September 20, 2012), counted and weighed (wet) to estimate yield. This is a standard method for estimating pre-harvest cranberry yields (e.g., Pozdnyakova et al. 2005). Wet weight was used as this can be easily converted to yield units used by cranberry growers (1 barrel [bbl] = 100 lbs = 45 kg; thus 1 barrel/acre = 111 kg/ha). In order to understand the level of pollination received in each treatment, the number of fully formed seeds from 20 berries from each plot were counted. The number of seeds is proportional to the amount of pollen deposited on the stigma and therefore represents an indication of pollination success (Cane & Schiffhauer 2003).

To test for cage effects we measured several environmental variables both inside and outside of the cages. An additional 15 treatment plots (5 each "open", "wind", and "closed", treatments as described above) were established and within each of these plots, we measured temperature, light intensity, soil moisture, and insect abundance. Temperature and light intensity were measured every 30 minutes for the duration of the cage study using HOBO data loggers (Onset Computer Corporation, Bourne, Massachusetts) hung inside inverted Styrofoam cups as sun shields. Percent soil moisture content was measured using a TDR 300 soil moisture meter (Spectrum Technologies, Aurora, Illinois) twice during the growing season (June 14 and July 9, 2012). Four measurements were taken per cage. The insect community within each treatment type (i.e., "open", "wind", and "closed") was measured continuously during bloom (May 29 - June 28, 2012) using one yellow sticky strip (10 × 15 cm, Great Lakes IPM 025-SS-35) per treatment plot replaced three times during bloom and three pan traps (blue, yellow, and white, ACE Brand Fluorescent paint) per treatment plot containing soapy water (Dawn™ blue dish soap).

To examine differences among treatments for yield, weight per berry, and berries per plot we used a one-way mixed model ANOVA with row and column locations as random effects. After fitting ANOVA models to the field data, a visual examination of the residuals determined that the assumptions of normality were met and no transformations were required. Differences among treatments were determined using Tukey's Honestly Significant Difference (HSD) test (Hsu 1996). Statistical analyses were performed using JMP Pro 10 (SAS Institute Inc. 2007).

Greenhouse experiment

Sixty dormant cranberry plants with visible buds, thirty individuals each of the "HyRed" and "Stevens" cultivars, were dug from a commercial cranberry marsh in late March of 2012. "Stevens" was chosen because it is the most commonly grown cultivar in Wisconsin and "HyRed" is a recently developed hybrid that has an earlier bloom and harvest time than "Stevens". Plants were rinsed thoroughly to remove all sand and possible pests from the roots and planted into 15 cm pots of moist peat moss. Pots were arranged randomly in a greenhouse set at 22°C with a 16 hour photoperiod. Approximately one month after potting, uprights were thinned to 5-6 flowering uprights per pot in order to reduce the total number of flowers and make hand pollination manageable. All but the first four flowers to bloom per upright were trimmed off as they opened since cranberry plants are more likely to set fruit on the lower, earlier flowers than the upper, later flowers (Birrenkott & Stang 1989).

Three treatments were established in the greenhouse: (1) "hand" pollination, which represents the biotic movement of pollen between flowers (mimicking bee visitation), (2) "agitation", which represents the physical movement of plants by wind, and (3) an undisturbed control, which provided a measure of autogamous self-pollination. To assess the potential for biotic pollination, flowers with a receptive stigma (i.e., those that were moist and protruding from the stamens) were hand pollinated daily during bloom by gently dipping the stigma into a small accumulation of pollen that had been collected from younger flowers into the cap of a micro-centrifuge tube. To assess the potential for abiotic pollination, plants in the "agitation" treatment were gently jostled daily during bloom by moving the palm of the hand across the vegetative top of the uprights for approximately 3 seconds, causing the plants to bump against each other. This action simulated the physical movement of plants as may be caused by wind while excluding the possibility of wind pollination *per se* in which pollen is moved through the air. Plants in the control treatment were left undisturbed throughout the study to assess whether fruit would be produced in the absence of either biotic or abiotic factors (i.e., autogamous self-pollination). For each treatment we established 10 replicates (i.e., pots) per cultivar. Pots were placed in two parallel rows ~0.2 m apart and spanning 3 - 3.5 m on both sides of a single aisle of greenhouse tables (one cultivar per side). Experimental treatments were initiated as soon as bloom began (April 24) and were continued daily until all flowers were done blooming (June 1). Berries were harvested approximately two months after the start of bloom when fruits began to turn red (June 20). The number of berries per upright were counted and each berry was weighed (wet weight, g). The product of these two variables (i.e., berry weight per upright) was used as a proxy for yield. Although the number of berries per area is an important variable in determining yield (Devetter 2013), the area in our experiment was limited by pot size and was therefore not included in our calculation of yield. The number of fully formed seeds was counted for each berry as a proxy for the amount of pollen reaching the stigma. Averages of berry weight, berries per upright, and seeds per berry were taken for each pot for a total of 10 replicates per treatment.

To examine differences among experimental treatments ("hand pollination", "agitation", undisturbed control) and cultivar ("Stevens", "HyRed"), we used a fully factorial two-way analysis of variance (ANOVA) to compare yield (berry weight per upright), berries per upright, and weight per berry. After fitting ANOVA models to the greenhouse data, a visual examination of the residuals suggested that a transformation (log x+1) was necessary to meet linearity assumptions for berry weight per upright and berries per upright. Each of these response variables indicates some form of pollination success. Differences among treatments by cultivar were determined using Tukey's HSD test. Statistical analyses were performed using JMP Pro 10 (SAS Institute Inc. 2007).

RESULTS

Field experiment

The field cages successfully excluded bees while maintaining comparable environmental conditions within each cage type (Appendix II). The cages blocked out all but the tiniest insects: the majority of those that made it into the cages were thrips (Thysanoptera). No bees at all were found in either cage type. The bridal veil mesh used in the "wind" treatment did not result in any difference in average daytime (09:00 - 15:00) temperature as compared to open plots. The floating row cover used in the "closed" and "agitation" treatments resulted in an 8% (3°C) increase in average daytime temperature as compared to the open plots ($F_{2,211.5} = 14.5$, $P < 0.0001$). The bridal veil resulted in decreased average daytime light levels of 7% ($F_{2,276.3} = 4.7$, $P < 0.0103$) in the "wind" treatment. Soil moisture did not vary among treatments (15.3 ± 0.6, $F_{2,23} = 0.17$, $P = 0.84$). Although some statistically significant differences in light levels and temperature were found, this variation is unlikely to be biologically relevant as light levels from all cage types were well within the range of average daylight intensity (10,000-25,000 lum/m²) and temperatures were within the normal growing range for cranberries (Roper 2006).

In the field, cranberry yields in the "open" treatments where highest (mean 367 bbl/ac), followed by "wind" (230 bbl/ac) and "agitation" (200 bbl/ac) and lowest for the "closed" treatment (108 bbl/ac, $F_{3,23.1} = 70.5$, $P < 0.0001$, Fig. 1A). Thus, yields in the treatments from which bees alone were excluded (i.e., "wind", "agitation") were on average about 59% that of the fully open plots, but double those of plots from which wind, agitation and bees were excluded. The variation in yield observed across treatments was the result of differences in both berry weight (Fig. 1B) and total number of berries produced per area (Fig. 1C). Weight per berry was significantly different among treatments with the heaviest berries in the "open" plots, followed by "agitation", and lowest in the "wind" treatment and "closed" control ($F_{3,24.2} = 36.6$, $P < 0.0001$, Fig. 1B). The total number of berries per plot also varied significantly among treatments with the most berries produced in the "open" and "wind" plots, followed equally by "agitation"

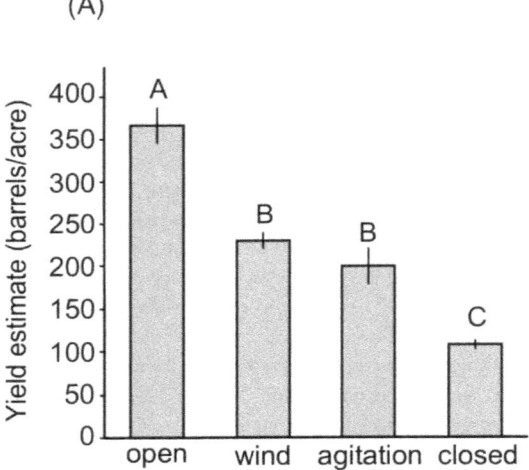

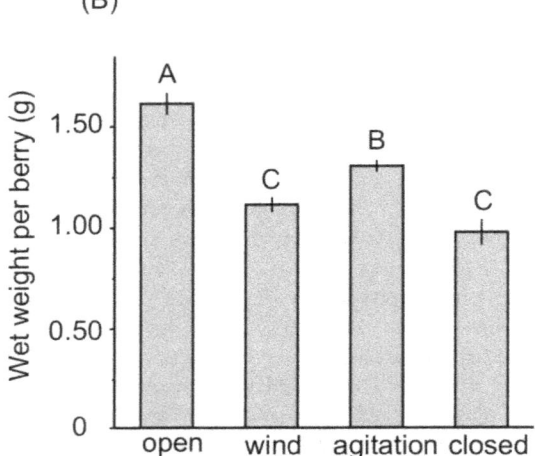

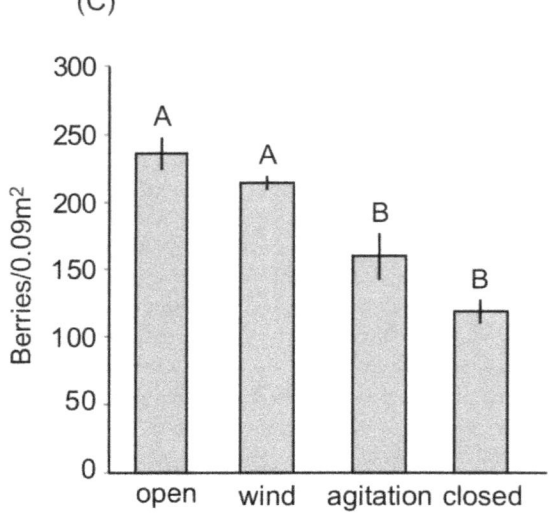

FIGURE I: Metrics of pollination success for the cranberry cultivar "Stevens" as measured in the field: (A) yield estimate (mean ± SE), (B) wet weight per berry (mean ± SE), and (C) number of berries per 0.09 m² plot (mean ± SE). Letters represent significant differences ($P < 0.05$) between treatments. Note that I barrel/acre = 111 kg/ha.

and the "closed" control (treatment, $F_{3,23.03} = 23.5$, $P < 0.0001$, Fig. IC).

Greenhouse experiment

In the greenhouse, considerable fruit set occurred on plants in both the biotic ("hand") and abiotic ("agitation") treatments but not in the undisturbed control. Ninety-eight percent of "HyRed" and 92% of "Stevens" uprights in the "hand" pollination treatment produced fruit and 52% of the "HyRed" and 30% of the "Stevens" uprights in the "agitation" treatment produced fruit. In contrast, only 2% of "HyRed" and 5% of "Stevens" uprights in the undisturbed control produced fruit.

Yield (represented as the product of weight per berry × berries per upright) was higher in the "hand" pollination treatment than in either of the other treatments and higher in the "agitation" treatment than the undisturbed control (treatment, $F_{2,58} = 203.8$, $P < 0.0001$, Fig. 2A). The number of berries per upright varied significantly among treatments in both cultivars (treatment, $F_{2,58} = 282.0$, $P < 0.0001$, Fig. 2B). More berries per upright were produced in the "hand" pollination treatment than in the "agitation" treatment, and more in the "agitation" treatment than in the undisturbed control. However, the difference in berries per upright between treatments was greater for "HyRed" than "Stevens" (treatment × cultivar, $F_{2,58} = 3.3$, $P = 0.046$). Berry weight did not vary across treatments ($F_{2,41} = 2.4$, $P = 0.11$, Fig. 2C) or cultivar ($F_{1,41} = 0.0021$, $P = 0.96$).

Furthermore, we found a relationship between berry weight and seeds per berry for "Stevens" in both the greenhouse ($R^2 = 0.42$, $P = 0.0011$, Fig. 3A) and the field ($R^2 = 0.83$, $P < 0.0001$, Fig. 3A) but not for "HyRed" (greenhouse only, $R^2 = 0.04$, $P = 0.39$, Fig. 3B).

DISCUSSION

Previous research on cranberry pollination has generally concluded that bees are required to produce fruit (Hutson 1925; Farrar & Bain 1946; Marucci 1966; Marucci & Moulter 1977). However, most of these studies did not consider vectors other than bees that could contribute to pollination. In fact, in over one hundred years of research, only four studies have considered the contribution of non-bee factors including wind and mechanical agitation to cranberry pollination (Roberts & Struckmeyer 1942; Filmer 1949; Papke et al. 1980; MacKenzie 1994), with the results providing mixed evidence that non-bee factors are relevant. In this study, we demonstrate that in the absence of bees, cranberry is still able to produce fruit. We found that plants from which bees alone were excluded but which were physically disturbed (i.e., by agitation, wind), produced a greater overall yield than plants from which both bees and disturbance were blocked. This result, combined with agitation treatments showing that physical movement of plants was sufficient to transfer enough pollen to produce full-sized, marketable fruit, challenges the notion that bees are the only way for cranberry plants to achieve fruit set, and suggests that both biotic and abiotic factors contribute to cranberry pollination.

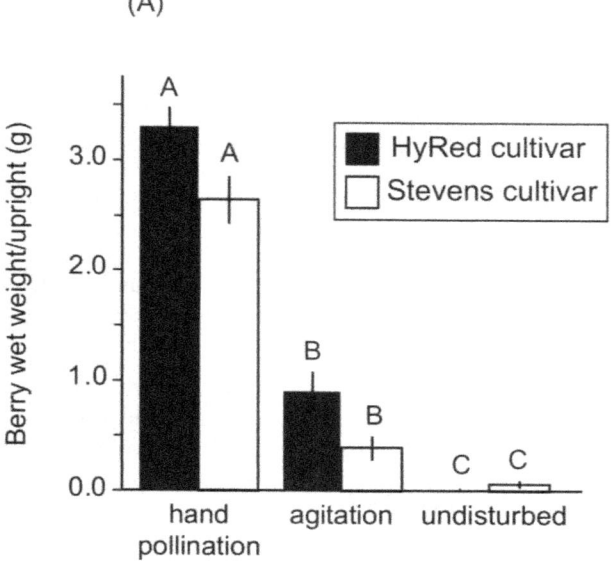

(A)

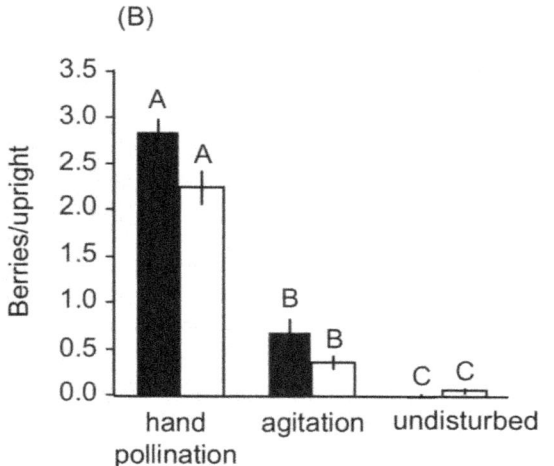

(B)

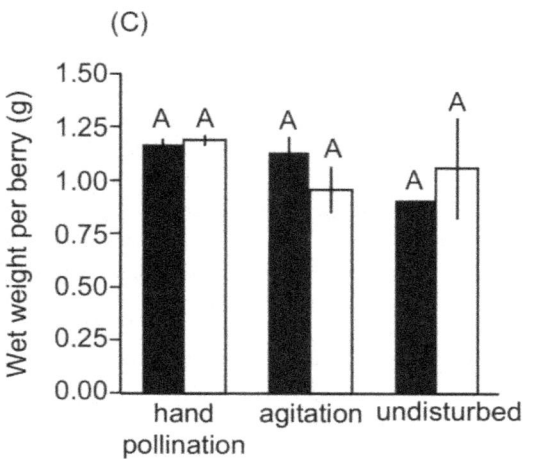

(C)

FIGURE 2: Metrics of pollination success for the cranberry cultivar "HyRed" (black bars) and "Stevens" (white bars) as measured in the greenhouse: (A) yield estimate (mean wet weight per upright ± SE), (B) number of berries per upright (mean ± SE), and (C) wet weight per berry (mean ± SE). Letters represent significant differences ($P < 0.05$) between treatments.

One possible reason that our results differ from previous studies on cranberry pollination may be differences in study design and scope. Often, studies that examine the importance of pollinators to fruit production in cranberry are designed only to test the importance of bees for pollination (e.g., Kevan et al. 1983; Cane & Schiffhauer 2003; Phillips 2011). In many of these experiments individual cranberry flowers are either isolated, testing their capacity to self-pollinate, or exposed to bees, demonstrating the effectiveness of bees to pollinate cranberry. Under normal growing conditions, however, flowers are not isolated and pollen may move between flowers through multiple mechanisms. Interestingly, several studies that used field cages to exclude bees from cranberry plants found that berries were produced within their cages, although at lower levels than open plots (Hutson 1925; Farrar & Bain 1946; Phillips 2011). These authors either provide no explanation of how berries formed without bees or interpret their findings as being the result of faulty cages that must have allowed bees to enter, rather than considering alternative mechanisms of pollination. In our study, careful cage construction (e.g., weighing down cage material to ground level, Appendix I) and insect monitoring within cages (Appendix II) makes us confident that bees were not contributing to pollination within the field cages and that other mechanisms are indeed contributing to cranberry pollination.

Differences in the relative contribution of biotic and abiotic factors between field and greenhouse results are likely due to uncontrolled factors in the field that were absent in the greenhouse. The greenhouse experiment suggests that cranberries do not autogamously self-pollinate as there was only ~1% fruit production in the undisturbed controls compared to the hand pollination treatment. In the field, the treatment with fully closed cages was meant to mimic the greenhouse undisturbed control. Yet, under field conditions, plots had yields about 30% (108 bbl/ac) that of open field conditions (367 bbl/ac). One possible explanation for this difference is the presence of non-bee insects contributing to pollination. Specifically, thrips (Thysanoptera) were found in all treatments in the field (Appendix II) but not observed in the greenhouse. These tiny pollen-eating insects have been shown to be effective pollinators in other systems, including ericaceous plants which are in the same family as cranberry (Hagerup & Hagerup 1953; Kirk 1988; Baker & Cruden 1991; Ananthakrishnan 1993). It is notable that although thrips are ubiquitous and abundant in cranberry marshes, their potential influence on cranberry pollination remains a hypothesis to be critically examined (but see Gaines-Day 2013). Differences between the agitation treatment in the field and greenhouse may also be due to differences in density of flowering uprights. In the greenhouse, flowering uprights were thinned to a low density (equivalent to about 279 flowering uprights m⁻²), whereas in the field, flowering uprights were on average six times as dense (1,705 flowering uprights m⁻²). We hypothesize that the higher relative effects of agitation in the field (107 bbl/ac greater than "closed", or 29% that of open plots) compared to the greenhouse (20% of hand pollination) is at least partly due to a higher density of flowering uprights, where pollen is more abundant, flowers are closer together, and there is a higher likelihood

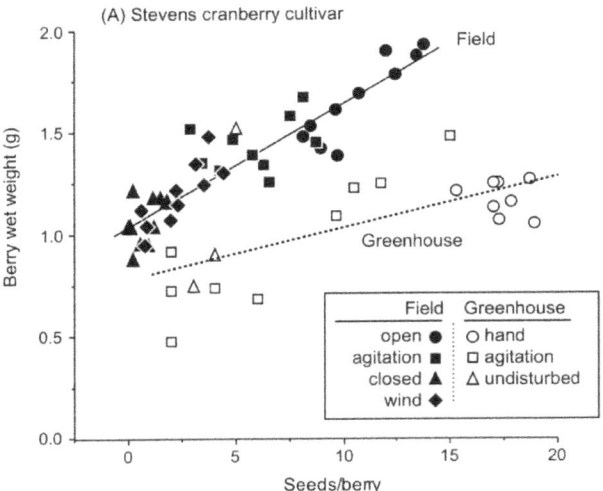

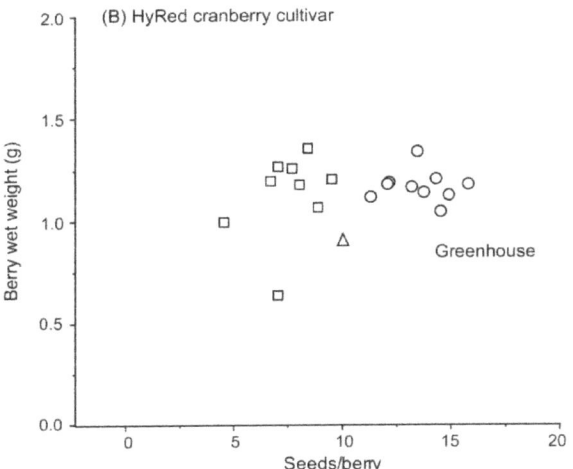

FIGURE 3: Relationship between berry weight (g wet weight) and seeds per berry for the cranberry cultivars (A) "Stevens" and (B) "HyRed", for field (solid symbols) and greenhouse (hollow symbols) experiments.

that pollen could be transferred between flowers with even slight movement. Upright density may also be one factor explaining differences between this study and Filmer (1949) which found no effect of manual agitation on yield in a field setting; in that study upright density was about 2.5 times lower (646 flowering uprights/ m²) than our field densities. Finally, another aspect that could have also contributed to differences in yield among treatments is the proportion of flowers setting fruit. This component was not evaluated in this study but could give additional information on pollination success across the treatments.

The increase in seed number and berry weight in the Stevens cultivar as treatments go from the most undisturbed (undisturbed and closed) to those with progressively more potential for pollen movement (manual agitation and wind) suggests that the amount of pollen deposited onto the stigma is also increasing. Previous studies have also found a strong correlation between seed number and berry weight (Hall & Aalders 1965; Rigby & Dana 1971). Although the berries with the most seeds were found in the hand pollination

treatment in the greenhouse, the heaviest berries were found in the open treatment in the field. Thus, other factors, such as water and nutrient availability, likely play a role in berry weight. Cane and Schiffhauer (2003) found that in the Stevens cultivar, a berry reaches a maximum weight when the flower receives 8 tetrads of pollen resulting in about 15 seeds - a level we achieved in the hand-pollinated treatments but not in the field. In the field, the seed number per berry never reached 15 suggesting that even with bees and abiotic factors contributing to pollination, pollen may be limiting. The lack of relationship between berry weight and seed number in the HyRed cultivar provides evidence that differences also exist among cultivars in their response to pollen availability.

Our study was specifically designed to test the co-occurrence of pollination mechanisms including bees (presence/absence), wind, mechanical action, and autogamous self-pollination. We found that mechanical agitation, by wind or by hand can move pollen between cranberry flowers resulting in significantly greater yields than when these factors are absent. In small plots that were caged with fine mesh and obstructed to prevent wind, yields were 30% (108 bbl/ac) that of plots which received open pollination (367 bbl/ac). The addition of wind or physical agitation resulted in nearly a 100% increase in yield (+107 bbl/ac) contributing an additional 29% of yield (59% of open/bee pollination). The difference between open plots and caged plots suggests that bees alone are responsible for 41% of cranberry yield. This number is consistent with, but on the low end of the range of those reviewed in Delaplane and Mayer (2000). Furthermore, the estimated yield due to non-bee factors (215 bbl/ac) is comparable with average yields observed in commercial operations across North America. For example average cranberry yields in 2012 for Wisconsin were 245 bbl/ac, but yields were lower in other parts of the country, ranging from 81-140 bbl/ac on the west coast of the USA (Washington and Oregon) to 163-183 bbl/ac on the east coast (Massachussets and New Jersey, USDA 2012 Fruit Summary). If the conditions created in our small experimental plots are similar to those observed at farm field scales (i.e., comparable wind or physical agitation regimes, flowering upright densities, minute insects), then farmers may be realizing more than half of their crop yield from non-bee factors. Whether non-bee related pollination by itself is sufficient for a farmer to achieve an economically viable yield however will depend on additional economic factors such as the market price of the commodity and additional input costs. Although bees clearly contribute to cranberry pollination, results from prior studies may have understated other pathways of pollination (movement of pollen by abiotic factors or tiny insects) that could also result in viable fruit production.

ACKNOWLEDGEMENTS

We thank C.J. Searles Cranberry Inc. for allowing us to conduct our field cage study on their marsh and Gaynor Cranberry for providing plant material for the greenhouse study. We thank Eric Zeldin for his assistance digging plant material and providing guidance on establishing the plants in the greenhouse. We also thank the field and lab technicians who assisted with the project:

Rachel Mallinger, Amanda Rudie, Collin Schwantes, Eric Wiesman, Christopher Watson, Sacha Horn, and Scott Lee. Shawn Steffan, Neal Williams, and Dan Cariveau reviewed earlier versions of this manuscript. Funding for this research was provided by a University of Wisconsin Formula Fund Hatch Grant (WIS01415).

APPENDICES

Additional supporting information may be found in the online version of this article:

Appendix I. Cage construction.
Appendix II. Environmental data collected within field cages.

REFERENCES

Ananthakrishnan TN (1993) The role of thrips in pollination. Current Science 65:262-264.

Baker JD, Cruden RW (1991) Thrips-mediated self-pollination of two facultatively xenogamous wetland species. American Journal of Botany 78:959-963.

Birrenkott BA, Stang EJ (1989) Pollination and pollen tube growth in relation to cranberry fruit development. Journal of the American Society for Horticultural Science 114:733-737.

Cane JH, Schiffhauer D (2003) Dose-response relationships between pollination and fruiting refine pollinator comparisons for cranberry (*Vaccinium macrocarpon* [Ericaceae]). American Journal of Botany 90:1425-1432.

Dana MN, Steinmann S, Goben L (1989) Pollen source and fruit set of cranberry. Cranberries 53:10-14.

Delaplane KS, Mayer DF (2000) Crop pollination by bees. CABI Publishing, New York.

Devetter LW (2013) Understanding yield of cranberry: Bud development, carbohydrate allocation, and yield component analysis. (Doctoral Dissertation). University of Wisconsin, Madison.

Dott RH, Attig JW (2004) Roadside Geology of Wisconsin, 1st edition. Mountain Press, Missoula, Mont.

Eck P (1986) Cranberry. In: Monselise, SP (ed) CRC handbook of fruit set and development. CRC Press, Boca Raton, FL, pp 109-117.

Eck P (1990) The American cranberry. Rutgers University Press, New Brunswick, NJ.

Farrar CL, Bain HF (1946) Honeybees as pollinators of the cranberry. American Bee Journal 86:503-504.

Filmer RS (1949) Cranberry pollination studies. In: Proceedings of the 80th Annual Meeting of the American Cranberry Growers Association. pp 14-22.

Free JB (1968) The pollination of strawberries by honey-bees. Journal of Horticultural Science 48:107-11.

Free JB (1993) Insect pollination of crops, 2nd edn. Academic Press Inc., London.

Gaines-Day H (2013) Do bees matter to cranberry? The effect of bees, landscape, and local management on cranberry yield. (Doctoral Dissertation). University of Wisconsin, Madison.

Greenleaf SS, Kremen C (2006) Wild bee species increase tomato production and respond differently to surrounding land use in Northern California. Biological Conservation 133:81-87.

Hagerup E, Hagerup O (1953) Thrips pollination of *Erica tetralix*. New Phytologist 52:1-7.

Hall IV, Aalders LE (1965) The relation between seed number and berry weight in the cranberry. Canadian Journal of Plant Science 45:292.

Hsu JC (1996) Multiple comparisons: theory and methods. Chapman & Hall, London.

Hutson R (1925) The honey bee as an agent in the pollination of pears, apples and cranberries. Journal of Economic Entomology 18:387-391.

Jaycox ER (1970) Pollination of strawberries. American Bee Journal:176-177.

Kevan P, Gadawski R, Kevan S, Gadawski S (1983) Pollination of cranberries, *Vaccinium macrocarpon*, on cultivated marshes. Proceedings of the Entomological Society of Ontario 114:45-53.

Kirk WDJ (1988) Thrips and pollination biology. In: Ananthakrishnan, T.N., Raman, A. (eds) Dynamics of insect-plant interaction: Recent advances and future trends. Oxford & IBH Publishing Company, New Delhi, pp 129-135.

Klein A-M, Vaissiere BE, Cane JH, Steffan-Dewenter I, Cunningham SA, Kremen C, Tscharntke T (2007) Importance of pollinators in changing landscapes for world crops. Proceedings of the Royal Society B-Biological Sciences 274:303-313.

MacKenzie KE (1994) The foraging behaviour of honey bees (*Apis mellifera* L) and bumble bees (*Bombus spp*) on cranberry (*Vaccinium macrocarpon* Ait). Apidologie 25:375-383.

Marucci PE (1966) Cranberry pollination. Cranberries 30:11-13.

Marucci PE, Moulter HJ (1977) Cranberry pollination in New Jersey. Acta Horticulturae (ISHS) 61:217-222.

McGregor SE (1976) Insect pollination of cultivated crop plants. Agriculture Handbook No. 496. Agricultural Research Service, US Department of Agriculture, Washington, D.C.

Mohr NA, Kevan PG (1987) Pollinators and pollination requirements of lowbush blueberry (*Vaccinium angustifolium* Ait. and *V. myrtilloides* Michx.) and cranberry (*V. macrocarpon* Ait.) in Ontario with notes on highbush blueberry (*V. corymbosum* L.) and lingonberry (*V. vitis-ideae* L.). Proceedings of the Entomological Society of Ontario 118:149-154.

Papke AM, Eaton GW, Bowen PA (1980) Airborne pollen above a cranberry bog. HortScience 15:756.

Phillips KN (2011) A comparison of bumble bees (*Bombus spp.*) and honey bees (*Apis mellifera*) for the pollination of Oregon cranberries (Ericaceae: *Vaccinium macrocarpon*). (Master's Thesis). Oregon State University, Corvallis, OR

Pozdnyakova L, Giménez D, Oudemans PV (2005) Spatial Analysis of Cranberry Yield at Three Scales. Agronomy Journal 97:49-57.

Reader RJ (1977) Bog ericad flowers: self-compatibility and relative attractiveness to bees. Canadian Journal of Botany 55:2279-2287.

Rigby B, Dana MN (1971) Seed number and berry volume in cranberry. HortScience 6:495-496.

Rigby B, Dana MN (1972) Flower opening, pollen shedding, stigma receptivity, and pollen tube growth in the cranberry. HortScience 7:84-85.

Roberts RH, Struckmeyer BE (1942) Growth and fruiting of the cranberry. Proceedings of the American Society for Horticultural Science 40:373-379.

Roper, TR (2006) The physiology of cranberry yield. Wisconsin Cranberry Crop Management Newsletter, Volume XIX, Madison, WI.

Sarracino JM, Vorsa N (1991) Self and cross fertility in cranberry. Euphytica 58:129-136.

SAS Institute Inc. (2007) JMP Pro 10. Cary, NC.

Southwick EE, Southwick L (1992) Estimating the economic value of honey bees (Hymenoptera: Apidae) as agricultural pollinators in the United States. Journal of Economic Entomology 85:621-633.

USDA NASS (2013) Wisconsin - 2012 Fruit Summary. Madison, WI.

USDA NASS WASS (2006) 2005 Cranberry production and pollination survey. Madison, WI

Williams I (1994) The dependence of crop production within the European Union on pollination by honey bees. Agricultural Zoology Reviews 6:229-257.

Wisconsin State Climatology Office [online] URL: http://www.aos.wisc.edu/~sco/seasons/summer.html

9

Bee diversity and floral resources along a disturbance gradient in Kaya Muhaka Forest and surrounding farmlands of coastal Kenya

David O. Chiawo*[1], Callistus K.P.O. Ogol[2], Esther N. Kioko[3], Verrah A. Otiende[4], Mary W. Gikungu[3]

[1]Strathmore University, P. O Box 59857-00200 Nairobi, Kenya,
[2]African Union Commission, Roosevelt Street, P. O. Box 3243, Addis Ababa, Ethiopia
[3]National Museums of Kenya, P.O. Box 40658-00100 Nairobi, Kenya
[4]Pan African University, P. O. Box 62000-00200 Nairobi, Kenya

Abstract—Bees provide important pollination services that maintain native plant populations and ecosystem resilience, which is critical to the conservation of the rich and endemic biodiversity of Kaya forests along the Kenyan Coast. This study examined bee composition and floral resources from the forest core to the surrounding farmlands around Kaya Muhaka forest. In total, 755 individual bees, representing 41 species from three families were recorded: Apidae, Halictidae and Megachilidae. Overall, Apidae were the most abundant with a proportion of 76% of the total bee individuals, Halictidae at 14% and Megachilidae at 10%. Bee composition was similar between forest edge and crop fields as compared to forest core and fallow farmlands. We found a significant decrease in bee diversity with increasing distance from the forest to the surrounding farming area. A high abundance of bees was recorded in fallow farmland, which could be explained by the high abundance of floral resources in the habitat. We found floral resources richness to significantly affect bee species richness. These findings are important for understanding the effects of land use change on insect pollinators and their degree of resilience in disturbed habitats.

Keywords: Bee diversity; conservation; habitat disturbance; Kaya Muhaka forest; pollinator abundance

Introduction

There is evidence that pollinators are declining in some parts of the world (Keams et al. 1998; Kremen & Ricketts 2000), due to habitat destruction and land use intensification (Steffan-Dewenter & Westphal 2008). A strategic pollinator conservation plan should include their associated floral resources because the community structure of insect pollinators is related to their host plants (Potts et al. 2003; São Paulo Declaration 1999). Past studies have revealed positive relationships between bee abundance and floral abundance as well as between bee diversity and floral diversity (Banaszak 1996; Banaszak 2000; Potts et al. 2003). Kremen et al. (2007) explains that pollination services are provided by a variety of wild, free-living organisms but chiefly bees. Bees are the primary pollinators of rare and endangered plants, maintaining the biodiversity of most terrestrial eco-systems (LaSalle & Gould 1993; Stubbs et al. 1997). However, pollination services by wild bees are likely reduced in many areas and pollination-related problems within natural and agricultural ecosystems are becoming more common (Baude et al. 2016; Koh et al. 2016). Habitat fragmentation and isolation due to land use may reduce bee species richness and abundance and change their foraging behaviour (Didham et al. 1996). The decline may be

aggravated at the Kenyan coast due to agricultural encroachment, timber extraction and charcoal production (CEPF 2005). Moreover, commercially managed colonies of *Apis mellifera* have also declined in many parts of the world (Kremen et al. 2007). However, some bee communities appear to have some degree of resilience to land-use change (Banaszak 1992). Approaching such issues by documenting which species are involved is a key step to facilitating their conservation and management (Danks 1994).

Despite the ongoing concerns and controversy, there is little information on the response of bees to land-use change (Brosi et al. 2008) and only a few studies in Kenya have been published (Eardley et al. 2009; Gikungu 2006; Gikungu 2002; Gikungu et al. 2011). In agricultural regions, bees (Hymenoptera: Apoidea) are vital for successful fruit production (O'Toole 1993; Sheffield et al. 2003). Data on their relative abundance and diversity gives an indication of pollinator activities (Kevan 1999). Such data are missing in many forest ecosystems in Africa especially along the coastal region of East Africa. This study documents for the first time bees and their floral resources in Kaya Muhaka Forest and surrounding farmland in the coastal region of Kenya.

Materials and Methods

Study area

The study was conducted at Kaya Muhaka forest (KMF) on the coastal plains of Kenya (Fig. I), East Africa at a geographical location of 04° 18' S; 39° 33' E to 04° 38' S;

*Corresponding author: dchiawo@strathmore.edu,
chiawo2006@gmail.com

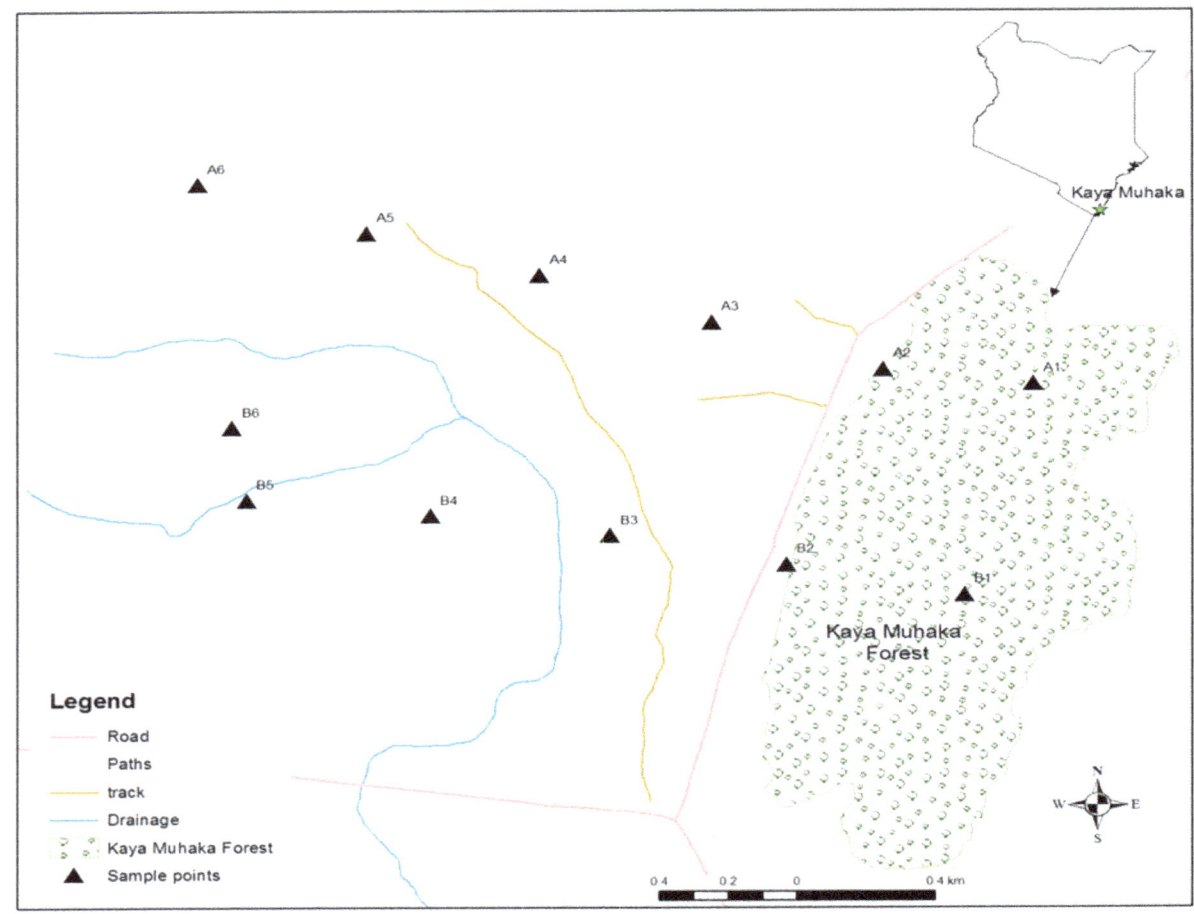

FIGURE 1:   Study area and satellite map of sampling points.

39° 53' E and surrounding farmlands. Kaya forests are residual patches of once extensive diverse lowland forest of Eastern Africa. It is a protected area and managed by Coastal Forest Conservation Unit (CFCU) of National Museums of Kenya (NMK) in conjunction with the local community. KMF covers about 130-150 ha and is located 32 km South of Mombasa town at an altitude of 20 - 40 m ASL.

Biodiversity of the coastal forests

The coastal forest stretches from Kenya to Tanzania and Islands of Zanzibar and Pemba, hosting more than 4,500 plant species and 1,050 plant genera with around 3,000 species and 750 genera occurring in the forest. At least 400 plant species are endemic to the forest patches and about another 500 are endemic to the intervening habitats that make up 99 percent of the eco-region area (WWF-US, 2003). They are botanically diverse with a high conservation value, consisting of regionally endemic climbers, shrubs, herbs, grasses and sedges (Burgess et al. 2000). The coastal forests are also known for high endemism of invertebrate groups such as millipedes, molluscs and forest butterflies (Burgess et al. 2000). KMF is known for a high Lepidoptera diversity and endemism (Lehmann & Kioko 2005). More than half of Kenya's rare plants occur in the coastal region, many in the Kayas. The flora of the forest is either vulnerable or endangered (TFCG 2007).

Farmlands

The surrounding farmlands are characterised by small-scale farming of subsistence crops, mainly cassava, cowpea, maize and rice. Also pigeon pea is sparsely distributed in these farms. Major commercial crops include coconut, citrus, cashew nut and mangoes. Cashew nuts and mangoes occupy major parts of the fallow farmlands. The latter is characterised by a mix of open grasslands, shrubs, mango and cashew nut trees. Farmlands close to settlements are dominated by coconut plantations.

Data collection

Data was collected in four habitat types, forest core, forest edge, fallow farmlands and crop fields. The forest core of Kaya Muhaka Forest was characterised by dense tree cover and a thick canopy. Forest edge was the transition from forest vegetation to fallow farmland. Fallow farmland was characterized by uncultivated land with wild herbaceous plants, stands of mango trees and cashew nuts while farmlands were cultivated areas. Two main transects, each 2.5 km long, were established from the forest core through the forest edge, fallow farmland to crop fields. Increasing distance from the forest to crop field was characterised by increased disturbance including uncontrolled habitat burning, human settlement, un-planned access routes and

TABLE I: GPS coordinates of sampling points of bees in Kaya Muhaka Forest and surrounding farmlands.

	N	E		N	E
A1	4.1972°	39.3159°	B1	4.2016°	39.3144°
A2	4.1968°	39.3141°	B2	4.2008°	39.3130°
A3	4.1965°	39.3117°	B3	4.2002°	39.3102°
A4	4.1957°	39.3091°	B4	4.1997°	39.3075°
A5	4.1948°	39.3064°	B5	4.1993°	39.3045°
A6	4.1940°	39.3036°	B6	4.1990°	39.3018°

planting of some crops with no floral benefit to the pollinators. Six sampling points were located along each transect at intervals of 0.5 km. A set of 3 parallel belt transects, 50 m long and 2 m wide (50 m × 2 m) were laid at each sampling point across the main transects. Belt transects are most effective active sampling methods for bees (Banaszak 1996). A total of 12 sampling points (Tab. I and Fig. I) and 36 belt transects were established and surveyed. Each belt transect was surveyed three times per month from April to September 2010 covering the wet and the dry period. To adequately sample species with different diurnal patterns, sampling was done between 8.30 a.m. – 12.30 a.m. and 2.00 p.m. – 4.00 p.m. during sunny and partly cloudy days.

All foraging bees encountered along the 50 m × 2 m belt transects were collected using sweep nets (hand netting) within a standard 20 minute sampling time per belt for diversity and abundance data as described by Potts et al. (2003) and Banaszak (1996). The total number of bee individuals collected during the sampling period was considered as an estimate of bee abundance at each sampling point (Diego & Simberloff 2002). Individual bee samples were coded to be able to associate them with their habitats and floral resources. Bee collections for each day were pinned and later identified at NMK. To assess floral richness at each site, the number of understory plant species with open flowers was recorded at each sampling point. Samples of plants visited by bees were pressed and given the same code as the corresponding bee to correctly document bee-plant association (Gikungu 2006; Gikungu et al. 2011). Plant materials were taken to NMK Herbarium for expert identification.

Data analysis

Diversity was determined based on number of individual species; α Shannon's diversity index. Evenness index (J) was used to measure the relative abundance of bees in the study area. Renyi diversity profiles were used to visually compare bee diversity of the habitats, a higher profile along the entire range from (alpha = 0) to (alpha = Inf.) is considered to be more diverse (Kindt & Coe 2005). Renyi evenness profiles were used to visually compare bee evenness of the habitats, a profile higher than others along the entire range is considered more even.

Cluster analysis was used to analyse the ecological distance among the habitats to depict their similarity in bee species composition. One-way ANOVA was used to compare the diversity and relative abundance of insects among habitats (forest core, forest edge, fallow farmland and crop fields). The relationship between bee species richness and floral resources richness was tested using simple linear regression analysis. Simple linear regression was also used to test the effect of increasing distance from forest core to crop fields on bee species diversity, richness and abundance. Analysis was done using R (R Development Core Team 2011).

Results

Bee richness and abundance

A total of 755 bee individuals were collected on 60 days of the 6 months sampling period. About 41 bee species from 3 families (Apidae 76%, Halictidae 14% and Megachilidae 10%) were recorded (Tab. 2). Bee species richness decreased gradually from crop fields, forest edge, fallow farmland, to forest core (Fig. 2). Fallow farmlands had the highest overall bee abundance followed by crop fields then forest edge. Forest core recorded the least number of bee individuals (Fig. 3).

Effect of increasing distance from forest core on bee diversity and abundance

We recorded high bee species diversity at 0.5 km and 1.5 km from the centre of the forest. Bee diversity increased gradually with decreasing distance to forest core which was considered a no disturbance area. Bee diversity was measured by Shannon diversity index (H'). Bee diversity in fallow farmland at 1 km distance from the forest core was H' = 2.673, while at 1.5 km it was H' = 2.668, at 2 km H = 2.300, and at 2.5 km in crop field it was H' = 1.883. The lowest diversity was recorded at 0 km (forest core), with H' = 1.571. Distance away from forest had significant effect on bee diversity ($F_{1,4} = 10.705$, $P < 0.05$). The highest bee abundance was recorded at 0.5 km (forest edge) and 1 km (fallow farmland) from the forest core. There was a marked reduction in bee abundance beyond 1 km from the forest core. Increasing distance from forest core had no significant effect on total bee abundance ($F_{1,4} = 0.389$, $P > 0.05$).

Effect of habitat and floral resources on bee composition

Crop fields and forest edge had similar bee species' composition. Fallow farmland shared more species with crop fields and forest edge than forest core, as shown by the cluster analysis (Fig. 4). Several floral resources, common in fallow farmlands and crop fields were visited by many bee species and were considered important flora for bees in the area, e.g. *Agathisanthemum bojeri*, *Crotalaria emarginata*, *Truimfetta rhomboidea*, *Cajanus cajan*, *Rhynchosia velutina*, *Julbernardia magnistipulata*, *Hyptis suaveolens*, *Eriosema glomeratum* and *Waltheria indica* (Tab. 2). The most abundant floral resource was *Agathisanthemum bojeri*; it was abundant at the forest edge. Floral richness declined gradually from the forest edge to crop fields. The highest richness was recorded at the forest edge followed by fallow farmlands then crop fields. The lowest floral resource

TABLE 2: Bee species and associated floral resources in Kaya Muhaka Forest and surrounding farmland collected in the period April 2010 to September 2010.

Bee Family	Bee species	Number of individuals encountered	Floral resources
Apidae	*Amegilla mimadvena*	4	*Hibiscus surattensis*
	Amegilla sp. 1	39	*Agathisanthemum bojeri*
			Julbernardia magnistipulata
			Rhynchosia velutina
			Vernonia cinerea
	Apis mellifera	59	*Abutilon zanzibaricum*
			Agathisanthemum bojeri
			Julbernardia magnistipulata
			Ludwigia sp.
			Nesaea radicans
			Sorindeia madagascariensis
			Tridax procumbens
	Braunsapis sp.	33	*Cocos nucifera*
			Crotalaria emarginata Benth
			Hoslundia opposita
			Paulinia pinata
	Ceratina sp. 1	22	*Allophylus rubifolius*
			Hoslundia opposita
	Ceratina sp. 2	7	*Agathisanthemum bojeri*
			Tridax procumbens
	Ceratina sp. 3	163	*Allophylus rubifolius*
			Eriosema glomeratum
			Gossypioides kirkii
			Paulinia pinata
			Waltheria indica
	Ceratina sp. 4	19	*Agathisanthemum bojeri*
	Ceratina sp. 5	1	*Waltheria indica*
	Ceratina sp. 6	4	*Agathisanthemum bojeri*
	Ceratina sp. 7	15	*Agathisanthemum bojeri*
			Waltheria indica
	Dactylurina schmidti	19	*Cajanus cajan*
			Urena lobata
	Hypotrigona sp. 1	15	*Cajanus cajan*
	Hypotrigona sp. 2	4	*Cajanus cajan*
	Macrogalea candida	30	*Agathisanthemum bojeri*
			Hewittia malabarica
			Waltheria indica
	Meliponula ferruginea	8	*Agathisanthemum bojeri*
			Cocos nucifera
	Xylocopa caffra	33	*Abutilon zanzibaricum*
			Agathisanthemum bojeri
			Cajanus cajan
			Rhynchosia velutina
			Rhynchosia velutina
			Vernonia cinerea
	Xylocopa flavicollis	38	*Abutilon zanzibaricum*
			Rhynchosia velutina
			Vernonia cinerea
	Xylocopa hottentota	20	*Julbernardia magnistipulata*
			Rhynchosia velutina
			Vernonia cinerea
			Waltheria indica
	Xylocopa nigrita	4	*Cajanus cajan*

TABLE 2 continued

Bee Family	Bee species	Number of individuals encountered	Floral resources
	Xylocopa scioensis	21	*Rhynchosia velutina*
Halictidae	*Lasioglosum* sp.	4	*Allophylus rubifolius*
			Eriosema glomeratum
			Truimfetta rhomboidea
	Lipotriches sp. 1	39	*Pupalia lappalea*
	Lipotriches sp. 2	6	*Pupalia lappalea*
	Lipotriches sp. 3	12	*Agathisanthemum bojeri*
	Lipotriches sp. 4	11	*Hoslundia opposita*
	Nomia sp.	4	*Julbernardia magnistipulata*
	Pseudapis sp.	24	*Agathisanthemum bojeri*
			Allophylus rubifolius
			Chamaeerista mimosoides
			Eriosema glomeratum
			Pupalia lappalea
	Pseudapis sp. 2	8	*Chamaeerista mimosoides*
	Steganomus sp.	11	*Crotalaria emarginata* Benth
Megachilidae	*Euaspis abdominalis*	3	*Paulinia pinata*
	Heriades sp.	18	*Truimfetta rhomboidea*
	Megachile discolour	4	*Crotalaria emarginata* Benth
			Crotalaria emarginata Benth
	Megachile felina	7	*Crotalaria emarginata* Benth
	Megachile sp. 2	19	*Cajanus cajan*
			Crotalaria emarginata Benth
			Hyptis suaveolens
			Indigofera paniculata
			Indigofera paniculata
			Julbernardia magnistipulata
			Tephrosia villosa
			Truimfetta rhomboidea
	Megachile sp. 3	10	*Hyptis suaveolens*
			Truimfetta rhomboidea
	Megachile sp. 5	1	*Hyptis suaveolens*
	Megachile sp. 7	4	*Philenoptera bussei*
	Megachile sp. 8	4	*Hyptis suaveolens*
	Megachille sp. 6	1	*Agathisanthemum bojeri*
	Pachyanthidium sp.	7	*Rhynchosia velutina*

richness was noticed in the forest core at 0 km. Increasing distance from forest core had no significant effect on floral richness ($F_{1.4} = 0.0005$, $P = 0.983$). Floral richness had a significant positive effect on bee species richness ($F_{1.10} = 34.5$, $P = < 0.0002$, $R^2 = 0.775$) (Fig. 5).

Bee relative abundance

Overall, there was uneven distribution of bee species in the survey area, the evenness index was $J = 0.427$. However, evenness was generally higher in the forest core than in other habitats followed by fallow farmlands. Forest edge and crop fields were largely uneven as indicated by their low profiles (Fig. 6). High abundance of *Ceratina* sp. 3, *Apis mellifera*, *Amegilla* sp. 1, *Lipotriches* sp. 1, *Xlocopa flavicollis*, *Braunsapis* sp., *Macrogalea candida*, and *Xylocopa caffra*, affected mainly the evenness of bee distribution.

DISCUSSION

Effect of habitat type on bee diversity and abundance

Overall bee abundance is a positive function of the abundance of flowers in a habitat (Banaszak 1996; Potts et al. 2003). This fact could explain the high overall diversity and abundance of bees in fallow farmland recorded in the study area. The key floral resources supporting high bee diversity in fallow farmlands were mainly annual plants. Bee diversity is known to be strongly correlated with the species

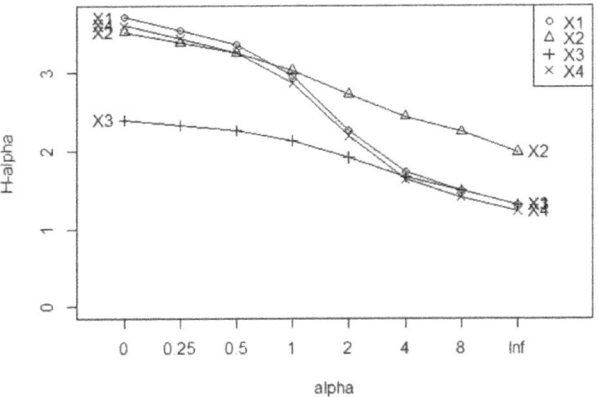

FIGURE 2: Rényi diversity profiles for separate habitats; alpha = 0 indicates species richness; marked points along each profile represent randomly selected sampling points for each habitat: X1-Crop fields, X2-Fallow farmland, X3-Forest core, X4-Forest edge.

richness of annuals (Potts et al. 2003) and overall floral diversity (Banaszak 1996). Our results finding agree with Gikungu et al. (2011) in Kakamega forest, where the overall bee abundance was high in fallow farmlands as well. Apart from a high richness of floral resources, fallow farmland was constantly very heterogeneous, consisting of woody and herbaceous plants offering nesting and feeding requirements for a large diversity of bee species. The heterogeneity attributed to a mix of large cashew nut trees, mango trees and associated woody shrubs, annual flowering plants and grassland patches.

At the forest edge, high abundance of *Agathesanthemum bojeri, Tridax procumbens* and *Waltheria indica* probably attracted foraging bees contributing to an important bee abundance. Bee abundance in crop fields was similar to that of the forest edge due to low-level land use and limited or no use of agrochemicals. After crop harvest, abundant weedy plants flowered on the fields attracting foraging bee species. Crop fields were also characterised by unmanaged hedgerows, which appeared advantageous for the survival of wild flowers that attracted bees. Also Potts et al. (2003) showed that open habitats with abundant floral resources attract numerous foraging bee species. Such disturbed habitats have favourable environmental conditions that are correlated with high bee abundance including temperature, light intensity and humidity (Liow et al. 2001). The low bee diversity and abundance observed in the forest core could thus be attributed to low temperatures, higher humidity and low light intensity associated with a closed forest canopy. Furthermore, the forest core had the lowest abundance and richness of floral resources, explaining the relatively low bee diversity we recorded. However, we considered only the understory community where very few plants were flowering.

Surprisingly, crop fields had the greatest absolute bee species richness. Besides floral resource abundance this could further be attributed to the high attraction of 'tourist' bee species to such disturbed habitats. These do not reside within them and have potentially large foraging ranges like *Amegilla* and *Xylocopa* sp. (Liow et al. 2001). If managed properly, crop fields may thus offer supplementary conservation sites for bee species. Carefully designed wild flower and crop mixtures in crop fields could supply important floral resources to bee species in farmlands and support bee conservation in farmlands. Highly abundant bee species included *Ceratina* sp, *Apis mellifera, Lipotriches* sp,

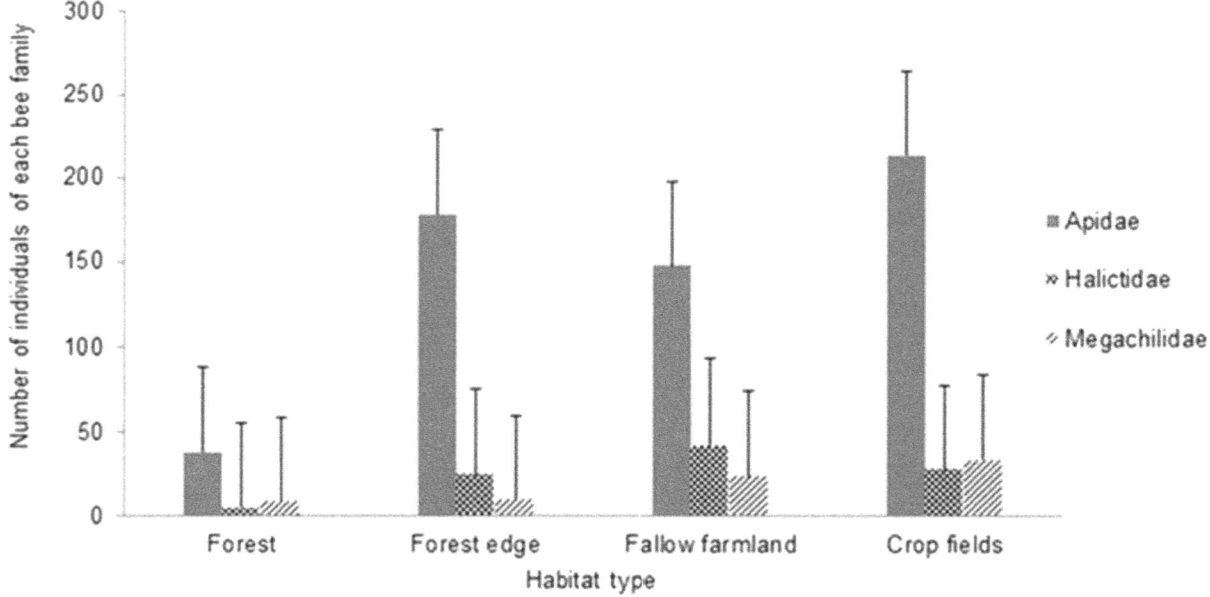

FIGURE 3: Abundance of each bee family per habitat. Bars represent number of individuals with standard error.

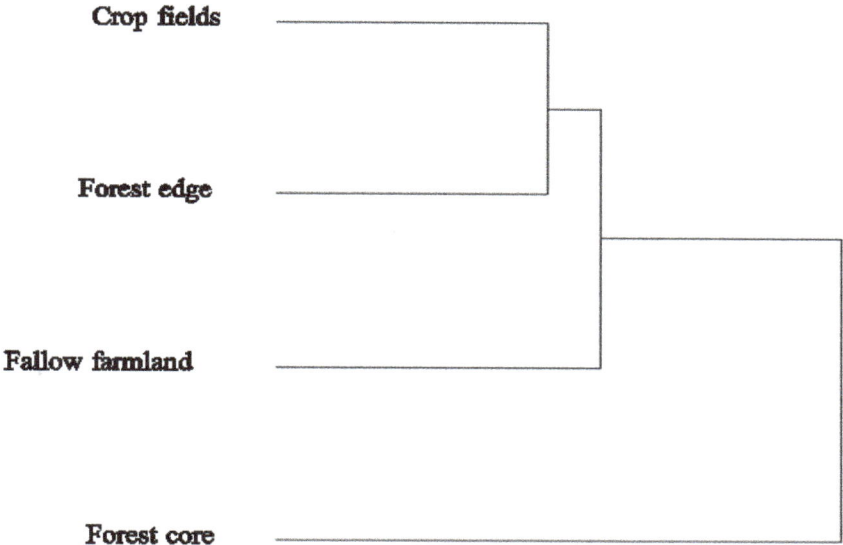

FIGURE 4: Cluster analysis of bee species composition in Kaya Muhaka Forest and surrounding farmland: X1-Crop fields, X2-Fallow farmland, X3-Forest core, X4-Forest edge.

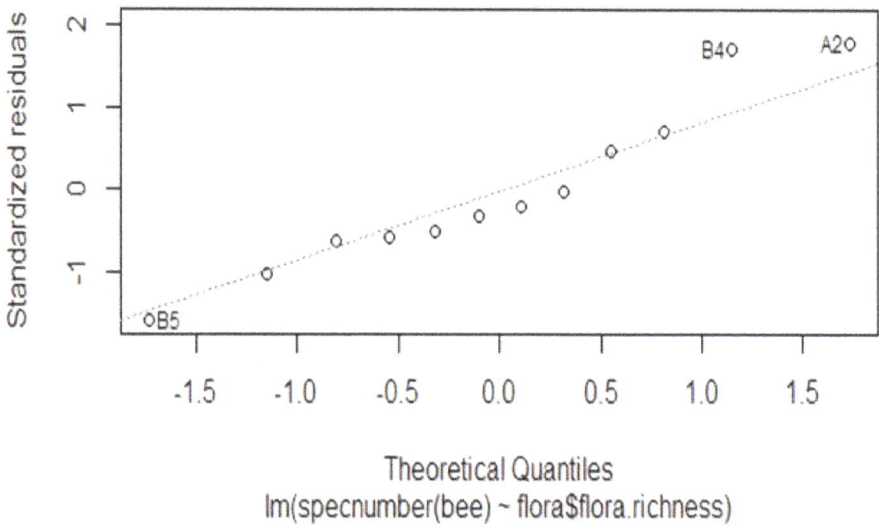

FIGURE 5: Effect of floral resources richness on bee species richness.

Xylocopa flavicollis, Braunsapis sp, *Macrogalea candida* and *Xylocopa caffra.* However, low abundance of some bee species, especially solitary bees, could be attributed to a limitation or absence of their preferred host plants. Possibly rare bee species, which were recorded in low numbers were i.a. *Euaspis* sp., *Lasioglosum* sp., *Amegilla mimadvena,* and *Xylocopa nigrita.* Future studies should focus on such rare species and their habitat requirements.

Halictidae were most abundant in fallow farmland where relatively stable habitat conditions with a mix of grasses and shrubs in this habitat could provide favourable nesting sites. Apidae was the dominant bee family across all habitats, similar to a study at Mt. Carmel (Potts et al. 2003). Most Apidae species are long distance foragers and can explore diverse nectariferous flowers across many habitats. They are less affected by differences in habitats and abundant in diverse habitat types. High abundance of *Ceratina* sp. and honey bees (*Apis mellifera*) contributed largely to the dominance of the family.

Megachilidae were more common in crop fields, attributable to the presence of their important floral resources, Leguminaceae plants e.g *Cajanus cajan* and *Crotalaria emarginata.* The distribution of Megachilidae appeared to be linked to pollen resources as in the case of leguminous plants, see also Potts et al. (2003). It is evident that different groups of bees show contrasting responses to land use change, likely driven by differences in their foraging and nesting biology (Brosi et al. 2008).

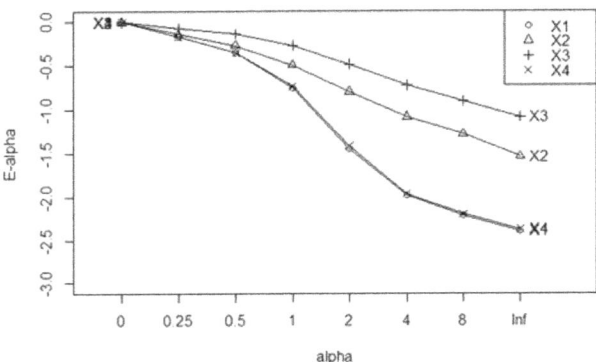

FIGURE 6: Rényi evenness profiles of bees for separate habitats; marked points along each profile represent randomly selected sampling points for each habitat: X1-Crop fields, X2-Fallow farmland, X3-Forest core, X4-Forest edge

Effect of increasing distance from forest on bee diversity and abundance

In contrast to Brosi et al. 2007 and Klein et al. (2007), we found a significant negative effect of distance from the forest core on bee diversity. The results provide evidence that distance from natural habitats may strongly determine spatial distribution of bees. This finding conforms to the pattern found by Rickets et al. (2008) which reported that native pollinator visitation rate drops to 50 % away from natural habitats. The result is also consistent with the findings of similar research (e.g., Morandin & Winson 2005; Chacoff & Aizen 2006; Bailey et al. 2014). However, overall bee abundance did not vary significantly with distance from the forest implying that most bees are dependent on habitat quality rather than proximity to primary forests and some occur in high abundance away from the natural forest. Kaya Muhaka forest and forest edge probably act as buffers for conservation of bees where they may seek refuge for nesting and foraging when the farmlands are extensively impoverished and indiscriminately disturbed. Though the level of habitat heterogeneity in the habitats was not the same, it was observed across all habitats. This could have been the most important factor influencing the diversity and abundance of bees. Regional habitat heterogeneity could be a more important factor than farming practice in determining the diversity and abundance of pollinators in agricultural landscapes (Brosi et al. 2008).

This study reveals that habitat heterogeneity, presence of natural habitats and land use practices are key factors in determining bee diversity and abundance in a given region. Natural habitats and sites with high heterogeneity have the highest capacity to provide diverse ecological requirements for insect pollinators including shelter, foraging, mating and breeding sites (Kremen et al. 2007). Open and heterogeneous habitats support high bee diversity and overall pollinator abundance. Forest core, forest edge, fallow farmlands and crop fields are important and complement each other in the conservation of insect pollinators. Contrary to traditional opinion, farmlands can be important conservation areas when properly managed to maintain habitat quality, a fact emphasised by Klein et al. (2007). Kaya Muhaka forest and surrounding farmlands have the potential to support diverse bee communities. However, current human activities may lead to habitat change and degradation and will need to be controlled to further threaten bee populations. Forest edge is an important foraging site for insect pollinators and needs to be conserved. An ecosystem approach to farming in the Muhaka area, along with careful management including wildflowers and crop mixtures, could help to make the farmlands important conservation sites for bees and other pollinators. Our findings echo the need for an ecosystem approach for the management of agro-ecosystems to support sustainable pollination services and contribute to our understanding of the effects of land use change on insect pollinators and their degree of resilience in disturbed habitats.

ACKNOWLEDGEMENTS

We acknowledge the National Council for Science and Technology (NCST) for funding the study. National Museums of Kenya (NMK) management for hosting the work at Centre for Bee Biology and Pollination (CBBP). CBBP supported the work with taxonomic skills, field materials and equipment; it also provided working space and resources that were useful for bee identification. We thank Kaya elders and Kaya Muhaka community. Staff of herbarium, NMK for floral resources identification and CFCU staff at Ukunda for logistical assistance.

REFERENCES

Allen-Wardell G, Berhardt T, Bitner R (1998) The potential consequences of pollinator declines on the conservation of biodiversity and stability of food crop yields. Conservation Biology 12:8-17.

Bailey S, Requier F, Nusillard B, Stuart PMR, Potts SG, Bouget C (2014) Distance from forest edge affects bee pollinators in oilseed rape fields. Ecology and Evolution 4:370-380.

Banaszak J (1996) Ecological bases of conservation of wild bees. In The conservation of bees, Linnean Society Symposium series 18:55-62.

Banaszak J (2000) Effect of habitat heterogeneity on the diversity and density of pollinating insects. In: Ekom B, Irwin ME, Robert Y (eds) Interchanges of Insects between Agricultural and Surrounding Landscapes. Kluwer Academic Publishers, Dordrecht, pp 123–140.

Banaszak J (1992) Strategy for conservation of wild bees in an agricultural landscape. Agriculture Ecosystems and Environment 40:179-192.

Baude M, Kunin WE, Boatman ND, Conyers S, Davies N, Gillespie MAK, Morton RD, Smart SM, Memmott J (2016). Historical nectar assessment reveals the fall and rise of floral resources in Britain. Nature 530: 85-88.

Biesmeijer JC, Roberts SPM, Reemer M, Ohlemuller R, Edwards M, Peeters T (2006) Parallel declines in pollinators and insect-pollinated plants in Britain and the Netherlands. Science 313:351-354.

Brosi BJ, Daily G, Ehrlich PR (2007) Bee community shifts with landscape context in a tropical countryside. Ecological Applications 17:418-430.

Brosi BJ, Daily GC, Shih TM, Oviedo F, Durán G (2008) The effects of forest fragmentation on bee communities in tropical countryside. Journal of Applied Ecology 45:773-783.

Burgess ND, Clarke, GP, Madgewick, J, Robertson SA, Dickinsen A (2000) Distribution and status. In: Burgess N, Clarke G (eds).

The Coastal Forests of Eastern Africa. IUCN, Gland and Cambridge, pp 71-81.

CEPF (2003) Ecosystem Profile: Eastern Arc Mountains and Coastal Forests of Tanzania and Kenya Biodiversity Hotspot. Critical Ecosystems Partnership Fund, Washington DC. http://www.cepf.net/Documents/final.easternarc.ep.pdf (accessed April 2014).

Chacoff NP, Aizen MA, (2006) Edge effects on flower-visiting insects in grapefruit plantations bordering premontane subtropical forest. Journal of Applied Ecology 43:18-27.

Danks D (1994) Regional diversity of insects in North America. American Entomologist 40:50-55.

Didham RK, Ghazoul J, Stork NE, Davis AJ (1996) Insects in fragmented forests: a functional approach. Trends in Ecology and Evolution 11:255-260.

Diego P, Simberloff D (2002) Ecological specialization and susceptibility to disturbance: Conjectures and refutations. The American Naturalist 159:606-623.

Driscoll D (2005) Is the matrix a sea? Habitat specificity in a naturally fragmented landscape. Ecological Entomology 30:8-16.

Eardley C, Gikungu M, Schwarz M (2009) Bee conservation in Sub-Saharan Africa and Madagascar: diversity, status and threats. Apidologie 40:355-366.

Gikungu M, Wittmann D, Irungu D, Kraemer M (2011) Bee diversity along a forest regeneration gradient in Western Kenya. Journal of Apicultural Research 50:22-34.

Gikungu M (2006) Bee Diversity and some Aspects of their Ecological Interactions with Plants in a Successional Tropical Community. PhD dissertation, University of Bonn.

Gikungu M (2002) Studies on bee population and some aspects of their foraging behaviour in Mt. Kenya forest. Msc thesis, University of Nairobi.

Hartmann I (2004) "No Tree, No Bee – No Honey, No Money": The Management of Resources and Marginalisation in Beekeeping Societies of South West Ethiopia. Proceedings of the conference: Bridge Scales and Epistemologies. Alexandria, Egypt.

Keams C, Inouye D, Waser N (1998) Endangered mutualisms: the conservation of plant–pollinator interactions. Annual Review of Ecology and Systematics 29:83-112.

Kevan GP (1999) Pollinators as bioindicators of state of the environment: species activity and diversity. Agriculture Ecosystems and Environment 74:373-393.

Kindt R, Coe R (2005) Tree diversity analysis. A manual and software for common statistical methods for ecological and biodiversity studies, ICRAF: Nairobi.

Klein AM, Vaissiere BE, Cane JH, Steffan-Dewenter I, Cunningham SA, Kremen C, Tscharntke T (2007) Importance of Pollinators in Changing Landscapes for World Crops. Proceedings of the Royal Society B 274:303-313.

Koh I, Lonsdorfa EV, Williams NM, Brittain C, Isaacs R, Gibbs J, Ricketts TH (2016) Modeling the status, trends, and impacts of wild bee abundance in the United States. PNAS 113:140-145.

Kremen C, Ricketts T (2000) Global perspectives on pollination disruptions. Conservation Biology 14:1226-1228.

Kremen C, Williams N, Thorp R (2002) Crop pollination from native bees at risk from agricultural intensification. In: Proceedings of the National Academy of Sciences 99:16812-16816.

Kremen C, Williams NM, Aizen MA, Gemmill-Herren B, LeBuhn G, Minckley R, Packer L (2007) Pollination and other ecosystem services produced by mobile organisms: a conceptual framework for the effects of land-use change. Ecology Letters 10:299-314.

LaSalle J, Gould ID (1993) Hymenoptera: Their diversity and their impact on the diversity of other organisms. In: LaSalle J, Gould ID (eds) Hymenoptera and biodiversity. CAB International, Wallingford, pp 1-26.

Lehmann I, Kioko E (2005) Lepidoptera diversity, floristic composition and structure of three kaya forests on the south coast of Kenya. Journal of East African Natural History 94:121-161.

Liow HL, Sodhi NS, Elmqvist T (2001) Bee Diversity along a Disturbance Gradient in Tropical Lowland Forests of South-East Asia. Journal of Applied Ecology 38:180-192.

Minckley R, Cane J, Kervin L (2000) Origins and ecological consequences of pollen specialization among desert bees. Proceedings of the Royal Society of London 267:265-271.

Morandin LA, Winston ML (2006) Pollinators provide economic incentive to preserve natural land in agroecosystems. Agriculture, Ecosystems & Environment 116:289-292.

O'Toole C (1993) Diversity of native bees in agro-ecosystems. In: LaSalle J, Gauld ID (eds). Hymenoptera and biodiversity. CAB International, Wallingford, pp 169-196.

Pauw A (2007) Collapse of a pollination web in small conservation areas. Ecology 88:1759-1769.

Potts SG, Vulliamy B, Dafni A, Ne'eman G, Willmer P (2003) Linking bees and flowers: How do floral communities structure pollinator communities? Ecology 84:2628-2642.

R Development Core Team (2011) R: A language and environment for statistical computing. The R Foundation for Statistical Computing, Vienna: http://www.r-project.org/ (accessed April 2014).

São Paulo Declaration (1999) Report on the recommendations of the workshop on the conservation and sustainable use of pollinators in agriculture with emphasis on bees. Brazilian Ministry of the Environment, Brasilia.

Sheffield CS, Kevan GP, Smith RF (2003) Bee Species of Nova Scotia, Canada, with New Records and Notes on Bionomics and Floral Relations (Hymenoptera:Apoidea). Journal of the Kansas Entomological Society 76:357-384.

Steffan-Dewenter I, Westphal C (2008) The interplay of pollinator diversity, pollination services and landscape change. Journal of Applied Ecology 45:737-741.

Stubbs S, Drummond A, Allard LS (1997) Bee Conservation and Increasing *Osmia sp.* in Maine Low bush Blueberry Fields. Northeastern Naturalist 4:133-144.

TFCG (2007) CEPF'S investment in the Eastern Arc and coastal forests of Tanzania and Kenya briefing book. CEPF, Washington DC.

Tylianakis J, Klein A, Tscharntke T (2005) Spatiotemporal variation in the diversity of hymenoptera across a tropical habitat gradient. Ecology 86:3296-3302.

Wadley L, Colfer I (2004) Sacred forest, hunting, and conservation in West Kalimantan, Indonesia. Human Ecology 32:313-338.

Winfree R, Griswold T, Kremen C (2007) Effect of human disturbance on bee communities in a forested ecosystem. Conservation Biology 21:213-223.

WWF-US (2003) The ecosystem profile: Eastern Arc Mountains and Coastal Forests of Tanzania and Kenya biodiversity hotspot. WWF Eastern Africa Regional Programme, Washington, DC. http://www.cbd.int/database/attachment/?id=717 (accessed May 2014).

10

Aggressive Displacement of Carpenter Bees *Xylocopa nigrita* from Flowers of *Lagenaria sphaerica* (Cucurbitaceae) by Territorial Male Eastern Olive Sunbirds (*Cyanomitra olivacea*) in Tanzania

Jeff Ollerton [1,*] and Clive Nuttman [2]

[1]*School of Science and Technology, Newton Building, Avenue Campus, University of Northampton, Northampton, NN2 6JD, U.K*
[2]*The Tropical Biology Association, Department of Zoology, University of Cambridge, Downing Street, Cambridge, CB2 3EJ, U.K*

Abstract— Male Eastern Olive Sunbirds (*Cyanomitra olivacea*) and *Xylocopa nigrita* carpenter bees in Tanzania both utilise the flowers of male plants of *Lagenaria sphaerica* (Cucurbitaceae) as a source of nectar. The sunbirds set up territories defending this nectar resource. Observations of interactions between the sunbirds and the carpenter bees show that the bees are aggressively displaced from flowers when spotted by the birds. Only the bees can be considered as legitimate pollinators as the birds do not contact the anthers of the male flowers and were never seen visiting nectarless female flowers of *Lagenaria sphaerica*. Such territory defence may have implications for the frequency of movement and composition of pollen being transferred from male to female flowers which warrants further research.

Keywords: *Africa, Bee pollination, Birds, Mutualism, Territoriality, Tropical ecology*

INTRODUCTION

Many animals are territorial and actively defend patches of habitat from individuals of the same and/or different species. The motivation for this aggression is varied and can include defence of mates and offspring, monopolisation of food resources, or combinations of these (e.g. Marler 1976). Although most pollinating animals are not strictly territorial, studies have shown that intra- and interspecific territorial aggression occurs in a wide range of flower visiting animals and that this may have implications for rates of pollen movement and reproductive success in the plants that they pollinate. Such territoriality has been observed in bees (e.g. Wirtz et al. 1988; Willmer et al. 1994; Johnson & Steiner 1994; Jürgens et al. 2009) and bats (Elmqvist et al. 1992), and there is an especially long and rich literature on hummingbirds (e.g. Stiles & Wolf 1970; Primack & Howe 1975; Stiles 1975; Boyden 1978; Carpenter 1979; Cotton 1998; Franceschinelli & Bawa 2000; Canela & Sazima 2003; Rocca & Sazima 2006; Jacobi & Antonini 2008; Lara et al. 2009).

In contrast there are relatively few published observations of such territoriality in Old World sunbirds and their relatives (Gill & Wolf 1977; Frost & Frost 1980; Akinpelu 1989; Lott & Lott 1991; Evans & Hatchwell 1992; Burd 1995; Larsson & Hemborg 1995; Symes et al. 2008; Geerts & Pauw 2009) and particularly scarce are examples of interspecific aggression, the only ones of which we are aware being Akinpelu (1989), Tropek et al. (in press) and Nuttman (unpublished data 2000) – see Conclusions. Territorial defence of floral resources can be considered a form of interference competition, as distinct from exploitative competition, both of which have been noted in bee-bird-flower systems (e.g. Hansen et al. 2002, Geerts & Pauw 2011).

During a period in the field in Tanzania we noticed that male Eastern Olive Sunbirds (*Cyanomitra olivacea* - Nectariniidae) feeding on flowers of *Lagenaria sphaerica* (Cucurbitaceae) would occasionally chase carpenter bees (*Xylocopa nigrita*) from those flowers. In this short communication we quantify the frequency of this interaction, and discuss resource use by the flower visitors and the potential negative impacts on pollination rate and seed set to *L. sphaerica*.

MATERIALS AND METHODS

Observations were made in secondary rainforest around the Amani Nature Reserve headquarters in the East Usambara Mountains, Tanzania (5° 6' 3.95" S, 38° 37' 45.26" E) between 26th July and 24th August 2011. *Lagenaria sphaerica* E. Mey. (Cucurbitaceae) is a climbing herb of forest edges with a distribution spanning Somalia to South Africa (Blundell 1987). At Amani, the plant is infrequently encountered at the edges of clearings, ponds and

*Corresponding author:
Jeff.Ollerton@northampton.ac.uk

rivers, clambering up trees to a height of around 10 m. In common with many of the Cucurbitaceae the species is dioecious. We estimated the local population of *L. sphaerica* at Amani to comprise three male and three female plants within an area of approximately 2 ha; the only other plant observed was a fourth female individual some 1 km from the Amani Nature Reserve offices. The genders were spatially clustered but this was probably coincidental.

Eastern Olive Sunbirds *Cyanomitra olivacea* are resident endemics of East Africa and occur from Kenya east of the Rift Valley and southern Somalia, down to the Eastern Cape, westwards to Malawi and southeast Zambia, eastwards as far as the Tanzanian islands of Zanzibar, Pemba and Mafia (Fry & Keith 2000). They are common birds in the undergrowth and at higher levels of both mature and secondary forest, and may be encountered as individuals, in pairs, or as groups of 4 to 5 birds, with larger aggregations occurring in trees that are mass flowering. These sunbirds have a mixed diet of small fruit, invertebrates and nectar, visiting a wide range of both native and introduced species. At Amani we observed them feeding on flowers of African tulip tree *Spathodea campanulata* (Bignoniaceae), bananas *Musa* var. (Musaceae), *Thunbergia grandiflora* (Acanthaceae), *Syzygium* sp. and *Callistemon* sp. (Myrtaceae), and other species.

Male Eastern Olive Sunbirds are polygynous and territorial. Dominant individuals set up territories around plants in flower that provide sufficient nectar, aggressively displacing all other sunbirds except females who copulate with the resident male. Mated females then make nests within the male's territory (Fry & Keith 2000, Cheke et al. 2001). Birds are relatively sedentary with the longest recovery distance of a ringed bird being only 4 km (Cheke et al. 2001).

The carpenter bees that we observed have been identified as female *Xylocopa nigrita* (Fabricius, 1775) (Apidae, Xylocopinae). A larger, reddish brown *Xylocopa* sp., observed only once, may be the male of this sexually dimorphic species. The females were only observed visiting *L. sphaerica* and a second species of Cucurbitaceae, as well as the non-native *Thunbergia grandiflora* (Acanthaceae) at Amani. We did not otherwise encounter it within the surrounding forest.

Observations of interactions between the flowers of *L. sphaerica*, Eastern Olive Sunbirds *C. olivacea* (A. Smith, 1840) and carpenter bees *Xylocopa nigrita* were carried out during daylight hours on a single, large male plant which was the most easily accessible of those that we located. We recorded the time, number of open flowers on the plant and the number of flowers foraged, as well as residence times of animals on flowers. Total observation time for the male plant was 1630 minutes over 18 days. Female plants in the vicinity were observed for flower visitors for a total of 480 minutes over 9 mornings, all between 0620 and 0915.

We also recorded the opening times of flowers and measured their dimensions. Nectar characteristics were assessed as standing crop from unbagged flowers at two time periods (0900-100 and 1500-1600). Volume was measured using microcapillary tubes and concentration assessed using sugar refractometers (Kearns & Inouye 1993; Dafni et al. 2005). Data from the two time periods were not statistically significantly different and were therefore pooled. Statistical analyses were carried out using SPSS 17.0.

FIGURE 1: (a) Male Eastern Olive Sunbird in its territory within a patch of male *Lagenaria sphaerica*. (b) Male Eastern Olive Sunbird feeding on nectar in flowers of male *Lagenaria sphaerica*. Note that the head and body of the bird is not coming into close contact with the anther cone of the flower. (c) & (d) Female *Xylocopa nigrita* visiting flowers of a male *Lagenaria sphaerica*. Note that the underside of the body of the bee is coming into close contact with the anther cone of the flower. Photographs by Anna Rausch.

FIGURE 2 – Male (left – scale bar = 15 mm) and female (right – scale bar = 25 mm) flowers of *Lagenaria sphaerica* (Cucurbitaceae). Note the nectar chamber formed by the filament bases in male flowers; this is absent in the rewardless female flowers.

RESULTS AND DISCUSSION

Male sunbirds were observed to establish territories only around male *L. sphaerica* plants (Fig. 1a). This is explained by the fact that whilst male flowers produce significant quantities of rather concentrated nectar, female flowers do not produce nectar (Tab. 1, Fig. 2). Flower dimensions of males and females were similar, though highly variable within gender; the largest female flowers observed had much a wider corolla diameter than the largest males, though overall there was no statistically significant difference, perhaps because of the smaller sample size of female flowers (independent samples t-test for unequal variances: t = -1.68, df = 8.4, P = 0.13). However mean diameter of the androecium was larger than that of the gynoecium (independent samples t-test: t = 7.37, df = 36, P < 0.001; Tab. 1). Both male and female flowers were scented with a similar, fresh, sweet odour, though our perception was that female flower odour contained a citrusy component that was absent from the male flowers (see Ashman 2009 for a discussion of gender scent differences in dioecious species). The overall similarity of the two genders suggests that unrewarding female flowers are mimicking rewarding male flowers and relying on occasional visits by pollinators which are deceived into expecting nectar to be present. Deceit pollination by rewardless female flowers is known from other species (e.g. Bawa 1980; Willson & Agren 1989; Armstrong 1997) and may explain why female floral display was small compared to male display, as is expected in model-mimic systems (female range = 0 to 5 flowers per plant per day; male range = 12 to 36 flowers per plant per day). However during the review process one anonymous referee noted that "This is common in dioecious species since female plants need to invest more in eventually producing fruits. Common for example in *Leucadendron* in South Africa." The small sample size and the slightly different growing conditions of male and female plants in our study make any conclusions tentative and would require testing using a common garden experiment.

Male flowers opened in the early morning (before 0700) and closed in the early evening at dusk, then reopened once more the next day, before finally closing for good the second evening. Female flowers also opened in the early morning; in contrast, however, they always closed prior to midday and did not reopen.

Nectar in male flowers was exploited by a diversity of flower visitors of varying abundance (Tab. 2). The most common flower visitors were female carpenter bees (*Xylocopa nigrita* – Figs. 1c and d) followed by Eastern Olive Sunbirds (*Cyanomitra olivacea* - Figs. 1a and b) who collectively accounted for over 87% of all visits. The only observed visitor to a female flower was a single female individual of *Xylocopa nigrita* (Tab. 2). The remainder of this paper will focus on these two most abundant flower visitors.

Both *Cyanomitra* sunbirds and *Xylocopa* bees actively visited male flowers throughout the day (Fig. 3). If average visitation rates during four time periods (0700-0959; 1000-1259; 1300-1559; 1600-1859) are considered, average flower visitation across the day was relatively constant in sunbirds (Kruskal Wallis Test, χ^2 = 2.3, df = 3, P = 0.52), whilst *Xylocopa* foraging was significantly higher in the morning (Kruskal Wallis Test, χ^2 = 16.9, df = 3, P = 0.001). The two most frequent visitors also differed in their residence times on flowers, carpenter bees spending on average more than 60% longer than sunbirds (2.9 ± 1.1s versus 1.8 ± 0.9s; independent samples t-test: t = -3.1, df = 31, P = 0.004). During five-minute foraging bouts, *Xylocopa* individuals on average visited more flowers than sunbirds (10.7 versus 7.3 flowers per 5 minutes; two sample t test: t = 2.9, df = 179, P = 0.004).

TABLE 1: Flower sizes and nectar characteristics of male and female flowers of *Lagenaria sphaerica*. Samples sizes (number of flowers assessed) are given in brackets.

	Corolla diameter (mm)		Androecium/gynoecium diameter (mm)		Nectar volume (µl)	Nectar concentration (%)
	Mean ±SD	Range	Mean ± SD	Range	Mean ± SD	Mean ± SD
Male	70.1 ± 5.5 (29)	60 – 79	16.2 ± 1.7	14 – 22	0.63 ± 0.46 (22)	42.4 ± 4.1 (20)
Female	82.9 ± 21.3 (9)	55 – 115	11.5 ± 1.7	9 – 14	0.0 (9)	-

	Male flowers	Female flowers
Carpenter bees (*Xylocopa nigrita*)	69.4	100
Eastern Olive Sunbirds (*Cyanomitra olivacea*)	18.1	0
Other bees (including some different *Xylocopa* spp.)	6.9	0
Lepidoptera (including Hesperiidae)	3.2	0
Other insects	2.4	0
Total number of animals observed	216	1

TABLE 2: Proportional (%) abundance of the different visitors to male and female flowers of *Lagenaria sphaerica*

The sunbirds and the carpenter bees interacted aggressively throughout the day (Fig. 3) but at a low frequency: of 134 *Xylocopa* foraging bouts observed, only 21 (almost 16%) resulted in aggressive behaviour by sunbirds. In all cases the birds chased the bees from flowers and actively pursued them into the forest; the sound of their beaks striking the bees' bodies could sometimes be heard. Intraspecific aggression was less common; female bees chased female bees only twice, whilst the one observation of a putative male *Xylocopa nigrita* resulted in it being chased off by two female bees that were foraging in the same patch. The male sunbird was aggressive to another male twice in 49 foraging bouts (i.e. just over 4% of observations). Other birds and mammals that entered the sunbird's territory but did not feed on the *Lagenaria* flowers (e.g. Green Barbet *Stactolaema olivacea*, Square Tailed Drongo *Dicrurus ludwigii*, bush squirrel *Paraxerus* sp.) were ignored.

Although we did not explicitly test for pollinator effectiveness we are confident that the most important pollinators of *Lagenaria sphaerica* during our period of observation at this site were the female *Xylocopa* bees. We draw this conclusion because: (1) bee visitation rate to male flowers was more than three times that of the sunbirds (Tab. 2); (2) when visiting male flowers, sunbirds appear to pick up little or no pollen on their bill or feathers, due to their long bills probing beneath the position of pollen release from the anthers (see Figs. 1 and 2) compared to the bees which grasp the androecium firmly with their legs and pick up pollen on their ventral surface (Figs. 1b, c and d); see also Janeček et al. (2007) for another very similar example; (3) crucially, bees are the only visitors to be observed on female flowers. The territoriality of the male sunbirds makes it unlikely that they would visit the female flowers and the very occasional visits by *Xylocopa* bees are probably by individuals moving between other nectar sources and testing these flowers to check if they contain nectar or pollen (though no obvious pollen collecting behaviour was observed by these bees on *L. sphaerica*).

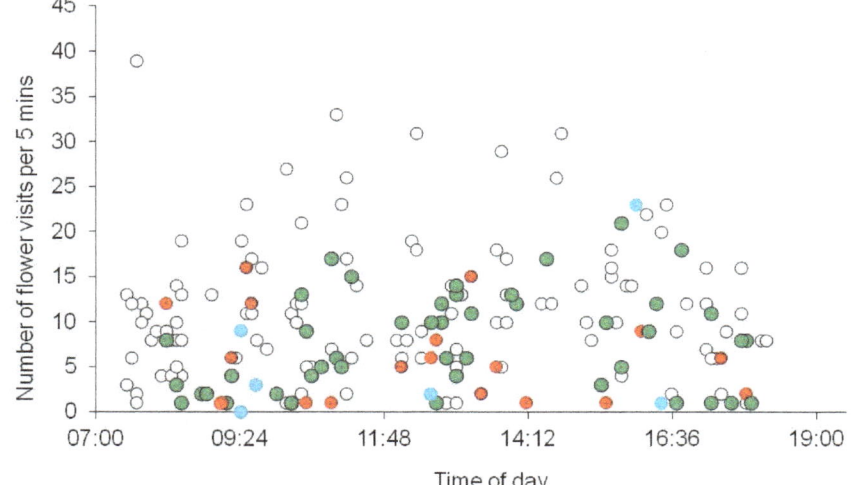

FIGURE 3: Rate of visitation to male *Lagenaria sphaerica* flowers by Eastern Olive Sunbirds (green markers) and carpenter bees (white markers). Intraspecific aggressive interactions between bees are coloured blue. Interspecific aggressive interactions between birds and bees are coloured red and have a nominal value of 1 when bees were chased away by birds before they could land on a flower. Where markers with different colours overlapped completely they have been slightly offset. Note that periods with zero visitation have been removed from the graph to aid interpretation but are included within the statistical analyses (see text).

Future work on this system should include pollinator exclusion and pollen addition experiments that specifically test for the effectiveness of these different flower visitors. A number of recent studies have demonstrated that plants with apparently mixed bird-bee pollination systems are in fact functionally specialized for bee pollination (e.g. Janeček et al. 2007, Watts et al. 2012, Padyšáková et al. 2013). However more such studies are required to assess the generality of these findings.

CONCLUSIONS

Only three other studies that we are aware of have noted interactions between sunbirds and bees on flower patches, but with conflicting results. In Nigeria, Akinpelu (1989) showed that honeybees (*Apis mellifera*) displace Western Olive Sunbirds (*Cyanomitra obscura* syn. *Nectarina olivacea*) from flowers of *Tecoma stans* (Bignoniaceae). However it is unclear whether these were male birds defending territories. In contrast, Nuttman (unpublished data 2000) observed aggressive interactions between Palestine Sunbirds (*Nectarinia osea*) and *Xylocopa pubescens*. The birds defended flowers of an *Erythrina* sp. (Fabaceae) and drove away bees that tried to access floral resources, as was found in our present study. Finally Tropek (in press) recently documented observations of *Cynniris* spp. sunbirds attacking *Xylocopa* spp. carpenter bees in patches of *Hypoestes aristata* (Acanthaceae) in Cameroon, with striking parallels to the present study. It seems likely that this phenomenon is widespread but unreported in the Old World.

Territorial defence of flowers by sunbirds may affect the frequency of movement of pollinators and the composition of pollen (in terms of paternal genotypes) that they carry. This could be particularly significant for dioecious species such as *L. sphaerica* in which the female flowers are unrewarding mimics of the males. If the male sunbirds are indirectly improving the likelihood of female flowers being pollinated by chasing away pollinators from male flowers and forcing them to explore the unrewarding female flowers, they could be engaging in a mutualistic relationship with *L. sphaerica* despite being ineffective pollinators. This would be an interesting question to follow up and warrants further research.

ACKNOWLEDGEMENTS

The framework of this short paper was written at Emau Hill Campsite on Sunday 14th August 2011. Accompanying the trill and cry of the local bird fauna was the constant whine of a chain saw and the occasional crash of another falling tree as land is cleared for a sugar field or *shamba*. The East Usambaras are an area of incredible, unique biodiversity. But they have been, and continue to be, changed faster than they can be conserved or studied. People require food and food production needs land. But people also require the forest for a range of ecosystem services. Conservation of the whole of the Eastern Arc Mountains, including the East Usambaras, should be a Tanzanian (indeed an African) priority and needs to be carefully balanced against the requirements of local people who need the forest as well as land for crops. We say *asante sane* to the people of the East Usambaras for their warmth and hospitality, in the sincere hope that they can achieve such a balance. Thanks also to the staff and students of the Amani 2011 Tropical Biology Association field course for their help, encouragement and discussions, particularly Anna Rausch for taking the photographs. We thank Duncan McCollin (University of Northampton), André Rodrigo Rech (University of Northampton and Universidade Estadual de Campinas) and two anonymous referees for comments on the manuscript; and K.-D. Dijkstra and Connal Eardley (ARC, Pretoria) for help with species identifications. JO is particularly grateful to Karin, Oli, Patrick and James for their companionship and help with data collection and input during our time in Tanzania.

REFERENCES

Akinpelu AI (1989) Competition for the nectar of *Tecoma stans* flowers between Olive Sunbird (*Nectarina olivacea*) and insects. Malimbus 11:3-6.

Armstrong JE (1997) Pollination by deceit in nutmeg (*Myristica insipid*, Myristicaceae): Floral displays and beetle activity at male and female trees. American Journal of Botany 84:1266-1274.

Ashman TL (2009) Sniffing out patterns of sexual dimorphism in floral scent. Functional Ecology 23:852-862.

Bawa KS (1980) Mimicry of male by female flowers and intrasexual competition for pollinators in *Jacaratia dolichaula* (D. Smith) Woodson (Caricaceae). Evolution 34:467-474.

Blundell M (1987) Wild Flowers of East Africa. Collins, London.

Boyden TC (1978) Territorial defense against hummingbirds and insects by tropical hummingbirds. Condor 80:216-221.

Burd M (1995) Pollinator behavioural responses to reward size in *Lobelia deckenii*: no escape from pollen limitation of seed set. Journal of Ecology 83:865-872.

Canela MBF, Sazima M (2003) *Aechmea pectinata*: a hummingbird-dependent bromeliad with inconspicuous flowers from the rainforest in south-eastern Brazil. Annals of Botany 92:731-737.

Carpenter FL (1979) Competition between hummingbirds and insects for nectar. American Zoologist 19: 1105-1114.

Cheke RA, Mann C, Allen R (2001) Sunbirds: a guide to the sunbirds, spiderhunters, sugarbirds and flowerpeckers of the world. Christopher Helm, London.

Cotton PA (1998) Temporal partitioning of a floral resource by territorial hummingbirds. Ibis 140:647-653.

Dafni A, Kevan PG, Husband BC (2005) Practical pollination biology. Enviroquest, Cambridge, Ontario.

Elmqvist T, Cox PA, Rainey WE, Pierson ED (1992) Restricted pollination on oceanic islands - pollination of *Ceiba pentandra* by flying foxes in Samoa. Biotropica 24:15-23.

Evans MR, Hatchwell BJ (1992) An experimental study of male adornment in the Scarlet-Tufted Malachite Sunbird. I. The role of pectoral tufts in territorial defense. Behavioral Ecology and Sociobiology 29:413-419.

Franceschinelli EV, Bawa KS (2000) The effect of ecological factors on the mating system of a South American shrub species (*Helicteres brevispira*). Heredity 84:116-123.

Frost SK, Frost PGH (1980) Territoriality and changes in resource use by sunbirds at *Leonotis leonurus* (Labiatae). Oecologia 45:109-116.

Fry CH, Keith S (2000) The Birds of Africa, vol. 6. Academic Press, London.

Geerts S, Pauw A (2009) Hyper-specialization for long-billed bird pollination in a guild of South African plants: the Malachite Sunbird pollination syndrome. South African Journal of Botany 75:699-706.

Geerts S, Pauw A (2011) Farming with native bees (*Apis mellifera* subsp. *capensis* Esch.) has varied effects on nectar-feeding bird communities in South African fynbos vegetation. Population Ecology 53: 333-339.

Gill FB, Wolf LL (1977) Non-random foraging by sunbirds in a patchy environment. Ecology 58: 1284-1296.

Hansen DM, Olesen JM, Jones CG (2002) Trees, birds and bees in Mauritius: exploitative competition between introduced honey bees and endemic nectarivorous birds? Journal of Biogeography 29:721-34.

Jacobi CM, Antonini Y (2008) Pollinators and defence of *Stachytarpheta glabra* (Verbenaceae) nectar resources by the hummingbird *Colibri serrirostris* (Trochilidae) on ironstone outcrops in south-east Brazil. Journal of Tropical Ecology 24:301-308.

Janeček Š, Hrázský Z, Bartoš M, Brom J, Reif J, Hořák D, Bystřická D, Riegert J, Sedláček O, Pešata M (2007) Importance of big pollinators for the reproduction of two *Hypericum* species in Cameroon, West Africa. African Journal of Ecology 45: 607-613.

Johnson SD, Steiner KE (1994) Pollination by megachilid bees and determinants of fruit-set in the Cape orchid *Disa teuifolia*. Nordic Journal of Botany 14:481-485.

Jürgens A, Bosch SR, Webber AC, Witt T, Frame D, Gottsberger G (2009) Pollination biology of *Eulophia alta* (Orchidaceae) in Amazonia: effects of pollinator composition on reproductive success in different populations. Annals of Botany 104:897-912.

Kearns CA, Inouye DW (1993) Techniques for pollination biologists. University Press of Colorado, Niwot.

Lara C, Lumbreras K, Gonzalez M (2009) Niche partitioning among hummingbirds foraging on *Penstemon roseus* (Plantaginaceae) in central Mexico. Ornitologica Neotropical 20:81-91.

Larsson C, Hemborg AM (1995) Sunbirds (*Nectarinia*) prefer to forage in dense vegetation. Journal of Avian Biology 26:85-87.

Lott DF, Lott DY (1991) Bronzy sunbirds (*Nectarinia kilimensis*) relax territoriality in response to internal changes. Ornis Scandinavica 22:303-307.

Marler PR (1976) On animal aggression: The roles of strangeness and familiarity. American Psychologist 31:239-246.

Padyšáková E, Bartoš M, Tropek R, Janeček Š (2013) Generalization versus specialization in pollination systems: Visitors, thieves, and pollinators of *Hypoestes aristata* (Acanthaceae). PLoS ONE e59299. doi:10.1371/journal.pone.0059299.

Primack RB, Howe HF (1975) Interference competition between hummingbird (*Amazilia tzacatl*) and skipper butterflies (Hesperiidae). Biotropica 7:55-58.

Rocca MA, Sazima M (2006) The dioecious, sphingophilous species *Citharexylum myrianthum* (Verbenaceae): Pollination and visitor diversity. Flora 201:440-450.

Stiles FG (1975) Ecology, flowering phenology, and hummingbird pollination of some Costa Rican *Heliconia* species. Ecology 56:285-301.

Stiles FG, Wolf LL (1970) Hummingbird territoriality at a tropical flowering tree. Auk 87:467.

Symes CT, Nicolson SW, McKechnie AE (2008) Response of avian nectarivores to the flowering of *Aloe marlothii*: a nectar oasis during dry South African winters. Journal of Ornithology 149:13-22.

Tropek, R, Bartoš, M, Padyšáková E, Janeček Š (in press) Interference competition between sunbirds and carpenter bees for the nectar of *Hypoestes aristata*. African Zoology

Watts S, Huamán Ovalle D, Moreno Herrera M, Ollerton J. (2012) Pollinator effectiveness of native and non-native flower visitors to an apparently generalist Andean shrub, *Duranta mandonii* (Verbenaceae). Plant Species Biology 27:147–158.

Willmer P, Gilbert F, Ghazoul J, Zalat S, Semida F (1994) A novel form of territoriality - daily paternal investment in an anthophorid bee. Animal Behaviour 48: 535-549.

Willson MF, Agren J (1989) Differential floral rewards and pollination by deceit in unisexual flowers. Oikos 55:23-29.

Wirtz P, Szabados M, Pethig H, Plant J (1988) An extreme case of interspecific territoriality: male *Anthidium manicatum* (Hymenoptera, Megachilidae) wound and kill intruders. Ethology 78:159-176.

11

DIVERSITY AND POLLINATION VALUE OF INSECTS VISITING THE FLOWERS OF A RARE BUCKWHEAT (*ERIOGONUM PELINOPHILUM*: POLYGONACEAE) IN DISTURBED AND "NATURAL" AREAS

V. J. Tepedino[1]*, W. R. Bowlin[2] and T. L. Griswold

USDA ARS Bee Biology & Systematics Lab, Department of Biology, Utah State University, Logan UT 84322-5310

Abstract—We compared flower-visitors of the endangered plant *Eriogonum pelinophilum*, at relatively undisturbed and highly disturbed sites. We found no difference between sites in flower visitation rate or species richness of flower-visitors; species diversity of flower-visitors was higher at disturbed than at undisturbed sites but there was no difference in equitability. We found significant differences in total *E. pelinophilum* pollen carried on the body among 14 abundant bee species; eight abundant wasp species; and 12 abundant fly species. Both bee and wasp species carried significantly more pollen on the ventral compared to dorsal segments of the body; pollen on the body of fly species was more equally distributed across body surfaces. Total pollen carried on flower-visitor bodies was significantly related to visitor length, suggesting that larger visitors were more effective pollinators. Total Pollination Value, a measure combining both visitor abundance and body pollen was greater at the disturbed site than the undisturbed site, further suggesting that pollination in fragments of this rare species is not a major concern. We conclude that the high diversity of insect flower-visitors and the generalized nature of *E. pelinophilum* flowers make a special management programme to conserve pollinators unnecessary. Conservation of this buckwheat is best achieved by simple habitat preservation, together with a program to enlist private citizens to include buckwheat plants in their backyard gardens.

Keywords: pollinators, pollen placement, bees, flies, wasps, degraded remnant population, conservation

INTRODUCTION

Foremost among the important causes of plant rarity are habitat loss, modification and fragmentation (Ehrlich 1988; Wilson 1988; McNeeley et al. 1990; Gentry 1996). Such deterioration of habitat can also adversely affect the pollinators (Vinson et al. 1993; Gess & Gess 1993; Westrich 1996) that many rare plants in the western United States depend upon (Tepedino 2000), and thereby further impair their seed production and recruitment. Thus, rare plant declines may be accelerated by a reduction in the number and kinds of animals that visit and pollinate their flowers (Sipes & Tepedino 1995; Kearns et al. 1998).

The effects of pollinator loss are not distributed equally across plant species because not all flowering plants require visitation by pollen vectors to set seed. A recent estimate of the percent of species whose reproduction is aided by pollinators is 85 – 90% (Ollerton et al. 2011), but vulnerability to pollinator loss varies even among pollinator-requiring species. Tepedino (1979), Bond (1994), Kearns et al. (1998) and others have noted that self-incompatible plant species and those that have evolved specialized associations with a few selected pollinator species are more vulnerable to pollinator loss than are self-compatible species and those whose unspecialised flowers are used by many generalised

flower visitors. Such unspecialised plant species are forecast to be less prone to the indirect effects of habitat change on their pollinators.

Tests of such hypotheses come mostly from studies of non-threatened plants in fragmented habitats. Rare plants have much in common with plants in habitat fragments (relatively few individuals, reduced habitat area, isolation from other populations) and results of studies of the effects of fragmentation on plant reproductive success should have application to the management of rare plants. Thus far, results have been equivocal. An early review (Aizen et al. 2002), found no support for these hypotheses but an expanded meta-analysis by the same group (Aguilar et al. 2006) found a strong negative effect of fragmentation on plant reproduction; some subsequent studies disagree, e.g., González-Varo et al. (2009). As predicted (Aizen et al. 2002), this effect was found for self-incompatible species but not for self-compatible species. However, counter to expectations, plants with specialised pollination systems were no more vulnerable than those with generalised pollination systems (ibid.).

Such findings make desirable additional information on the pollination biology of rare plant species. A dependency on pollinators for successful reproduction generally means that land managers must plan both for the protection of plant populations and for their pollinators. On the other hand, in a world of limited time and funding, eliminating a need for pollinator management frees up money and effort for other conservation concerns (Schemske et al. 1994). To make such

---

*Corresponding author; email: andrena@biology.usu.edu
[1]Federal Collaborator
[2]Present Address: 2380 East 11000 North, Richmond UT 84333

decisions, conservationists must know a plant's pollination requirements, and the identity and resource requirements of its pollinators.

One rare species that may be less vulnerable to specific pollinator loss is clay-loving wild buckwheat, *Eriogonum pelinophilum* Reveal (Polygonaceae), a narrow endemic listed in 1984 as endangered under the U. S. Endangered Species Act; *E. pelinophilum* occurs in an area of rapid residential and commercial growth in west central Colorado (USA). Although *Eriogonum* is one of the two largest plant genera native to North America (Reveal 2005), the pollination biology of few species is known. All such studies report at least partial self-compatibility (Bowlin et al. 1993, *E. pelinophilum*; Kan 1993, *E. umbellatum* v. *torreyanum;* Duff 1996, *E. argophyllum*; Archibald et al. 2001, Tepedino et al. 2002, *E. ovalifolium* v. *williamsiae;* Neel et al. 2001, *E. ovalifolium* v. *vineum*) and low seed set (see also Kaye et al. 1990, *E. crosbyae*). Buckwheat flowers are visited by many bees, flies, wasps, butterflies and beetles (Kaye et al. 1990; Kan 1993; Archibald et al. 2001; Tepedino et al. 2002; Neel et al. 2001; Neel & Ellstrand 2003) suggesting that few, if any, insects are morphologically excluded from harvesting the small, easily obtained, nectar and/or pollen rewards.

To examine the general vulnerability of *E. pelinophilum* to pollinator loss, we compared a large, relatively undisturbed site with a much smaller, highly disturbed and fragmented site for number and identity of insects visiting the flowers, species richness, and species diversity and equitability of flower visitors. We also evaluated flower-visitors for their potential as pollinators using indirect (Spears 1983) quantitative and qualitative criteria (Herrera 1987, 1989). Our quantitative measures were visitation-rate and frequency on the flowers. Because important pollinators must accumulate pollen on relevant parts of the body (e.g., Lamborn & Ollerton 2000), vector pollen load (Inouye et al. 1994) was used as the qualitative measure. We also asked if there was a connection between total vector pollen load and size of flower-visitor (Kandori 2002). Quantitative and qualitative measures were then combined for abundant species in major taxa to obtain a measure of pollination value. Finally, we relate our findings to efforts to conserve *E. pelinophilum*.

METHODS

The Plant

Eriogonum pelinophilum is a rounded, sometimes spreading (8 – 30 cm diam, 10 – 20 cm hgt), sub-shrub,

endemic to heavy clay soils derived from Mancos Shale in Delta and Montrose counties in west central Colorado USA at an elevation of 1580 – 1935 m. Undisturbed populations dominate local shrub-steppe rangelands; disturbed populations occur in small remnant patches surrounded by farmland. Currently there are fourteen populations (not counting known extirpations or historical populations that have not been visited in 20 years) ranging in size from less than 100 individuals to more than 10,000 (U. S. Department of the Interior 2009).

Eriogonum pelinophilum blooms from late May to early September. Individual plants produce many small (3 – 5mm) flowers for extended periods (3 – 6 weeks); each plant may produce thousands of protandrous flowers. Flowers contain 6 – 9 stamens, three styles and produce small amounts of nectar at the base of the ovary (see Bowlin et al. 1993 for further details).

Study Sites

We compared flower-visitors at two sites, the relatively undisturbed Wacker's Ranch (UN; EO#018), and the highly disturbed North Mesa site (DI; EO#006) about 5 km away (site EO #s refer to U. S. Department of the Interior (DOI) 2009a). UN, 8 km southeast of Montrose CO is administered by DOI, Bureau of Land Management (BLM); it is approximately 146 hectares of predominantly native shrubs and forbs, elevation about 1875 m. *Eriogonum pelinophilum* was the dominant species in bloom with over 10,000 generally robust, large plants with many flowers. The site was ungrazed for at least three years prior to our study in 1990 (J. Ferguson, BLM, pers. comm.). It was surrounded by mostly private lands of native and improved rangeland, with a predominance of species of *Artemesia* and *Atriplex*. During the study, surrounding areas were moderately to heavily grazed by livestock, mostly cattle.

DI, was just north of Montrose, about 4 hectares in size, at approximately 1735 m elevation. There were approximately 200 *E. pelinophilum* plants in a private pasture/livestock holding area with much bare, heavily compacted ground. Blooming alfalfa fields occurred to the north and south, and a road right-of-way planted to grass, but with many weedy forbs, especially *Melilotus officinalis*, *Centaurea* sp. and *Convolulus arvensis*, bordered on the west. To the east were several corrals, and a storage area for farm equipment. *E. pelinophilum* plants here were significantly smaller than those at UN (unpublished data), and had little or no new vegetation and few leaves (they had been grazed

TABLE 1. The number of bee, wasp, ant and fly species captured from the flowers of *E. pelinophilum* at the undisturbed (UN), disturbed (DI) and Lawhead Gulch (LG) sites. SH is the number of species shared between UN and DI, $\Sigma\Sigma$ is the total number of unique species recorded in the study at UN and DI. LGUN is the number of species shared only between LG and UN, LGDI only between LG and DI, and LGUD at all three sites.

	UN	DI	SH	$\Sigma\Sigma$	LG	LGUN	LGDI	LGUD
Bees	18	20	10	28	14	3	2	6
Wasps	22	27	9	40	10	2	1	3
Ants	7	5	4	8	1	0	0	1
Flies	23	18	11	30	17	5	2	7
Totals	70	70	34	106	42	10	5	17

earlier the year by both cattle and sheep; Tepedino and Bowlin, unpublished information). The cattle had been off the site for four weeks when the study started; their return prompted the study's end. Despite having been grazed, plants produced abundant flowers for the entire period (starting a few days later than at the undisturbed site).

We also conducted limited collecting at Lawhead Gulch (LG; EO#001), the type locality for *E. pelinophilum*. LG is north of Montrose, about 5 km NE of Austin, and was about 40 hectares at 1600 m elevation; several thousand plants shared the area with species of *Artemesia* and *Atriplex*. Many *M. officinalis* plants were in bloom in surrounding areas. Parts of LG had been heavily grazed in past years and were being lightly grazed during our collections. Overall, LG was intermediate to UN and DI in disturbance impact; we confined our collections to a lightly disturbed section.

Flower Visitor Diversity

Systematic collections of insects visiting *E. pelinophilum* flowers were begun in 1990 at each site soon after flowering commenced and continued every seven days, except during inclement weather. The sole exception was the last collection date which was two weeks later than the penultimate one. Collection days were: UN – June 12, 19, 26, July 2, 9, 18, Aug 2; DI – June 13, 20, 28, July 3, 10, 19, Aug 3; LG – June 5 (AM), June 11 (PM), June 19, 20 (all day). Collecting ceased when the DI site became unavailable due to the return of cattle; flowering also had declined greatly at both sites. Insect flower-visitors were collected, usually by two collectors (at LG only one collector was active) repeatedly traversing different parts of the sites. Because flower-visitors usually change over a day, we employed four one-hour collecting periods each collection day: 0800 – 0900, 1100 – 1200, 1400 – 1500, and 1700 – 1800 hrs. Rare visitors, not seen on *E. pelinophilum* during systematic collections, were collected during additional short collecting bouts conducted at each site throughout the study. Insects were collected with butterfly nets, and immediately placed in cyanide kill-jars. Later, they were pinned, labeled and stored for identification. Since more time was spent collecting at UN (10 hrs/wk) than at DI (9 hrs/wk), all survey data is presented as insects or species per person-hour per week. Weekly results were compared between sites by taxon using the Wilcoxon Matched-Pairs Signed Ranks Test (Daniel 1990).

To measure diversity (*D*) and equitability (*E*) at UN and DI, Simpson's *D* (Begon et al. 1986), and accompanying *E*, were used. *D* is calculated as $[1/\Sigma (n_i/N)^2]$, where n = the number of individuals of species *i*, and N = the total number of individuals. *D* varies between 1.0 and the total number of species (*S*) in the collection. Equitability (*E*) is then *D*/S. *D* and *E* were calculated for each week's collection and compared between sites using the Wilcoxon Matched-Pairs Signed Rank Test.

Pollen placement

The amount of pollen on the body of abundant (≥ 5 individuals) hymenopterans (bees, wasps, ants) and dipterans (flies) was scored under a binocular microscope at 160X. For each insect, we confirmed that *Eriogonum* pollen was present

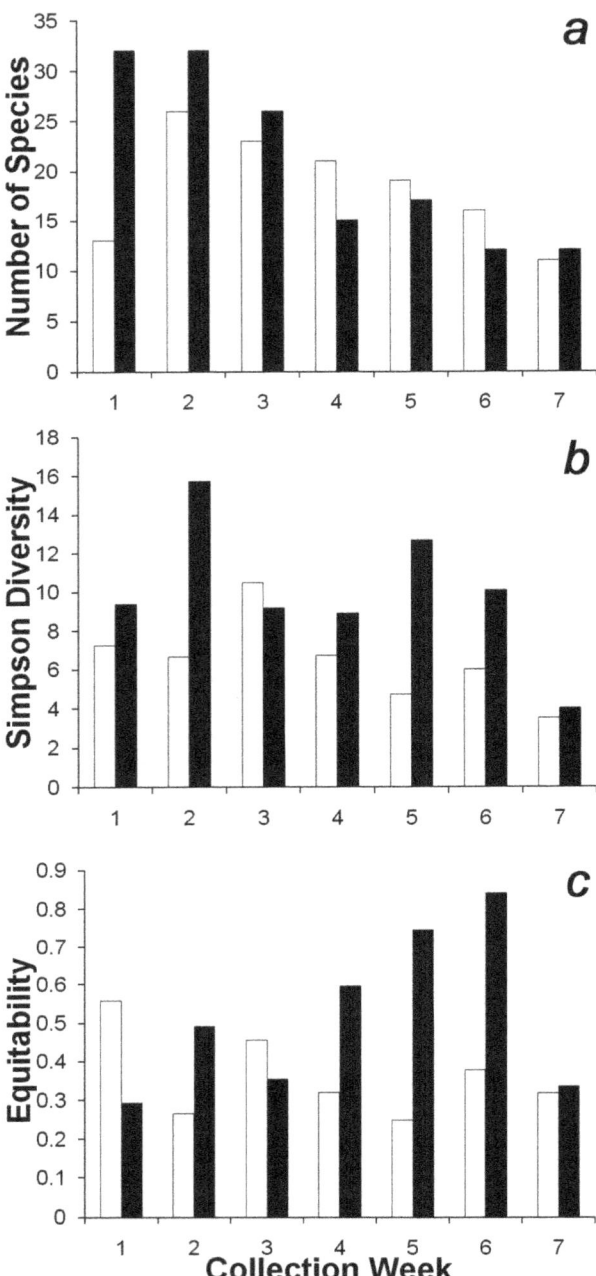

FIG. 1. Three estimates of diversity of insects captured each week from the flowers of *Eriogonum pelinophilum* at undisturbed (open bars) and disturbed (solid bars) sites. Top panel (a) shows total species richness (ants, bees, flies, wasps), mid panel (b) shows species diversity (SD), lower panel (c) shows equitability (EQ).

by its bright yellow hue, size (~ 30μ) and tricolporate, ellipsoidal shape, and that it comprised the majority of pollen types, although we did not estimate the percentage of buckwheat pollen on any insect. We examined six body parts for pollen (dorsal and ventral head, thorax, abdomen; note that the legs, where most bees collect pollen for transport to the nest, were excluded). When pollen was scored, the length of the insect was measured to the nearest half mm.

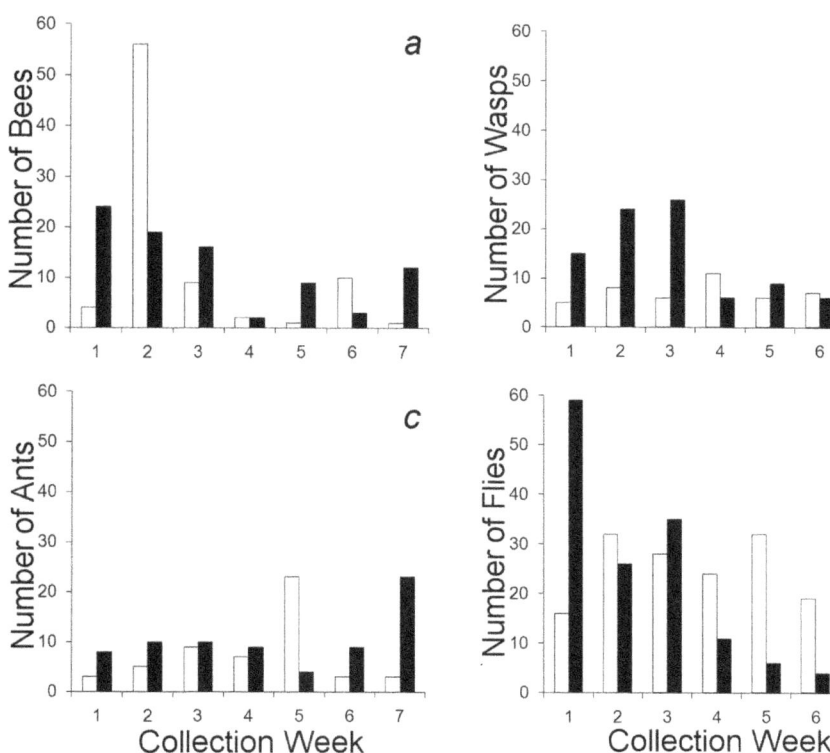

FIG. 2. The total number of a) bee, b) wasp, c) ant and d) fly individuals collected per week from flowers of *Eriogonum pelinophilum* at undisturbed (open bars) and disturbed (solid bars) sites (unadjusted for the 10% lower collecting effort at disturbed).

For each insect examined, pollen on each of the six body parts was scored as either 0 (no pollen present); 1 (1 – 10 grains), 2 (11 – 25 grains), 3 (26 – 50 grains), 4 (> 50 grains). To determine if pollen distribution varied with location on the body, we compared these pollen ratings for each abundant species (≥ 5 individuals) of bee, wasp, ant, fly using the Friedman nonparametric two-way AOV test (Daniel 1990). The three dorsal and three ventral body segments were then combined and compared for each abundant species using the Wilcoxon Matched-Pairs Signed Rank test. We also compared pollen ratings separately for each of the six body segments among abundant species of bees, wasps and flies using the Kruskal-Wallis test (with expertimentwise corrections). Total pollen load among abundant species within each major taxon (bees, wasps, flies) was compared using the Kruskal-Wallis test (with expertimentwise corrections). Finally, linear regressions were performed using median body length as the independent variable and total pollen load as the dependent variable. Regressions were performed separately for bees, wasps, flies, and total insects.

Total Pollination Value (TPV)

We combined quantitative and qualitative estimates into a more realistic representation of the "pollination value" (TPV) of *E. pelinophilum* flower visitors. Our objective was to express weekly TPV for each major taxon (bees, wasps, ants, flies) in terms of total TPV calculated for the entire blooming season at each site. For each week we obtained the product of the number of each abundant species captured (uncommon species were ignored) and its total pollen load rating and summed these to obtain a total value for the week. Weekly sums were then added to obtain a total estimate of

TPV for the entire blooming season. Total TPV was then used as the denominator to estimate the pollination due to each major taxon for each week.

RESULTS

Flower Visitor Diversity and Abundance

With its canopy of open, accessible flowers produced over two months, *E. pelinophilum* attracted a wide variety of insects (Appendix I). Although the total number of species captured at UN and DI was identical (70 species, Tab. I), the sites shared only 32.1% of the combined 106 species. Only five species were abundant (≥ 5 individuals) at both sites (one species each of wasps and ants, three of flies); 14 species at DI and 11 at UN were abundant but present in low numbers or completely absent at the other site.

At LG, 42 species were recorded (Tab. I, Appendix I) even though collecting was limited to one collector for a few days early in the study. Ten species, all with a single individual, occurred only at LG. Most LG species (76.2%) were shared with at least one other site. A few more species were shared exclusively with UN (10) than with DI (5); 17 species were common to all three sites. Of the nine abundant species at LG (five bees, four flies), seven were also present at both UN and DI (Appendix I).

Weekly species richness comparisons showed that DI had more total species than UN for the first three weeks and the last week (Fig. 1a). Simpson's *D* was greater at DI than at UN for six of the seven weekly comparisons (Fig. 1b) and narrowly missed significance (Z = 1.94, P = 0.052; Wilcoxon matched-pairs Signed Rank Test). DI also had a more even apportionment of individuals into species than UN

Table 2. Mean number of individuals and species (± sd) captured per person hour per week at the flowers of *E. pelinophilum* at disturbed (DI) and undisturbed (UN) sites. Number of weeks sampled = 7 in all cases. P = Probability; Wilcoxon Matched-Pairs Signed Rank Tests.

	Individuals			Species		
	UN	DI	P	UN	DI	P
Bees	1.19 ± 2.0	1.33 ± 0.9	0.53	0.33 ± 0.3	0.53 ± 0.5	0.14
Wasps	0.73 ± 0.2	1.43 ± 1.0	0.27	0.60 ± 0.2	0.81 ± 0.3	0.35
Ants	0.76 ± 0.7	1.17 ± 0.7	0.20	0.24 ± 0.2	0.29 ± 0.1	0.59
Flies	2.39 ± 0.7	2.31 ± 2.3	0.45	0.67 ± 0.4	0.66 ± 0.3	0.67
Total	5.06 ± 2.5	6.31 ± 3.7	0.67	1.81 ± 0.5	2.36 ± 1.0	0.24

TABLE 3. Mean (± sd) of combined pollen ratings (see text) for three dorsal, three ventral and six total body sections for abundant species of bees, wasps, ants and flies (≥ five individuals at UN, DI and LG sites combined). Asterisks in Dorsal or Ventral column indicate significant differences for that species (* = ≤ 0.05, ** ≤ 0.01, *** ≤ 0.001; Wilcoxon Signed Rank test). Different superscripts in Total column indicate significant differences (Kruskal-Wallis and multiple comparison correction tests, P = 0.05) among species of bees, wasps and flies.

Taxa			Dorsal	Ventral	Total
Bees	Andrenidae	*Andrena hallii*	5.63 (0.74)	8.13 (1.64)*	13.75 (2.31)[a]
		Perdita calloleuca	0.32 (0.63)	1.52 (1.36)***	1.84 (1.75)[d]
		Perdita wilmattae	0.44 (0.70)	2.28 (2.47)***	2.72 (2.91)[cd]
	Apidae	*Apis mellifera*	6.90 (3.48)	9.00 (2.53)*	15.90 (5.74)[a]
		Bombus huntii	7.20 (1.92)	9.40 (2.41)	16.60 (4.10)[a]
		Ceratina nanula	1.36 (1.43)	4.91 (2.21)*	6.27 (2.97)[abcd]
	Colletidae	*Colletes phaceliae*	3.22 (2.11)	6.00 (2.69)*	9.22 (4.66)[abc]
		Hylaeus episcopalis	3.43 (1.45)	2.43 (1.22)	5.86 (1.79)[abcd]
	Halictidae	*Agapostemon femoratus*	4.63 (2.62)	7.88 (2.42)*	12.50 (4.75)[ab]
		Lasioglossum caducum	0.80 (1.10)	2.00 (2.83)	2.80 (3.90)[bcd]
		Halictus confusus	2.09 (1.70)	6.55 (2.70)**	8.64 (3.98)[abc]
		Halictus ligatus	3.50 (1.22)	8.17 (1.33)*	11.67 (2.16)[ab]
		Halictus tripartitus	1.58 (0.90)	6.75 (2.42)**	8.33 (2.87)[abc]
	Megachilidae	*Ashmeadiella aridula*	4.27 (2.41)	7.00 (3.49)*	11.27 (5.50)[ab]
Wasps	Eumenidae	*Euodynerus annulatus*	5.15 (1.79)	7.75 (2.40)***	12.90 (3.73)[ab]
		Euodynerus exoglyphus.	4.33 (1.21)	7.33 (2.88)*	11.67 (3.88)[abc]
		Stenodynerus apache	2.44 (1.42)	3.89 (2.31)*	6.33 (3.28)[c]
		Stenodynerus sp. 1	3.84 (2.09)	5.16 (2.31)**	9.00 (4.18)[bc]
	Pompilidae	*Anoplius sp.*	4.00 (2.24)	5.71 (2.56)*	9.71 (4.46)[abc]
	Sapygidae	*Sapyga sp.*	5.00 (1.82)	9.29 (2.21)*	14.29 (3.30)[ab]
	Sphecidae	*Cerceris sp.* 1	6.50 (2.83)	8.88 (1.64)*	15.25 (4.06)[a]
		Cerceris sp. 2	4.53 (2.13)	6.00 (2.10)*	10.53 (3.52)[abc]
Ants	Formicidae	*Formica obtusopilosa*	3.00 (2.15)	4.53 (2.49)***	7.54 (4.42)
		Leptothorax tricarinatus	0.11 (0.33)	1.22 (0.83)*	1.33 (1.00)
		Pogonomymex occidentalis	0.88 (0.64)	2.25 (1.28)*	3.13 (1.64)
Flies	Bombyliidae	*Anastoechus sp.*	3.37 (1.67)***	1.05 (0.91)	4.42 (2.17)[b]
		Aphoebantus sp.	0.64 (0.73)	3.86 (2.42)***	4.50 (2.65)[b]
		Chrysanthrax sp. 1	0.33 (0.82)	0.67 (0.82)	0.83 (0.98)[c]
		Phthiria sp.	0.82 (0.74)	0.60 (0.59)	1.42 (0.99)[c]
		Thyridanthrax pallida	6.44 (1.48)	8.15 (1.79)***	14.59 (2.66)[a]
		Villa sp. 1	1.67 (1.37)***	1.00 (0.63)	2.67 (1.75)[bc]
		Villa sp. 2	1.86 (1.28)	3.41 (1.77)	5.27 (2.69)[b]
	Milichiidae	*Leptometopa halteris*	0.69 (0.74)	1.27 (0.53)**	1.96 (1.00)[c]
	Muscidae	*Peleteria sp.*	3.60 (1.37)	6.68 (2.27)***	10.28 (3.40)[ab]
	Syrphidae	*Eupeodes volucris*	3.33 (1.07)	7.08 (2.31)**	10.42 (3.00)[ab]
		Paragus tibialis	2.00 (1.06)*	1.29 (1.00)	3.29 (1.60)[b]
	Stratiomyidae	*Hedriodiscus binotatus*	7.83 (1.60)	10.17 (2.40)	18.00 (3.74)[a]

on five of seven sampling dates (Fig. 1c), but there was no significant difference between them in E ($Z = 1.18$, $P > 0.20$).

UN and DI were similar in the rank order abundance of insect visitors over time (Fig. 2): flies were usually the most abundant taxon, with wasps vying for second position with ants (UN) or bees (DI). Exceptions were the second week at UN, when a large number of *Perdita wilmattae* (Andrenidae) were recorded, and the last two weeks at DI, when ants were most abundant.

Despite the lower collection effort, DI yielded more individuals of bees and wasps than did UN for four of the seven collection weeks and more ants for six of the seven weeks (Fig. 2). UN usually exceeded DI for flies (five of seven weeks). When collection effort was adjusted for person hours per week, the sites did not differ significantly in either the number of individuals or species collected by taxon or in total (Tab. 2).

For our quantitative estimate of pollinator importance, we used the mean number of insects captured on flowers at UN and DI combined (Tab. 2). This estimate gave a ranking of flies > bees > wasps > ants. Because the number of individuals of major taxa was quite variable from week-to-week (Fig. 2), and the important pollinators of buckwheat changed over the course of its extended blooming period, such a ranking, by itself, is suspect. A better estimate of pollinator importance would integrate both visitor abundance and pollen accumulation with flower phenology (see below).

Pollen placement on flower-visitor bodies

Pollen was unequally distributed across the six body parts of most abundant species (Appendices II - IV). Twelve of 14 bee species, seven of eight wasp species, all three ant species and nine of 12 fly species carried significantly more pollen on some body sections than on others (Friedman Tests). Because of the small, open buckwheat flowers with erect stamens and styles, and the scramble-like foraging behaviour of insects across the inflorescence, most abundant flower-visitors carried more pollen ventrally than dorsally (Tab. 3). Thirteen of 14 bee species, and all eight wasp and three ant species carried more buckwheat pollen ventrally than dorsally, and most of these comparisons were significant (Wilcoxon Signed-Rank test). The predominantly ventral distribution of pollen was less pronounced for flies: only eight of 12 fly species accumulated more pollen ventrally, and only five of these comparisons were significant. Conversely, three of four fly species had significantly more dorsal than ventral pollen.

There were also significant differences in total *Eriogonum* pollen grains carried on the body among abundant flower-visiting bee species (Kruskal-Wallis Test, $H = 101.6$, $P < 0.0001$); wasp species ($H = 28.7$, $P = 0.0002$); ant species ($H = 25.9$, $P < 0.0001$) and fly species ($H = 248.5$, $P < 0.0001$) (Tab. 3). The most abundant visitors tended to carry fewer pollen grains than less common visitors. For each major taxon, a Pearson product-moment correlation of number of visitors to the flowers and our pollen accumulation estimate was inverse for each group but was significant only for bees (bee $t_{13} = -2.74$, $P < 0.02$, $r^2 = 0.39$; fly $t_{11} = -0.48$, wasp $t_7 = -0.34$, both $P > 0.50$; no analysis was conducted for ants).

We searched for patterns in pollen placement among bees, wasps and flies by grouping their species and comparing them for pollen placement on each of the six body parts (Tab. 4). There were significant differences among visitors for mean pollen grains on the head ($P < 0.003$), face ($P < 0.03$), and lower abdomen ($P < 0.02$) but not for the upper or lower thorax or upper abdomen (all $P > 0.40$). Surprisingly, wasps accumulated more pollen on four of the six body segments than did the other groups; bees accumulated more on the lower abdomen. Flies had the fewest pollen grains on three of the six body segments. The ant *F. obtusopilosa* was usually intermediate in pollen grains but had many more than flies on the lower thorax and abdomen.

Body Size and Pollen Load

The variability in pollen accumulation among abundant species of flower-visitors can be partially explained by their size. The relationship of average length to pollen carried on the body was positive and highly significant for all abundant species ($t_{1.35} = 5.96$, $P < 0.0001$, $r^2 = 0.50$, Fig. 3). (Removing two outlier species, the flies *Villa* sp. 2, *Chrysanthrax* sp. 1, both present in relatively low numbers ($N = 6$), increased r^2 to 0.72.)

Not all groups of taxa contributed equally to the size-pollen accumulation association (Fig. 3). The relationship was significant for bees ($t_{13} = 6.62$, $P < 0.0001$, $r^2 = 0.79$) and flies ($t_{11} = 2.96$, $P < 0.02$, $r^2 = 0.44$), but not for wasps ($t_7 = 0.85$, $P > 0.40$); ants, with only three species were not analysed. (An identical analysis using only ventral pollen accumulation yielded similar results). However, wasps generally carried more pollen per unit length than did other taxa. Bees and flies accumulated pollen on their bodies at the same rate, but for a given size, bees carried more pollen than

	Bees (14)	Wasps (8)	Flies (12)	Ants (1)
Head	1.28 (0.89)[ab]	2.40 (0.53)[a]	0.82 (0.72)[b]	1.7
Face	1.90 (0.84)[ab]	2.73 (0.71)[a]	1.28 (1.30)[b]	1.26
Upper Thorax	1.01 (0.79)	1.12 (0.45)	1.14 (0.98)	0.67
Lower Thorax	1.97 (1.06)	2.28 (0.59)	1.71 (1.29)	2.11
Upper Abdomen	0.95 (0.74)	0.96 (0.43)	0.77 (0.76)	0.63
Lower Abdomen	1.99 (1.11)[a]	1.74 (0.75)[ab]	0.78 (0.96)[b]	1.16
Total	9.10 (4.80)	11.27 (3.03)	6.47 (5.57)	

TABLE 4. Mean ($\pm$ sd) rating for pollen scores for bee, wasp and fly species for the six body parts surveyed. Pollen rating follows that in text: scale 0 (no pollen) to 4 (> 50 grains). Numbers in parentheses adjacent to taxa denote number of species in each group. Different superscripts within a row denote significant differences between taxa ($P = 0.05$; Kruskal-Wallis tests, multiple comparison correction tests). Ant species *Formica obtusopilosa* shown only for comparison.

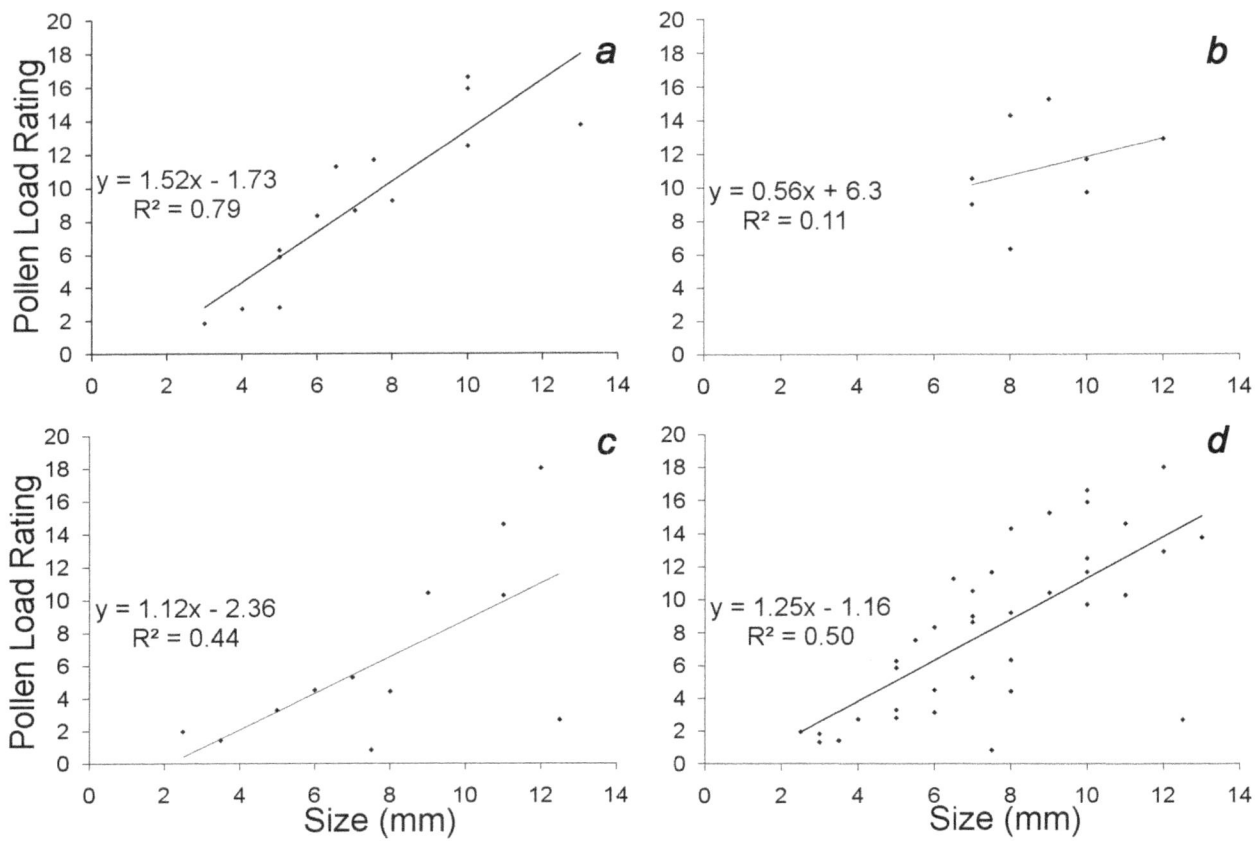

FIG.3. Median size of abundant insect species vs. rating for amount of pollen carried, by taxonomic group (a) bees, (b) wasps, (c) flies (d) total flower visitors.

did flies. A comparison of the regression lines for bees and flies was insignificant for slope ($F_{1,22}$ = 0.57, P > 0.40) but significant for elevation of the intercept ($F_{1,23}$ = 7.54, P = 0. 01). The common ant, *F. obtusopilosa*, also carried more pollen than did flies.

Quantitative and qualitative criteria: Total Pollination Value

Several intriguing findings emerged from our estimates of TPV (Fig. 4): 1) TPV was over 30% greater at DI than at UN (2083 – DI; 1587.5 – UN); 2) most pollinations at both sites were likely accomplished during the first three weeks of the flowering season when many more pollinators with high pollen accumulation ratings were present; 3) in general, all major taxa participated in the seasonal decline; 4) variability among major taxa in pollination value also declined with time; 5) summing pollination values for each major taxon across the seven weeks gave, for UN: bees – 30.8; wasps – 17.6; ants – 17.2; flies – 34.6; for DI: bees – 18.5; wasps – 27.9; ants – 20.3; flies – 33.1. Thus, bees were likely to be much more important pollinators at UN than at DI while wasps displayed the reverse pattern. Overall, neither ants nor flies differed much between sites.

DISCUSSION

In the United States, plants comprise 54.9% of the 1361 species listed under the Endangered Species Act, yet in 2009, the latest year for which statistics are available, the federal government spent only 3.7% of its total species budget on rare plant conservation (U. S. Fish & Wildlife Service 2009). As expenditures on animals are likely to remain disproportionately high for the foreseeable future, funding for rare plant management and research must be allocated effectively. Thus, it is important to help managers of listed plants such as clay-loving wild buckwheat to prioritise their list of concerns. If pollinator-limitation of reproduction were important, direct management intervention to encourage pollinator populations, such as nest site designation and preparation, might be warranted, particularly at highly disturbed sites such as DI.

Earlier studies provisionally suggested that management of pollinators to enable *E. pelinophilum* reproduction was unnecessary. Using pollen supplementation at the flower level (Knight et al. 2006) in a protocol with a strong bias towards finding pollinator limitation, Bowlin et al. (1993) found only occasional pollinator limitation of seed production at UN. But are pollinators more likely to be uncommon in disturbed areas? Our results suggest not. We addressed this question at

DI by comparing several indirect measures of pollinator effectiveness; each suggested that reproduction at DI was not being limited by inadequate pollination to a greater extent than at UN.

First, as with other species with small, open flowers and readily accessible rewards (e.g., Ramirez 2003; Fenster et al. 2004; Olesen et al. 2007; Zych 2007), buckwheat flowers at both sites were visited by a large assortment of non-specialized insects. We found no difference between UN and DI in species richness, diversity or equitability of flower-visitors (Tab. 2; Fig. I; App. I), suggesting that plants at the two sites were equally attractive, and that there was a comparable diversity of visitors to attract. Such measures are useful for describing the range of insect visitors and, thereby, the potential for pollinator species redundancy or insurance at a site (e.g., Winfree and Kremen 2009). They therefore suggest that *E. pelinophilum* plants at DI were as resistant to pollinator limitation as were plants at UN.

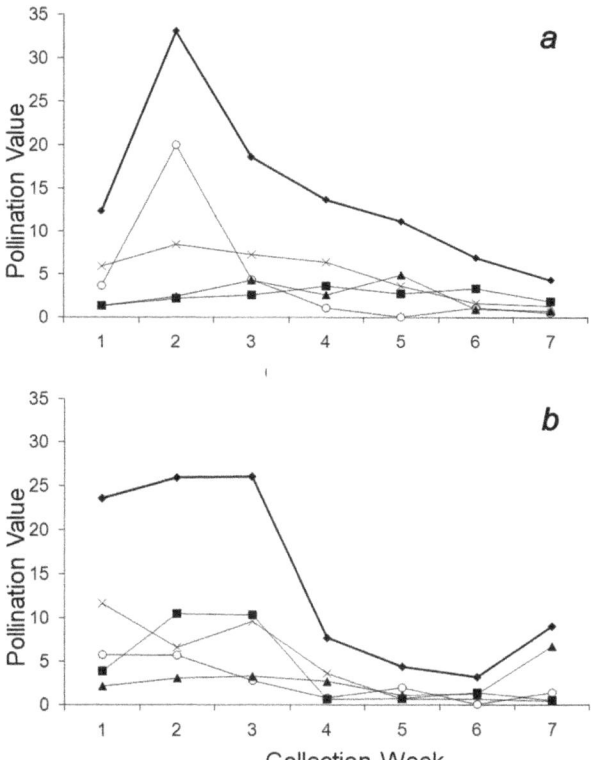

FIG. 4. Total Pollination Value (♦; see text for details), a measure combining abundance on the flowers with pollen accumulation on the body of bees (O), wasps (■), ants (▲) and flies (X) by collection week at UN (a) and DI (b).

Second, we found no significant differences between DI and UN in overall pollinator-visitation rates or in frequency on the flowers (Tab. 3, Fig. 2, App. II - IV). Thus, if Vazquez et al. (2005; see also Morris 2003; Sahli & Conner 2006), are correct in their provisional conclusion that visit-frequency is more valuable as an indicator of pollinator effectiveness than total vector pollen load or pollen deposition, then our sites were indistinguishable. However, unlike Vazquez et al. (2005), who based their conclusion on visitation rates that varied much more widely than did pollen

deposition, we found that abundant visitors in all major taxa varied greatly in vector pollen load (Tab. 3), and that commonness on the flowers and vector pollen load were inversely related for all comparisons. (We used pollen load because, like Zych (2007) with *Heracleum*, we found pollen deposition studies with *E. pelinophilum* infeasible. Usage of pollen load assumes it is positively related to pollen deposition, a reasonable assumption given the similar heights of dehiscing anthers and receptive stigmas of *E. pelinophilum*.) Some examples: common bees such as *Perdita calloleuca*, *P. wilmattae* and *Hylaeus episcopalis* and the flies *Phthiria* sp., *Villa* sp. I, and *Leptometopa halteris*, carried little pollen (App. II - IV) and, as a result, were likely to be inferior pollinators. Conversely, visitors overladen with pollen, such as the bees *Andrena hallii* and *Bombus huntii*, the wasps *Sapyga* sp. and *Cerceris* sp. I, and the fly *Hedriodiscus binotatus*, were uncommon and, therefore, unreliable pollinators.

Results of other studies also raise doubts that commonness on the flowers is always indicative of pollinator effectiveness: some common visitors actually lower plant reproduction by removing large amounts of pollen from flowers but depositing little (Thomson and Thomson 1992; Franzen and Larsson 2009; Hargreaves et al. 2009). More disquieting is the common disassociation between visitation rate and pollination, especially for simple flowers. Zych (2007), for example, working with a common umbellifer, found that 2-21% of all visits accounted for 70% of pollinations. Herrera (1989, 1990) reported similar findings for *Lavendula*, as did Kandori (2002) for *Geranium*. Thus, we doubt that some measure of flower-visitation rates alone is a sufficient estimate of pollinator effectiveness, at least for plants with simple pollen removal and deposition systems like buckwheats.

More realistic was our measure TPV, which like similar measures (e.g., Zych 2007) combined frequency on the flowers with pollen load. TPV was no greater at UN than at DI. Indeed, TPV estimated that pollination value was actually one-third greater at DI than at UN (Fig. 4). Thus, like our results using diversity measures and estimates of abundance and flower-visitation, TPV clearly indicated that disturbance had no negative effect on pollination.

It is not clear why TPV was higher at DI than at UN (Fig. 4). The most obvious difference between sites was a shift from bees as most common flower-visitors at UN to wasps at DI. However, such a change is not usually associated with an increase in pollination potential: wasps are usually thought to be inferior to bees as pollinators, primarily because their less hairy bodies accumulate less pollen (obviously not true here), and because they forage only for nectar and therefore visit fewer flowers than do bees which visit flowers for both nectar and pollen. Thus, it is unlikely that changes in species composition *per se* were responsible for differences in TPV (see also Kandori 2002). More suggestive was an increase in visitor size at DI compared to UN, and a significant relation between visitor size and pollen load (Fig. 3). Kandori (2002) also reported that larger visitors were more effective pollinators of *Geranium*. In addition to carrying more pollen grains, larger species have other potential advantages as pollinators: because they must visit more flowers to satisfy their greater nectar and pollen demands, they are likely to visit and pollinate more flowers per foraging bout and over their lifetimes.

Conservation

Although we uncovered no evidence that reproduction by *E. pelinophilum* was likely to be pollinator-limited, pollinator welfare cannot be removed from the priority list of management objectives, though it may be lowered. Others have also downplayed the importance of pollinators and the need for pollinator management for species in disturbed or fragmented areas. Donaldson et al. (2002) and Yates & Ladd (2005) found that species with generalised pollination systems had no problems reproducing in fragments and that pollinators were not a major concern. Aizen & Feinsinger (2003) thought that concerns other than pollination (e.g., recruitment, soil compaction, grazing, trampling) merited more attention, at least over the short term. Likewise, if seed production of *E. pelinophilum* was more limited at DI than at UN, it was more likely due to grazing or general habitat deterioration than to pollinators.

With their numerous niches and life histories, the diverse and unpredictable group of insects that visit and pollinate *E. pelinophilum* flowers limits and simplifies management options for pollinators. Management of particular pollinator species or groups of species is both futile and unnecessary because it is doubtful that there is a predictable suite of buckwheat pollinators. Insect flower-visitors commonly change from site-to-site and from year-to-year as has been shown in many other autecological and community studies (e.g., Tepedino & Stanton 1981; Herrera 1989, 2005; Aizen & Feinsinger 2003; Alarcon et al. 2008; Olesen et al. 2008; Petanidou et al. 2008) and this is especially expected with species like *E. pelinophilum* whose flowers are so readily exploitable. Indeed, evidence of such spatial variation in pollinator composition, and buckwheat's relative independence from the composition of the pollinator fauna, was present here (Tabs. 1,2; Figs. 1, 2). Even though our sites were only about 5 km apart, only five insect species were abundant at both disturbed and undisturbed sites; in general, frequency of particular species varied appreciably between sites. It is highly unlikely that any of these between site differences had, or is likely to have, any substantive effect on reproductive success of this rare plant.

To encourage such insect diversity, one can only strive for 1) general habitat preservation of as many separate areas as is possible, and 2) enrichment of areas adjacent to, and surrounding, buckwheat population fragments. The former policy is a truism; evidence of the importance of the latter is suggested by the likely contribution to insect abundance and diversity on buckwheat flowers by adjacent weed and alfalfa/lucerne fields. Such alternates to extirpated native plant species likely supported numerous generalist native flower-visitors which also included buckwheat in their foraging ambit. Others also have suggested that adjacent vegetation can help augment flower-visitors to plants within fragments (Steffan-Dewenter et al. 2002; Ricketts et al. 2008; Winfree et al. 2009).

Buckwheat reproductive biology enables another approach: institutional efforts to encourage small private plantings within both typical Mancos shale badlands habitat, and perhaps other edaphic substrates as well. For example, backyards of concerned, conservation-minded private citizens are being touted for preservation of biodiversity (Goddard et al. 2009) and as areas that improve connectivity between patches of endemic species. Such connectivity increases gene flow and dispersal among patches, and between patches and larger source areas (Rudd et al. 2002; Parker et al. 2008; Davies et al. 2009). One envisions a programme to distribute one-or two-year old plants that have been propagated as part of a community conservation project. Such efforts should be encouraged in areas like west central Colorado, where rapid residential development is impinging upon rare endemic plants like clay-loving wild buckwheat.

ACKNOWLEDGEMENTS

We thank the USDA ARS Grasshopper IPM program and the USDI-BLM (Melissa Siders) for funding; the Hughes folks for allowing us to work on their place; Robert Fitts for tireless field assistance; Jim Ferguson, USDI-BLM, for all sorts of help across two decades; Gina Glenne, USFWS, for assistance in relocating sites and encouragement; F. D. Parker (wasps) and W. Hanson (flies) for identifications; "H" Ikerd for aiding with details electronic; and Carol Kearns (Univ. of Colorado) and James Reveal (Cornell Univ.) for helping to untangle the manuscript.

APPENDICES

Additional supporting information may be found in the online version of this article:

Appendix I. Bee, wasp, ant and fly species captured on *E. pelinophilum* flowers at three Colorado sites.
Appendix II. Mean scores of *E. pelinophilum* pollen carried on six body parts of 14 bee species.
Appendix III. Mean scores of *E. pelinophilum* pollen carried on six body parts of eight wasp and three ant species.
Appendix IV. Mean scores of *E. pelinophilum* pollen carried on six body parts of 12 fly species.

REFERENCES

Aguilar R, Ashworth L, Galetto L, Aizen MA (2006) Plant reproductive susceptibility to habitat fragmentation: review and synthesis through a meta-analysis. Ecology Letters 9: 968-980.

Aizen MA, Ashworth L, Galetto L (2002) Reproductive success in fragmented habitats: do compatibility systems and pollination specialization matter? Journal of Vegetation Science 13: 885-892.

Aizen MA, Feinsinger P (2003) Bees not to be? Responses of insect pollinator faunas and flower pollination to habitat fragmentation. In: Bradshaw GA, Marquet PA (eds) How landscapes change. Springer-Verlag,,Berlin, pp 111-129.

Alarcon R, Waser NM, Ollerton J (2008) Year-to-year variation in the topology of a plant-pollinator interaction network. Oikos 117: 1796-1807.

Archibald JK, Wolf PG, Tepedino VJ, Bair J (2001) Genetic relationships and population structure of the endangered Steamboat Buckwheat (*Eriogonum ovalifolium* var. *williamsiae*) (Polygonaceae). American Journal of Botany 88: 608-615.

Begon M, Harper JL, Townsend CR (1986) Ecology. Sinauer Assoc., Sunderland MA.

Bond WJ (1994) Do mutualisms matter? Assessing the impact of pollinator and disperser disruption on plant extinction.

Philosophical Transactions of the Royal Society of London, B 344: 83-90.

Bowlin WR, Tepedino VJ, Griswold TL (1993) The reproductive biology of *Eriogonum pelinophilum* (Polygonaceae). In: Sivinski R, Lightfoot K (eds) Proceedings of the. Southwestern Rare and Endangered Plant Conference, New Mexico Forestry & Resources Conservation Division, Misc. Publ. #2, Santa Fe NM, pp 296-300.

Daniel WW (1990) Applied nonparametric statistics, 2nd Ed. PWS-Kent, Boston MA .

Davies ZG, Fuller RA, Loram A, Irvine KN, Sims V, Gaston KJ (2009) A national scale inventory of resource provision for biodiversity within domestic gardens. Biological Conservation 142: 761-771.

Donaldson J, Nänni I, Zachariades C, Kemper J (2002) Effects of habitat fragmentation on pollinator diversity and plant reproductive success in renosterveld shrublands of South Africa. Conservation Biology 16: 1267-1276.

Duff M (1996) The pollination biology and demography of *Eriogonum argophyllum*. Unpublished Report, The Nature Conservancy, Las Vegas NV .

Ehrlich PR (1988) The loss of diversity. In: Wilson EO (ed). Biodiversity. National Academy Press, Washington D. C., pp 21-27.

Fenster CB, Armbruster WS, Wilson P, Dudash MR, Thomson JD (2004) Pollination syndromes and floral specialization. Annual Review of Ecology and Systematics 35: 375-403.

Franzen, M, Larsson, M (2009) Seed set differs in relation to pollen and nectar foraging flower visitors in an insect-pollinated herb. Nordic Journal of Botany 27: 274-283.

Gentry AH (1996) Species expirations and current extinction rates: A review of the evidence. In: Szaro RC, Johnston DW (eds.) Biodiversity in managed landscapes. Oxford Univ. Press, Oxford, UK, pp 17-26.

Gess FW, Gess SK (1993) Effects of increasing land utilization on species representation and diversity of Aculeate wasps and bees in the semi-arid areas of southern Africa. In: LaSalle J, Gauld ID.(eds) Hymenoptera and biodiversity. C. A. B. International, Wallingford, UK, pp 83-113.

Goddard, MA, Dougill, AJ, Benton, TG (2009) Scaling up from gardens: biodiversity conservation in urban environments. Trends in Ecology and Evolution 25: 90-98.

González-Varo JP, Arroyo J, Aparicio A (2009) Effects of fragmentation of pollinator assemblage, pollen limitation and seed production of Mediterranean myrtle (*Myrtus communis*). Biological Conservation 142: 1058-1065 .

Hargreaves, AL, Harder, LD, Johnson, SD (2009) Consumptive emasculation: the ecological and evolutionary consequences of pollen theft. Biological Reviews: 84: 259-276.

Herrera CM (1987) Components of pollinator "quality": comparative analysis of a diverse insect assemblage. Oikos 50: 79-90.

Herrera CM (1989) Pollinator abundance, morphology, and flower visitation rate: analysis of the "quantity" component of a plant-pollinator system. Oecologia 80: 241-248.

Herrera CM (1990) Daily patterns of pollinator activity, differential pollinating effectiveness, and floral resource availability, in a summer-flowering Mediterranean shrub. Oikos 58: 277-288.

Herrera CM (2005) Plant generalization on pollinators: species property or local phenomenon? American Journal of Botany 92: 13-20.

Inouye DW, Gill DE, Dudash MR, Fenster CR (1994) A model and lexicon for pollen fate. American Journal of Botany 81: 1517-1530.

Kan T (1993) The distribution and ecology of the narrow endemic *Eriogonum umbellatum* var. *torreyana*. M. S. thesis, University of California, Davis CA.

Kandori I (2002) Diverse visitors with various pollinator importance and temporal change in the important pollinators of *Geranium thunbergii* (Geraniaceae). Ecological Research 17: 283-294.

Kaye T, Messinger W, Massey S, Kephart S, Flanagan D (1990) *Eriogonum crosbyae* and *E. prociduum* inventory, reproduction, and taxonomic assessment. Oregon Department of Agriculture, OSDA/BLM Challenge Cost Share Project No. 89-6. [online] http://www.centerforplantconservation.org/ASP/CPC_ViewProf ile.asp?CPCNum=1708 (accessed June 2011).

Kearns CA, Inouye DW, Waser NM (1998) Endangered mutualisms: the conservation of plant-pollinator interactions. Annual Review of Ecology and Systematics 29: 83-112.

Knight, TM, Steets, JA, Ashman, T-L (2006) A quantitative synthesis of pollen supplementation experiments highlights the contribution of resource reallocation to estimates of pollen limitation. American Journal of Botany 93: 271-277.

Lamborn E, Ollerton J (2000) Experimental assessment of the functional morphology of inflorescences of *Daucus carota* (Apiaceae): testing the "fly catcher effect". Functional Ecology 14: 445-454.

McNeely JA, Miller KR, Reid WV, Mittermeier RA, Werner TB (1990) Conserving the world's biological diversity. World Bank Publications, Philadelphia PA.

Morris WF (2003) Which mutualists are most essential? Buffering of plant reproduction against the extinction of pollinators. In: Kareiva P, Levin SA (eds) The importance of species. Princeton University Press, Princeton NJ, pp 260-280.

Neel MC, Ross-Ibarra J, Ellstrand NC (2001) Implications of mating patterns for conservation of the endangered plant *Eriogonum ovalifolium* var. *vineum* (Polygoneceae). American Journal of Botany 88: 1214-1222.

Neel MC, Ellstrand NC (2003) Conservation of genetic diversity in the endangered plant *Eriogonum ovalifolium* var. *vineum* (Polygonaceae). Conservation Genetics 4: 337-352.

Olesen JM, Dupont YL, Ehlers, BK, Hansen DM (2007) The openness of a flower and its number of flower-visitor species. Taxon 56: 729-736.

Olesen JM, Bascompte J, Elberling H, Jordano P (2008) Temporal dynamics in a pollination network. Ecology 89:1573-1582.

Ollerton J, Winfree R, Tarrant S (2011) How many flowering plants are pollinated by animals? Oikos 120: 321-326.

Parker K., Head L, Chisolm LA , Feneley N (2008) A conceptual model of ecological connectivity in the Shellharbour local government area, New South Wales, Australia. Landscape and Urban Planning 86:47-59.

Petanidou T, Kallimanis AS, Tzanopoulos J, Sgardelis SP, Pantis JD (2008) Long-term observation of a pollination network: fluctuation in species and interactions, relative invariance of network structure and implications for estimates of specialization. Ecology Letters 11: 564-575.

Ramirez N (2003) Floral specialization and pollination: a quantitative analysis and comparison of the Leppik and the Faegri and van der Pijl classification systems. Taxon 52: 687-700.

Reveal JL (2005) Flora of North America. Vol. 5: Polygonaceae, *Eriogonum* Michaux. [online] http://www.efloras.org/florataxon.aspx?flora_id=1&taxon_id=112045.

Rudd H, Vala J, Schaefer V (2002) Importance of backyard habitat in a comprehensive biodiversity conservation strategy: A connectivity analysis of urban green spaces. Restoration Ecology 10: 368-375.

Sahli HF, Conner JK (2006) Characterizing ecological generalization in plant-pollinators systems. Oecologia 148: 365-372.

Schemske DW, Husband BC, Ruckelshaus MH , Goodwillie C, Parker IM, Bishop JG (1994) Evaluating approaches to the conservation of rare and endangered plants. Ecology 75: 584-606.

Sipes SD, Tepedino VJ (1995) Reproductive biology of the rare orchid, *Spiranthes diluvialis*: Breeding system, pollination, and implications for conservation. Conservation Biology 9: 929-938.

Spears Jr EE (1983) A direct measure of pollinator effectiveness. Oecologia 57: 196-199.

Steffan-Dewenter, I et al. (2002) Scale-dependent effects of landscape context on three pollinator guilds. Ecology 83: 1421-1432.

Tepedino VJ (1979) The importance of bees and other insect pollinators in maintaining floral species composition. Great Basin Naturalist Memoirs 3:139-150.

Tepedino VJ, Stanton NL (1981) Diversity and competition in bee-plant communities on short-grass prairie. Oikos 36:35-44.

Tepedino VJ, Archibald JK, Bowlin WR (2002) Reproduction and pollination in the rare Steamboat Buckwheat (*Eriogonum ovalifolium* var. *williamsiae*). Unpublished Report, The Nature Conservancy, Las Vegas NV .

Thomson JD, Thomson BA (1992) Plant presentation and viability schedules in animal-pollinated plants: consequences for reproductive success. In: Wyatt R (ed) Ecology and evolution of plant reproduction: New approaches. Chapman & Hall, New York NY, pp 1-24.

U. S. Department of the Interior, Fish and Wildlife Service (2009) Endangered and threatened wildlife and plants; 90-day finding on a petition to revise critical habitat for *Eriogonum pelinophilum* (Clay-Loving Wild Buckwheat). Federal Register 74 (118): 29456-29461.

U. S. Fish and Wildlife Service (2009) Federal and state endangered and threatened species expenditures Fiscal Year 2009 [online] http://www.fws.gov/endangered/esa-library/pdf/ 2009_EXP_Report.pdf.

Vazquez DP, Morris WF, Jordano P (2005) Interaction frequency as a surrogate for the total effect of animal mutualists on plants. Ecology Letters 9: 1088-1094.

Vinson SB, Frankie GW, Barthell J (1993) Threats to the diversity of solitary bees in a Neotropical dry forest in central America. In: LaSalle J, Gauld ID (eds) Hymenoptera and biodiversity. C. A. B. International, Wallingford, UK, pp 53-81.

Westrich P (1996) Habitat requirements of central European bees and the problems of partial habitats. In: Matheson A, Buchmann SL, O'Toole C, Westrich P, Williams IH (eds) The conservation of bees. Academic press, London UK, pp 1-16.

Wilson EO (1988) The current state of biological diversity. In: Wilson EO (ed) Biodiversity. National Academy Press, Washington DC, pp 3-18 .

Winfree R, Kremen, C (2009) Are ecosystem services stabilized by differences among species? A test using crop pollination. Proceedings of the Royal Society, B 276: 229-237.

Winfree R, Aguilar R, Vazquez DP, LeBuhn G, Aizen MA (2009) A meta-analysis of bees' responses to anthropogenic disturbance. Ecology 90: 2068-2076.

Yates CJ, Ladd PG (2005) Relative importance of reproductive biology and establishment ecology for persistence of a rare shrub in a fragmented landscape. Conservation Biology 19: 239-249.

Zych M (2007) On flower visitors and true pollinators: The case of protandrous *Heracleum sphondylium* L. (Apiaceae). Plant Systematics and Evolution 263: 159-179.

12

POLLEN REMOVAL AND DEPOSITION BY POLLEN- AND NECTAR-COLLECTING SPECIALIST AND GENERALIST BEE VISITORS TO *ILIAMNA BAKERI* (MALVACEAE)

V. J. Tepedino*[1], Laura C. Arneson[1,2], and Susan L. Durham[3]

[1]*Department of Biology, Utah State University, Logan UT, USA 84322-5305*
[2]*present address: 333 North Quince St., Salt Lake City UT, USA 84103*
[3]*Ecology Center, Utah State University, Logan UT, USA 84322-5205*

Abstract—Up to 60% of the bee species of a region are oligolectic; they collect pollen only from a closely related group of plants though nectar-collecting choices are often broader. Bee specialists are expected to be superior to generalists in gathering pollen from their host plants and perhaps in transferring pollen to host stigmas. We used the oligolege *Diadasia nitidifrons* and its pollen-host *Iliamna bakeri* to ask if specialists 1) were more efficient than generalists as pollen-collectors; 2) deposited more pollen on stigmas than generalists; and 3) if pollen-collectors removed and deposited more pollen than did nectar-collectors. We found support for the first and third hypotheses. *Diadasia* pollen- and nectar-collectors removed more pollen per flower-visit than did their primary generalist competitors (*Agapostemon* spp.). The superior pollen-gathering efficiency of *Diadasia* exceeded differences that might be attributed to size: although *Agapostemon* females are, on average, 12.5% smaller than *Diadasia* females, pollen-collecting *Agapostemon* left 22.9% more pollen in flowers than did *Diadasia*. We found no difference between taxa in time spent foraging on a single flower. *Diadasia* and *Agapostemon* pollen-collectors deposited significantly more pollen on *I. bakeri* stigmas than did nectar-collectors; there was no difference between taxa in pollen deposition. *Diadasia* was superior to generalists as a pollinator in two ways: *Diadasia* was 1) a more reliable presence in *I. bakeri* populations; and 2) always most abundant at *I. bakeri* flowers. The association between *D. nitidifrons* and *I. bakeri* appears to be another example of a highly specialised bee affiliated with an unspecialised host-plant.

Keywords: *Pollen removal, Pollen deposition, Specialization, Oligolecty, Iliamna, Diadasia*

INTRODUCTION

Oligolecty describes an inherent preference by all members of a bee population or species for the pollen of a circumscribed taxon of plants (Minckley & Roulston 2006). Up to 60% of the bee species of a region may be oligolectic (Minckley & Roulston 2006). Why this fraction is so large has intrigued bee biologists and pollination ecologists for at least a century (Robertson 1914). The commonest explanation for oligolecty is related to one advanced by Darwin (1876) for flower constancy (Cane & Sipes 2006; Raine & Chittka 2007). Darwin thought that individuals that restricted their visits to a few closely related plants would learn from frequent use to collect pollen from those flowers more effectively than their generalist competitors. Over time, at least some of those learned foraging behaviors are presumed to have become instinctive.

Pollen is the main source of amino acids, protein, lipids, and starch that female bees supply to their progeny (Roulston & Cane 2000); nectar, by contrast, is primarily an energy source and is commonly collected from both the pollen host and various other species (Robertson 1914; Linsley 1958; Eickwort & Ginsberg 1980; Wcislo & Cane 1996). More efficient collection of specific host-plant pollen by female oligoleges than by polyleges is likely to be strongly selected for at least two reasons: 1) pollen is more likely to be a limiting trophic resource because, unlike nectar, it is not replenished within flowers (e.g., Percival 1955; Linsley 1978; Minckley et al. 1994; Schlindwein et al. 2005; Larsson 2005; Larsson & Franzén 2007; Carvalho & Schlindwein 2011); and 2) only host-plant pollen is available to oligoleges but polyleges forage on a variety of different floral species with diverse morphologies and resource dispensing mechanisms.

While there is some evidence to support the hypothesis that specialists are superior to generalists in collection of pollen grains/unit time from their host plants (e.g., Strickler 1979; Cane & Payne 1988; Laverty & Plowright 1988; Thostesen & Olsen 1996; Larsson 2005; reviewed in Minckley & Roulston 2006), not all studies concur (Harder & Barrett 1993; Castellanos et al. 2003). Thus, our first objective was to compare the pollen-collecting ability of a specialist, the mallow oligolege *Diadasia nitidifrons* Cockerell (Arneson 2004; Arneson et al. 2004; Sipes & Tepedino 2005) on its host plant, *Iliamna bakeri* (Jepson)

---

*Corresponding author: vince.tepedino@usu.edu

Wiggins (Malvaceae), with that of its generalist bee competitors.

Plant species such as *I. bakeri* that have been "adopted" by a specialised pollinator such as *D. nitidifrons* may, in turn, begin to adapt to or coevolve with those pollinators, particularly when their presence at the flowers is as reliable as is that of *D. nitidifrons* (Arneson et al. 2004). However, because *I. bakeri*, with its large, open flowers, and easily accessible pollen and nectar, is also an attractive prospect for generalised visitors, there may be little selective pressure to co-evolve with a specialist pollinator. Alternatively, floral adaptations can sometimes be quite subtle and unexpected (Armbruster 2006), and specialists may sometimes be superior to generalists as pollinators of host flowers (Thomson 2003; Williams & Thomson 2003; Fenster et al. 2004; Minckley & Roulston 2006) though sometimes they are not (Tepedino 1981; Sampson & Cane 2000; Mayfield et al. 2001; Castellanos et al. 2003; Larsson 2005). Thus, our second objective was to determine whether visits by *D. nitidifrons* to *I. bakeri* flowers result in more fruit or seed than visits by generalists.

Differences in the behaviour of pollen and nectar collectors on flowers may also lead to differences in pollen removal, deposition, or the likelihood of contact with reproductive parts (Tepedino & Parker 1982; Wilson & Thomson 1991; Williams & Thomson 2003; Castellanos et al. 2003; McIntosh 2005). For example, simple preferences for flowers in the male stage by pollen collectors may lead to ineffective pollination even by specialists (Tepedino & Parker 1982). Therefore, our third objective was to compare pollen removal and deposition between nectar and pollen foragers.

MATERIALS AND METHODS

Study plant

I. bakeri is a globally rare though locally abundant, short-lived ($\leq$ 10 yr), fire-following mallow limited to the volcanic soils of high-elevation arid shrublands and open forests of northeast California and southeast Oregon, USA (Wooley 2000). Flowers are large and open, are open throughout the day, last for approximately 24 hrs, and produce between 16 and 18 stigmas and about 140 stamens (range 116-160; Arneson et al. 2004). Flowers are hermaphroditic and protandrous; male and female phases each last up to 12 hrs.

Over 85% of anthers are fully dehisced by the time the flower is one-third open, and bees are common visitors at these early stages when pollen and nectar are readily obtainable (Arneson 2004). Thus, pollen can be depleted quickly if visitors are abundant. The styles gradually emerge from the malvaceous column after anther dehiscence. By the time the styles extend above the column, the central most anthers have begun to reflex away from the column thus making minimal contact with stigmatic surfaces (Arneson 2004).

Arneson (2004) found that most *I. bakeri* plants are self-incompatible and that fruit and viable seed production requires pollinators. Hand-pollination experiments showed limited fruit and seed production from geitonogamous treatments and no fruit production from autogamy. Outcrossing treatments yielded significantly higher fruit set than did geitonogamy treatments.

Pollen Remaining

Single-visit experiments were used to estimate the number of pollen grains removed from virgin flowers. From 9 – 25 July 2003, budded inflorescences on 10 *I. bakeri* plants at the Clark Valley Road population in NE California (see Arneson et al. 2004 for details) were bagged (10 × 10 × 30 cm) with white bridal veil material (mesh 1 mm^2) to exclude visitors. When flowers displayed receptive stigmas, bags were removed and the target flower was allowed one bee visit and then immediately rebagged. Stigmas were judged receptive when they had extended above the column and appeared moist under 10× magnification.

A visit was defined as an actual landing upon the flower followed by an obvious attempt to gather pollen and/or nectar. We sight-identified generalist and specialist flower-visitors to the lowest taxonomic level possible. Subsequent identification of collected specimens revealed generalists to be four species of bumblebees (Apidae: mostly *Bombus vosnesenskii* Radoszkowski, but also *B. centralis* Cresson, *B. huntii* Greene, *B. melanopygus* Nylander) and two species of green sweat bees (Halictidae: mostly *Agapostemon angelicus* Cockerell/*A. texanus* Cresson, but also *A. femoratus* Crawford). We noted whether foragers collected nectar, pollen or both. "Pollen" and "both" categories were subsequently combined because the time spent collecting nectar following pollen collection was always brief, usually lasting only a few seconds. As we had no control over the identity or abundance of visitors, we had to settle for the visits that occurred. Therefore, our sample sizes are unequal and, in some cases, small.

Unvisited control flowers selected for pollen counts were immediately adjacent to single-visit flowers and in the same bag. Depending on availability, the number of control flowers varied from 2 to 6 per plant (median = 3.5, N = 39 total flowers), and the number of visited flowers varied from 1 - 25 per plant (median = 8.5, N = 102 total flowers). There was no indication that the unequal numbers of control and visited flowers per plant biased our results; for example, control flowers from a plant with one experimental flower produced approximately the same number of pollen grains as did those from a plant with 25 experimental flowers (mean $\pm$ SE (# flowers): plants with one flower visited = 24,930 $\pm$ 9,981 (2); plants with 25 flowers visited = 25,568 $\pm$ 5,182 (5)).

We used pollen remaining in unvisited controls and in single-visited flowers as our dependent variable. The day after single-visit flowers and control flowers closed, whole corollas were detached from the pedicel, placed individually in glass vials, and stored in the freezer to prevent fungal infestation. Pollen grains remaining, including those that had fallen into the corolla or been dislodged from the anthers, were counted in the lab two months later.

All remaining pollen grains were removed from stamens and petals in each flower via sonication in filtered ethanol, then counted using a HIAC Royco 8-channel particle counter (Model 8000A). *I. bakeri* pollen grains are approximately 50 microns in diameter and were distinguished, by size, from all particles less than 40 or greater than 60 microns in diameter. Flowers that became infested with fungal mycelia were discarded because pollen grains could not be dis-aggregated.

Strickler (1979) proposed that a comparison of pollen-removal efficiency among bee species should incorporate adjustments for size differences and for the average duration of individual flower visits. We used body volume as our size metric using the formula for a cylinder (volume $= \pi r^2 h$; where r = distance between the wing bases (tegulae) and h = body length (abdomen unteloscoped)). Body size has been shown to be a reasonable substitute for offspring provision size (Strickler 1979; Müller et al. 2006; Neff 2008). We measured the inter-tegular distance and body length of 15 females each of *D. nitidifrons*, *B. vosnesenskii* (workers), and *A. angelicus/A. texanus*, computed volume for each individual, and averaged over individuals. Duration of most individual single-flower visits was recorded with a stopwatch. Precise time was recorded for visits ≤ 60 seconds; for longer visits (97% of visits) we rounded up or down to the nearest half minute.

Pollinator Quality

We used the number of pollen grains deposited on *I. bakeri* stigmas as our metric of pollinator quality. We estimated pollen deposition for single bee visits to an unbagged, virgin flower using the same flowers used to measure pollen removal. Five of the 16-18 stigmas were chosen for pollen grain counts from a distance at which presence of pollen grains could not be distinguished by the unaided eye (6–9 dm). We selected the central-most stigma and one each near the periphery in the four compass directions (N, S, E, W). Pollen grains on each stigma were counted under 10× magnification in the field. We used the average number of pollen grains on the five stigmas for each flower as an estimate both of pollinator quality and also to compare with fruit set and seed number. Fruit capsules were collected approximately 3-4 weeks later when mature, returned to the lab, recorded and dried, and seeds were counted.

Statistical Analyses

The number of pollen grains in control flowers was compared to the number of pollen grains remaining in single-visited flowers for five visitor groups using a one-way ANOVA in a completely randomised design; Dunnett's test was used to control Type I error for multiple comparisons of the control group to each visitor group. There were five visitor groups: *Diadasia* (specialist) pollen and nectar collectors, *Agapostemon* (generalist) pollen and nectar collectors, and *Bombus* (generalist) nectar collectors.

Differences due to species (generalist versus specialist) and purpose (nectar versus pollen) on number of remaining pollen grains were assessed using a two-way factorial ANOVA in a completely randomised design. As *Bombus*

workers did not "intentionally" collect pollen, they were excluded from this analysis.

Differences in number of remaining pollen grains among the three nectar-feeding groups (generalist, specialist, and *Bombus*) were assessed using a one-way ANOVA in a completely randomised design. Post-hoc multiple comparisons among the three groups were controlled for Type I error using the Tukey-Kramer method.

Differences in visit duration by species (generalist versus specialist) and purpose (nectar versus pollen) were assessed using a two-way factorial ANOVA in a completely randomised design. Because *Bombus* visit durations were clearly shorter than those of other bee species and because *Bombus* data had only two unique values, no statistical comparisons with this visitor group were made. To elucidate the nature of interaction between species and purpose, we compared simple effects (i.e., generalists to specialists for each level of purpose, and nectar to pollen for each level of species); reported *P* values are unadjusted.

We used similar two-way and one-way ANOVAs to analyse number of pollen grains deposited in single visits. A two-sample t-test was used to compare the number of pollen grains deposited on flowers that set fruit with those that aborted.

To better meet assumptions of normality and homogeneity of variance, number of remaining pollen grains and number of deposited pollen grains were square-root transformed, and visit duration was log-transformed prior to analysis. Data calculations were made using the GLIMMIX procedure in SAS/STAT 13.2 in the SAS System for Windows 9.4 (TS1M2).

RESULTS

Iliamna bakeri flowers were visited by bee taxa that differed in size and in purpose. *Agapostemon* females were smallest (volume mean ± SE = 99.1 ± 8.2 mm^3) and actively collected pollen in 20 of 30 (67%) visits; *Diadasia nitidifrons* females were larger (volume mean ± SE = 111.5 ± 7.2 mm^3) and collected pollen during 26 of 38 (68%) visits; *Bombus vosnesenskii* workers (volume mean ± SE = 590.4 ± 79.5 mm^3) were over five times *Diadasia*'s size and removed pollen only passively as they actively collected nectar. Pollen-collecting visitors of both *Diadasia* and *Agapostemon* tended to accumulate more pollen on body parts as they scrambled around and across the multi-stamen flowers whereas nectar collectors remained stationary for longer periods on flowers while 'drinking.'

Pollen Remaining

Pollen remaining in flowers differed among control and visitor groups ($F_{5,112}$ = 14.4, $P < 0.001$; Fig. 1A). Pairwise mean comparisons showed significantly more pollen remained in control flowers than in flowers visited by *Diadasia* and *Agapostemon* pollen collectors (DP, AP; $P < 0.001$) or by *Diadasia* and *Bombus* nectar collectors ($P = 0.002$ and < 0.001, respectively) but not in *Agapostemon* nectar collectors ($P = 0.545$).

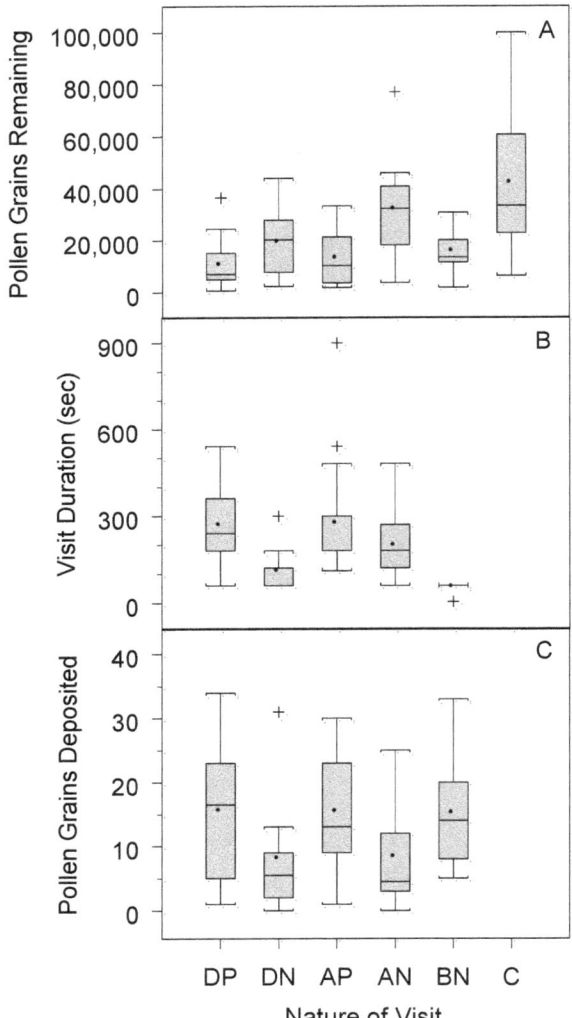

FIGURE 1. Box plots of results of single bee-visits to previously unvisited *Iliamna bakeri* flowers: A) number of pollen grains remaining; B) the duration of single flower-visits; C) the number of pollen grains deposited on stigmas. Shown are means (circles), maximum and minimum values, 25th and 75th percentiles, and outliers (+). Abbreviations: D (*Diadasia* – specialist), A (*Agapostemon* – generalist), B (*Bombus* – generalist), P (pollen-collectors), N (nectar-collectors), C (Control). Sample sizes: DP (26), DN (12), AP (20), AN (10), BN (11), C (39).

A two-factor analysis of pollen remaining showed that flowers visited by pollen-collectors had significantly less pollen remaining than those visited by nectar collectors ($F_{1.64}$ = 14.82, $P < 0.001$) and that flowers visited by *Diadasia* had fewer pollen grains remaining than those visited by *Agapostemon* ($F_{1.64}$ = 3.63, $P = 0.061$). The interaction term did not approach significance ($F_{1.64}$ =1.06, $P = 0.308$). There was no difference in pollen remaining among the three nectar-collecting taxa ($F_{2.30}$ = 2.35, $P = 0.113$).

We also incorporated a measure of visit duration into our assessment of foraging efficiency (Fig. 1B). The comparison of visit duration between specialist (*Diadasia*) and generalist (*Agapostemon*) by purpose (nectar vs. pollen-collection) yielded a significant interaction ($F_{1.82}$ = 6.36, $P = 0.014$). While there was no evidence of a difference between

specialist and generalist pollen collectors ($P = 0.832$), specialist nectar-collectors foraged more rapidly than did generalist nectar-collectors ($P = 0.004$). Specialist and generalist nectar-collectors also foraged more rapidly than did their pollen-collecting counterparts (P < 0.001; 0.063, respectively).

Pollinator Quality

We found no difference in pollen deposition between specialist (*Diadasia*) and generalist (*Agapostemon*) ($F_{1.85}$ = 0.12, $P = 0.729$; Fig. 1C). Pollen foragers deposited more pollen on stigmas than did nectar foragers ($F_{1.85}$ = 15.23, $P < 0.001$). There was no evidence of interaction ($F_{1.85}$ = 0.03, $P = 0.870$). A one-way ANOVA showed differences in pollen deposition among nectar collectors ($F_{2.40}$ = 3.93, $P = 0.028$): *Bombus* workers deposited more pollen on *Iliamna* stigmas than did *Diadasia* ($P = 0.029$) or *Agapostemon* ($P = 0.095$) nectar collectors. *Diadasia* and *Agapostemon* nectar collectors were not shown to be different ($P = 0.948$).

Overall, only 11 of 101 singly-visited flowers (10.9%) set fruit. Fruit-set was below 25% for all bee-purpose categories. Fruit set was too low to support comparisons between generalists and specialists or nectar and pollen collectors in fruit set. The 11 flowers that set fruit did not receive more pollen grains per stigmatic surface (mean = 15.3, S.E. = 2.9, $N = 11$) than did those flowers which did not set fruit (mean = 12.8, SE = 1.0, $N = 90$; $t_{99} = 0.82$, $P = 0.41$). Clearly, flowers must be visited more than once to realise the 70% level of fruit set reported by Arneson et al. (2004) for this population.

DISCUSSION

Recently, several enduring hypotheses of melittologists and pollination biologists have been questioned: 1) that specialist bees remove more pollen per flower visit from their preferred host plants than do generalist bees (Minckley & Roulston 2006); 2) that specialists deposit more pollen per flower visit to their host plants than do generalists and are thus better pollinators (Tepedino 1981; Thomson 2003; Williams & Thomson 2003; Franzén & Larsson 2009); and 3) that pollen-collecting bees collect and deposit more pollen per visit than do nectar-collectors (Williams & Thomson 2003). We found tentative support for the first and third hypotheses but not the second.

Specialist *Diadasia* foragers, whether collecting pollen or nectar, left fewer pollen grains in *I. bakeri* flowers after a single visit, i.e., they collected more pollen than did their generalist *Agapostemon* counterparts. Not only did *Diadasia* females collect more pollen than *Agapostemon* in an average visit but they collected disproportionately more as a percentage of body size. *Agapostemon* females are 12.5% smaller than *Diadasia* females but *Agapostemon* pollen-collectors left 22.9% more pollen grains in flowers (12,687) than did their specialist counterparts (10,322). The difference in amount of pollen left by nectar-collectors was even more dissimilar: *Agapostemon*-visited flowers averaged 65.6% more pollen grains remaining than did *Diadasia*-visited flowers (31,701 versus 19,134 respectively). Because body size and provision size tend to be positively correlated (Muller et al. 2006, Neff 2008), this suggests that *Diadasia* collected a greater proportion of each larval provision on each flower visit than did *Agapostemon*. Thus, the specialist *D. nitidifrons* appears to be superior to *Agapostemon* as a forager on *I. bakeri*, especially when size is factored in. However, this interpretation should be viewed with caution for several reasons: 1) the large variation in the number of pollen grains produced in unvisited flowers, and remaining in single-visited flowers (Fig. IA); 2) our inability to measure the size of individual foragers, and thus to statistically test the effects of size on pollen extraction differences; 3) potential complications arising from different proportions of pollen in the provisions of different bee taxa (Neff 2008); 4) our estimates of pollen removal are indicative only of freshly opened, unvisited flowers (see below).

The addition of time measures to our comparison of foraging efficiency (Strickler 1979) gave no reason to alter our assessment of *Diadasia* as a superior gatherer of *Iliamna* pollen. In removing greater numbers of pollen grains (Fig. IB), *Diadasia* foragers did not use significantly more time/flower than did *Agapostemon* foragers. Or, stated differently, *Agapostemon* foragers did not gain equivalence with *Diadasia* in pollen-harvesting efficiency by foraging faster on individual flowers. Additional time savings are likely to accrue to *Diadasia* foragers from their propensity to nest near their host plants. *Diadasia* species such as *D. nitidifrons* (personal observations) and several others (Schlising 1972; Neff et al. 1982; Ordway 1984, 1987) that nest near their host plants, likely enjoy a competitive advantage over generalists in accessing resources and also save time and expend less energy in doing so (Eickwort & Ginsberg 1980; Gathmann & Tscharntke 2002; Franzén et al. 2009). In contrast, generalists are not under any selective pressure to nest closer to *Iliamna* populations than to other plant species from which they forage. Indeed, we were unable to compare travel times amongst bee taxa because only *D. nitidifrons* nests were evident near *Iliamna* populations. The incorporation of travel time to resource patches would likely show significant time-saving for *Diadasia*.

A final complicating factor in this and related studies of foraging efficiency is that, to standardise our conditions, we were forced to time visits to virgin flowers with copious pollen. Thus, our estimates of visit duration are indicative of freshly opened, unvisited flowers and are of much longer duration than visits to previously visited, pollen- and nectar-exploited flowers (Arneson 2004). We conclude that measuring the time factor in foraging efficiency studies will require much more careful gathering and integration of data on flower-visit time, total foraging-trip time, and travel time at different times during the day, all against amounts of pollen and/or nectar removed, to reach convincing conclusions on the overall role of time in floral resource collection efficiency.

Our tentative results on pollen removal are in agreement with the few other studies that have compared specialist and generalist bees. In her pioneering study, Strickler (1979) found that, when adjusted for size and flower-handling time, the specialist bee, *Hoplitis anthocopoides*, removed more

pollen per *Echium vulgare* flower than did four species of generalists. Harder & Barrett (1993) compared two generalist *Bombus* species with the much smaller solitary specialist, *Melissodes apicata*, on the tristylous floral host *Pontederia cordata*. Although they did not report *M. apicata* to be a more efficient pollen harvester, when adjustment is made for the large differences in body size between bee species (*M. apicata* is 50-65% smaller than the bumblebees), *M. apicata* becomes a far more efficient pollen harvester than either of the generalists. Thostesen & Olesen (1996) and Larsson (2005) also showed that specialist bees removed a larger quantity of pollen than did generalists.

We found no indication that *D. nitidifrons* was superior to generalists as a pollinator when number of pollen grains deposited on receptive stigmas was used as the criterion of pollination success (Fig. IC). Other comparisons of pollen deposition between specialist and generalist bees are few. Thostesen & Olesen (1996) showed that only the specialist bumblebee *B. consobrinus* deposited pollen on host plant stigmas and was therefore superior to generalists as a pollinator. While not examining pollen deposition directly, McIntosh (2005) found that visits by specialists (species of *Diadasia*) were more effective than generalists at producing seeds of *Ferocactus*. Studies by Tepedino (1981) and Larsson (2005) illustrate how subtle and misleading simple comparisons of pollen deposition can be (reviewed in Ne'eman et al. 2010). Tepedino (1981) found that while the specialist bee *Peponapis pruinosa* deposited 33% more pollen grains on squash stigmas than did the generalist honeybee, honeybees improved their pollination value by preferring pistillate flowers while squash bees preferred staminate flowers. Similarly, Larsson (2005) found that the specialist bee, *Andrena hattorfiana*, deposited more pollen on stigmas of its host plant than did generalists but that it visited fewer stigma-presenting flowers and more pollen-presenting flowers than did generalists. Larsson (2005) concluded that *A. hattorfiana* was actually an inferior pollinator. Thus, the value of specialist bees to plant species with open flowers varies with the bee-plant association.

Other factors are at least as important in evaluating pollinator value as the number of pollen grains deposited per flower visit. Indeed, the total number of pollen grains deposited can be a misleading index of pollinator value (e.g., Cane & Schiffhauer 2003; Ne'eman et al. 2010), especially for mostly self-incompatible species like *I. bakeri* where geitonogamous pollinations are less effective than outcrossing at producing seeds. Arneson (2004) found that *Diadasia* typically visited only one flower per inflorescence before moving to other inflorescences on the same, or other, plants while generalists tended to visit multiple *Iliamna* flowers per inflorescence. One *Diadasia* visit/inflorescence is likely to translate to more inter-plant movements, and therefore more potential pollinating visits than by generalists which were more likely to move pollen between flowers within plants, thereby effecting unproductive selfing. Because of this likely difference in movement patterns between visitor types, *D. nitidifrons* visits are probably more effective at achieving pollinations on a per visit basis.

D. nitidifrons confers other advantages as a pollinator of its host plant: predictability of occurrence and numerical dominance. In their review, Vazquez et al. (2005) concluded that visitation frequency was the most valuable pollinator attribute to a plant. Arneson et al. (2004) found *D. nitidifrons* visiting *I. bakeri* flowers in all nine populations observed in northern California including those that were only two years old and > 25 km from the nearest *Iliamna* population, demonstrating high colonization potential. In addition, Tepedino (unpublished) found nesting populations of *D. nitidifrons* and high visitation rates to *Iliamna longisepala* (Torrey) Wiggins flowers at all nine sites surveyed in south central Washington, USA; *D. nitidifrons* was almost always the most abundant species visiting *Iliamna* flowers. Arneson et al. (2004) showed that at five of six *I. bakeri* sites studied most intensively, *D. nitidifrons* accounted for at least 60% of all *I. bakeri* visits and that high visitation rates were typically associated with high fruit production. Thus, the most important attribute of *D. nitidifrons* as a pollinator of *I. bakeri* may be its fidelity.

We found clear support for the third hypothesis, that pollen collectors removed and deposited more pollen per visit than did nectar collectors (Figs. 1A, C). This contrasts with some other studies which have shown that pollen collectors removed more pollen from flowers than did nectar collectors but that there was no difference in pollen deposition (e.g., Freitas & Paxton 1998; Thomson & Goodell 2001; Williams & Thomson 2003; Castellanos et al. 2003; Young et al. 2007). Pollen collectors sometimes deposited less pollen than nectar collectors (Thomson & Thomson 1992; Young et al. 2007) and sometimes more (Williams & Thomson 2003). In a related study, McIntosh (2005) found no difference in seed production by *Ferocactus* species with many stamens and open flowers between native bee collectors of nectar versus pollen.

The association between *D. nitidifrons* and *Iliamna* species appears to follow a general pattern for solitary bees and their host plants delimited by Waser et al. (1996), Wcislo & Cane (1996), and Minckley & Roulston (2006). First, oligolectic bees seem to become allied with host plants that are widespread, abundant, iteroparous, and have reliable and easily-exploited blooms. *Iliamna* species have all these characteristics except they are not widespread (indeed, *I. bakeri* and several congeners are globally rare). However, *Iliamna* species are usually locally abundant and predictable in that they are primary successional species that follow frequent fires (Wooley 2000, Arneson et al. 2004). A second emerging characteristic is that while the bee member of the association is highly specialised, the plant taxon is less so (Minckley & Roulston 2006). Indeed, *I. bakeri*, with ample patronage by both specialist and generalist bees, is not specialised in its pollinator requirements.

Acknowledgements

We thank Karrie Moredock for unstinting field assistance and companionship; and B. Corbin and M. Dolan of the BLM, and B. and I. Davidson, T. Fuentes, R. Posey, A. Sanger and R. Wooley of the Forest Service for meeting numerous needs throughout. Jim Cane and Faye Rutishauer instructed us in the use of the particle counter. Sheri Smith, Forest Service, Susanville CA, provided the impetus for this work along with monetary and logistical support and encouragement throughout. We thank Jim Cane, USDA ARS, PIRU-Logan and Jack Neff, CTMI-Austin, for their helpful review of an ancient version of the manuscript.

References

Armbruster WS (2006) Evolutionary and ecological aspects of specialized pollination: views from the arctic to the tropics. In: Waser NM, Ollerton J (eds) Plant-Pollinator interactions. University of Chicago Press, Chicago, pp 260-282.

Arneson LC (2004) Natural history of the rare, fire-following mallow, *Iliamna bakeri*, and its association with a specialist pollinator, *Diadaisa nitidifrons* (Hymenoptera: Apoidea) in northeastern California. MS thesis, Department of Biology, Utah State University, Logan, Utah, USA.

Arneson LC, Tepedino VJ, Smith SL (2004) Reproductive success of Baker's globe mallow and its association with a native specialist bee. Northwest Science 78:141-149.

Cane JH, Payne JA (1988) Foraging ecology of the bee *Habropoda laboriosa* (Hymenoptera: Anthophoridae), an oligolege of blueberries (Ericaceae: *Vaccinium*) in the southeastern United States. Annals of the Entomological Society of America 81:419-427.

Cane JH, Schiffhauer D (2003) Dose-response relationships between pollination and fruiting refine pollinator comparisons for cranberry (*Vaccinium macrocarpon* [Ericaceae]). American Journal of Botany 90:1425-1432.

Cane JH, Sipes S (2006) Characterizing floral specialization by bees: analytical methods and a revised lexicon for oligolecty. In: Waser NM, Ollerton J (eds) Plant-Pollinator interactions. University of Chicago Press, Chicago, pp 99-122.

Carvalho AT, Schlindwein C (2011) Obligate association of an oligolectic bee and a seasonal aquatic herb in semi-arid north-eastern Brazil. Biological Journal of the Linnean Society 102:355-368.

Castellanos MC, Wilson P, Thomson JD (2003) Pollen transfer by hummingbirds and bumblebees, and the divergence of pollination modes in *Penstemon*. Evolution 57:2742-2752.

Darwin C (1876) The effect of cross and self-fertilization in the vegetable kingdom. John Murray, London.

Eickwort GC, Ginsberg HS (1980) Foraging and mating behavior in Apoidea. Annual Review of Entomology 25:421-446.

Fenster CB, Armbruster WS, Wilson P, Dudash MR, Thomson JD (2004) Pollination syndromes and floral specialization. Annual Review of Ecology and Systematics 35:375-403.

Franzén M, Larsson M (2009) Seed set differs in relation to pollen and nectar foraging flower visitors in an insect-pollinated herb. Nordic Journal of Botany 27:274-283.

Franzén M, Larsson M, Nilsson SG (2009) Small local population sizes and high habitat patch fidelity in a specialized solitary bee. Journal of Insect Conservation 13:89-95.

Freitas BM, Paxton RJ (1998) A comparison of two pollinators: the introduced honey bee *Apis mellifera* and an indigenous bee *Centris tarsata* on cashew *Anacardium occidentale* in its native range of NE Brazil. Journal of Applied Ecology 35:109-121.

Gathmann A, Tscharntke T (2002) Foraging ranges of solitary bees. Journal of Animal Ecology 71:757-764.

Harder LD, Barrett SCH (1993) Influence of anther position on pollen removal by bees. Ecology 74:1059-1072.

Larsson M (2005) Higher pollinator effectiveness by specialist than generalist flower-visitors of unspecialized *Knautia arvensis* (Dipsacaceae). Oecologia 146:394-403.

Larsson M, Franzén M (2007) Critical resource levels of pollen for the declining bee *Andrena hattorfiana* (Hymenoptera, Andrenidae). Biological Conservation 134:405-414.

Laverty TM, Plowright RC (1988) Flower handling by bumblebees: A comparison of specialists and generalists. Animal Behavior 36:733-740.

Linsley EG (1958) The ecology of solitary bees. Hilgardia 27:543-599.

Linsley EG (1978) Temporal patterns of flower visitation by solitary bees, with particular reference to the southwestern United States. Journal of the Kansas Entomological Society 51:531-546.

Mayfield MM, Waser NM, Price MV (2001) Exploring the 'most effective pollinator principle' with complex flowers: bumblebees and *Ipomopsis aggregata*. Annals of Botany 88:591-596.

McIntosh ME (2005) Pollination of two species of *Ferocactus*: interactions between cactus-specialist bees and their host plants. Functional Ecology 19:727-734.

Minckley, RL, Wcislo, WT, Yanega, D, Buchmann SL (1994) Behavior and phenology of a specialist bee (Dieunomia) and sunflower (Helianthus) pollen availability. Ecology 75:1406-1419.

Minckley RL, Roulston TH (2006) Incidental mutualisms and pollen specialization among bees. In: Waser NM, Ollerton J (eds) Plant-Pollinator interactions. University of Chicago Press, Chicago, pp 69-98.

Müller A, Diener S, Schnyder S, Stutz K, Sedivy C, Dorn S (2006) Quantitative pollen requirements of solitary bees: implications for bee conservation and the evolution of bee-flower relationships. Biological Conservation 130:604-615.

Ne'eman GA, Jürgens A, Newstrom-Lloyd L, Potts SG, Dafni A (2010) A framework for comparing pollinator performance: effectiveness and efficiency. Biological Reviews 85:435-451.

Neff JL (2008) Components of nest provisioning behavior in solitary bees (Hymenoptera: Apoidea). Apidologie 39:30-45.

Neff JL, Simpson BB, Dorr LJ (1982) The nesting biology of *Diadasia afflicta* Cress. (Hymenoptera: Anthophoridae). Journal of the Kansas Entomological Society 55:499-518.

Ordway E (1984) Aspects of the nesting behavior and nest structure of *Diadasia opuntiae* Cockerell (Hymenoptera: Anthophoridae). Journal of the Kansas Entomological Society 57:216-230.

Ordway E (1987) The life history of *Diadasia rinconis* Cockerell (Hymenoptera: Anthophoridae). Journal of the Kansas Entomological Society 60:15-24.

Percival MS (1955) The presentation of pollen in certain Angiosperms and its collection by *Apis mellifera*. New Phytologist 54:353-368.

Raine NE, Chittka L (2007) Flower constancy and memory dynamics in bumblebees (Hymenoptera: Apidae: Bombus). Entomologia Generalis 29:179-199.

Robertson C (1914) Origins of oligotrophy of bees. Entomological News 25:67-73.

Roulston TH, Cane JH (2000) Pollen nutritional content and digestibility for animals. Plant Systematics and Evolution 222:187-209.

Sampson BJ, Cane JH (2000) Pollination efficiencies of three bee (Hymenoptera: Apoidea) species visiting rabbiteye blueberry. Journal of Economic Entomology 93:1726-1731.

Schlindwein C, Wittmann D, Feitosa Martins C, Hamm A, Alves Siqueira J, Schiffler D, Machado IC (2005) Pollination of Campanula rapunculus L. (Campanulaceae): How much pollen flows into pollination and into reproduction of oligolectic pollinators? Plant Systematics and Evolution 250: 147-156.

Schlising RA (1972) Foraging and nest provisioning behavior of the oligolectic bee, *Diadasia bituberculata* (Hymenoptera: Anthophoridae). Pan-Pacific Entomologist 48:175-188.

Sipes SD, Tepedino VJ (2005) Pollen-host specificity and evolutionary patterns of host switching in a clade of specialist bees (Apoidea: *Diadasia*). Biological Journal of the Linnean Society 86:487-505.

Strickler K (1979) Specialization and foraging efficiency of solitary bees. Ecology 60:998-1009.

Tepedino VJ (1981) The pollination efficiency of the squash bee (*Peponapis pruinosa*) and the honey bee (*Apis mellifera*) on summer squash (*Cucurbita pepo*). Journal of the Kansas Entomological Society 54:359-377.

Tepedino VJ, Parker FD (1982) Interspecific differences in the relative importance of pollen and nectar to bee species foraging on sunflowers. Environmental Entomology 11:246-250.

Thomson J (2003) When is it mutualism? American Naturalist 162 (Supplement):S1- S9.

Thomson JD, Goodell K (2001) Pollen removal and deposition by honeybee and bumblebee visitors to apple and almond flowers. Journal of Applied Ecology 38:1032-1044.

Thomson JD, Thomson BA (1992) Pollen presentation and viability schedules in animal-pollinated plants: consequences for reproductive success. In: Wyatt R (ed) Ecology and evolution of plant reproduction. Chapman & Hall, New York, pp 1-24.

Thostesen AM, Olesen JM (1996) Pollen removal and deposition by specialist and generalist bumblebees in *Aconitum septentrionale*. Oikos 77:77-84.

Vázquez DP, Morris WF, Jordano P (2005) Interaction frequency as a surrogate for the total effect of animal mutualists on plants. Ecology Letters 8:1088-1094.

Waser NM, Chittka L, Price MV, Williams NM, Ollerton J (1996) Generalization in pollination systems, and why it matters. Ecology 77:1043-1060.

Wcislo WT, Cane JH (1996) Floral resource utilization by solitary bees (Hymenoptera: Apoidea) and exploitation of their stored foods by natural enemies. Annual Review of Entomology 41:257-286.

Williams NW, Thomson JD (2003) Comparing pollinator quality of honey bees (Hymenoptera: Apidae) and native bees using pollen removal and deposition measures. In: Strickler K, Cane JH (eds) For nonnative crops, whence pollinators of the future? Thomas Say Publications, Entomological Society of America, pp 163-179.

Wilson P, Thomson JD (1991) Heterogeneity among floral visitors leads to discordance between removal and deposition of pollen. Ecology 72:1503-1507.

Wooley RL (2000) *Iliamna bakeri* (Jepson); Conservation Assessment. [online] URL: http://www.fs.usda.gov/detail/fremont-winema/learning/nature-science/?cid=fsbdev3_061888 (Accessed April 2016).

Young HJ, Dunning DW, von Hasseln KW (2007) Foraging behavior affects pollen removal and deposition in *Impatiens capensis* (Balsaminaceae). American Journal of Botany 94:1267-1271.

13

The behaviour of *Bombus impatiens* (Apidae, Bombini) on tomato (*Lycopersicon esculentum* Mill., Solanaceae) flowers: pollination and reward perception

Patrícia Nunes-Silva*[1], Michael Hnrcir[2], Les Shipp[3], Vera Lucia Imperatriz-Fonseca[2] & Peter G. Kevan[4]

[1]*Departamento de Biologia, Faculdade de Filosofia, Ciências e Letras de Ribeirão Preto, Universidade de São Paulo, Avenida Bandeirantes, 3900, 14040-901, Ribeirão Preto, Brazil.*
[2]*Departamento de Ciências Animais, Universidade Federal Rural do Semi-Árido, Avenida Francisco Mota, 572, 59.625-900, Mossoró, Brazil.*
[3]*Agriculture and Agri-Food Canada, 2585 County Road #20, Harrow, Ontario, Canada, N0R 1G0*
[4]*Canadian Pollination Initiative, School of Environmental Sciences, University of Guelph, ON N1G 2W1, Guelph, Canada*

Abstract—The foraging behaviour of pollinators can influence their efficiency in pollinating certain plant species. Improving our understanding of this behaviour can contribute to an improvement of management techniques to avoid pollination deficits. We investigated the relationship between the number of visits of bumble bees (*Bombus impatiens*) to tomato flowers (*Lycopersicon esculentum*) and two variables related to the quality of the resulting fruits (weight, number of seeds), as well as the relationship between foragers' thoracic weights, physical characteristics of thoracic vibrations (main frequency, velocity amplitude), amount of pollen removed from flowers, and the quality-related variables. In addition, we studied the capability of foragers to assess the availability of pollen in flowers. Tomato weight and seed number did not increase with the number of bee visits, neither were they correlated with the foragers' thorax weight. Thorax weight also did not correlate with the amount of pollen removed from the flowers nor with the physical characteristics of vibration. Vibration characteristics did not change in response to the amount of pollen available on tomato flowers. Instead, foragers adjusted the time spent visiting the flowers, spending fewer time on flowers from which some pollen had already been removed on previous visits. The quantity and the production-related variables of tomatoes are not dependent on the number of bee visits (usually one visit suffices for full pollination); bigger foragers are not more efficient in pollinating tomato flowers than smaller ones; and *B. impatiens* foragers are capable of evaluating the amount of pollen on a flower while foraging and during pollination.

Keywords: bumblebee, pollination, tomato, vibration

Introduction

Bumble bees (*Bombus* spp.) are highly efficient pollinators of tomato (*Lycopersicon esculentum* Miller) flowers and, for commercial purposes, yield far better results than honeybees, manual vibration, or self-pollination (Banda & Paxton 1991; Kevan et al. 1991; Dogterom et al. 1998; Morandin et al. 2001a, 2001b; Palma et al. 2008; Choi et al. 2009; Torres-Ruiz & Jones 2012). Today, approximately 95% of all bumble bee sales worldwide are destined for tomato production, with the estimated value of the bumble bee-pollinated crops reaching 12 billion Euros per year (Velthuis & van Doorn 2006).

Although tomato plants are self-compatible, the anthers need to be shaken to allow effective pollen release (Buchmann 1983). Many bee species, among them the bumble bees, generate thoracic vibrations when visiting tomato flowers therewith facilitating the release pollen from the anthers ("buzz-pollination"; Buchmann & Hurley 1978,

Buchmann 1983). However, whether and to which extent the physical characteristics of thoracic vibrations are correlated with fruit characteristics (e.g. weight, size, seed number) remains unknown.

Tomato fruit size depends, to a certain extent, on the amount of pollen transferred to the stigma (Morandin et al. 2001a). Even so, it has been suggested that the quality of tomatoes (weight, size, seed number) does not increase any further at flower visitation rates above one or two bumble bee visits (*Bombus impatiens*; Morandin et al. 2001a). In case, however, pollen is transferred inadequately to the stigma, seed production is impaired, therewith resulting in sub-optimal crop yields ("pollination deficit"; Vaissière et al. 2011).

The adequacy, efficiency, and quality of bee pollination are affected by many factors, such as the floral characteristics of a plant species that influence the behaviour of flower visitors (Lefebvre & Pierre 2006). Tomato flowers, for instance, produce certain chemicals (β-phellandrene and 2-carene) as part of their scent bouquet that reduce the visitation frequency of *B. impatiens* to the flowers, thus impeding bee pollination (Morse *et al.* 2012). The amount

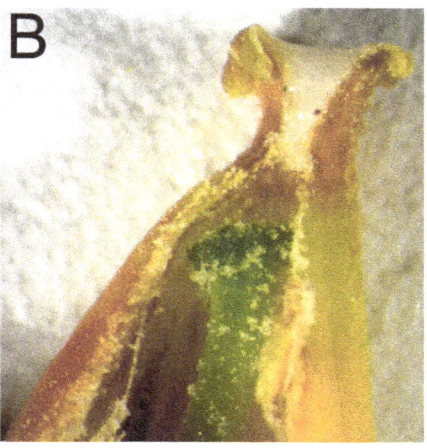

FIG. I. (A) The pore of the anthers' cone of tomato (*Lycopersicon esculentum*) flowers after being blocked with silicon. (B) By opening the cone it was possible to verify that the silicon was not removed by *Bombus impatiens* (Apidae) visitation and that pollen was deposited on stigma.

of these chemicals can be altered by different cultivation practices: vegetative plants produced less β-phellandrene and 2-carene and received more visits than generative plants (Morse 2009). Another important factor influencing pollinator visits is the presence of floral rewards.

In the case of tomato flowers, pollen is the only resource collected by bees, and its availability may affect the behaviour of these pollinators, yet only if the bees are capable of assessing the amount of pollen in flowers. Few studies have tackled this subject so far. But most of these investigations point to the ability of bees to evaluate the amount of pollen in flowers (Buchmann & Cane 1989, Harder 1990, Shelly et al. 2000, see however Hodges & Miller 1981).

Given the importance of bumble bees in tomato pollination and considering the putative relationship between the bees' behaviour and their efficiency as pollinators, we investigated the following questions: (1) Do fruit set and production-related parameters (weight and seed number) depend on the number of bumble bee (*Bombus impatiens*) visits? (2) Are big foragers more efficient in pollinating tomato flowers than small individuals? (3) Are *B. impatiens* foragers capable of evaluating the amount of pollen available in a flower during pollination?

MATERIALS AND METHODS

Study site and bee species

The present study was performed at the Greenhouse and Processing Crops Research Center of Agriculture and Agri-Food Canada (Harrow, Ontario, Canada) between April and June of 2010. The tomato plants (*Lycopersicon esculentum* Mill var. Clarance: Solanaceae) to be used in the experiments were grown and maintained in one compartment of the experimental greenhouse complex (plant compartment). A second compartment (bee compartment) contained two screened cages (2.5m x 5.3m x 2.3m), each of which contained one bumble bee colony (*Bombus impatiens* Cresson, Apidae) and 20 tomato plants, even during the experiments. The colonies were provided by Biobest Canada (Leamington, Ontario, Canada) and consisted of one queen and initially 15 workers. Due to the fact that tomato flowers

do not produce nectar, the hives had their own compartment containing a sugar solution as carbohydrate substitute.

Experiment 1

This first set of experiments was designed to evaluate whether and to which extent the number and duration of bee visits to tomato flowers influence posterior fruit set, weight and seed number. Additionally, it allowed us to investigate whether or not *B. impatiens* foragers are capable of assessing pollen reward.

We transferred tomato plants from the plant compartment to the bee compartment. Prior to the transfer, we covered the completely opened flowers to be used in the experiments with mesh bags to prevent bee visitation. In the bee compartment, the mesh bag of one flower was removed and only a single forager was released from the colony. The bumble bee nest was kept closed for the remainder of experiment. After the forager had visited an experimental flower for the desired number of times (see flower visit treatments), we covered the flower again with a mesh bag. Each forager was used for one to four subsequent flower visit treatments and then collected and killed by freezing. Immediately after death, the thorax was separated form head, abdomen, legs and wings and afterwards weighed on a precision scale (10^{-4} g).

The flower visit treatments were: C: control, no visit (n=18 flowers); IV: one bee visit (n=16); 2V: two visits (n=16); 3V: three visits (n=17); 4V: four visits (n=18); SV: several visits, plants from the plant compartment that had virgin flowers were kept for 8 hours in the bee compartment where the hive was opened and bees could forage freely (n=20); C2: control 2, a drop of silicon placed on the surface of the anther cone (n=16); and BP: anther pores blocked, one visit to flowers which had the pore of the anther cone blocked with silicon to stop pollen release (n=16, Fig. I). All visits were video-taped (JVC Everio GZ-MS 100V camcorder) for later analysis. After the visitation treatments, the plants were returned to the plant compartment, where they were kept until fruit ripening. Each flower and subsequent fruit was tagged for individual identification. After ripening, fruits were weighted (10^{-4} g) and their seeds counted.

FIG. 2. Bumble bee (*Bombus impatiens*) vibration recording. (A) The setup showing the Laser Doppler Vibrometer mounted on a small four-wheeled cart (1) and the flower fixed to a tripod pan handle with adhesive tape (2) being visited by a forager. (B) The red dot on the scutum of the bee is the laser beam of the vibrometer.

Bee visits were analysed by observing the video recordings concerning the total visit duration (visit duration = time between first landing and leaving for the colony or another flower). The number of buzzes made by the foragers could not be analysed from the videos because the noise of the ventilation system of the greenhouse interfered with detection of the bee sounds, thus compromising the accuracy of data. For evaluating the reward perception by bumble bee foragers, we compared the visit duration among the different treatments. As described by Buchmann & Cane (1989) it was expected that foragers spend more time visiting flowers with higher pollen reward.

Experiment 2

In a second set of experiments, we evaluated whether the amount of pollen removed from tomato flowers is related to the physical aspects of the thoracic vibrations generated by the bumble bee foragers during flower visits.

Individual virgin flowers were transferred to the bee compartment and fixed to a tripod pan handle with adhesive tape (Fig. 2A). Single bumble bee foragers were allowed to visit the flowers as described above. The thoracic vibrations generated by the foragers were recorded using a portable Laser Doppler Vibrometer (PDV-100, Polytec, Waldbronn, Germany; Fig. 2A), mounted on a small four-wheeled cart (for details see Hrncir *et al.* 2004) and positioned on a table right beneath the flower (Fig. 2A). The laser beam of the vibrometer was directed upwards via a diagonal mirror, and oriented perpendicular to the surface of the thorax as the bee hung inverted from the anther cone of the flower. Movements of the foragers could be followed by moving the cart. Thus, the laser beam aimed at the scutum of a forager

during the entire flower visit (Fig. 2B). The output of the vibrometer was fed into a notebook using the software Soundforge 7.0 (Sony Pictures Digital Inc., Madison, WI, USA). Vibration analyses were performed using the software SpectraPro 3.32 (Sound Technology Inc., Campbell, CA, USA). For each forager, we calculated the average main frequency (Hz) and the average velocity amplitude (mm/s) of its thoracic vibrations (average of 3 to 108 pulses). Statistical tests were performed using these individual averages.

The visitation treatments (1V: n=15; 4V: n=15; C2: n=14; BP: n=15) were the same as in experiment 1. In treatment 4V, bee vibrations were recorded only during the first and the last visit, based on our observations in the first experiment that revealed a great difference in behaviour between the first and the fourth visit (see results).

After the respective treatment, the anther cones were carefully removed from flowers and stored individually in tubes containing 1 ml of alcohol 70%. Afterwards, the anthers (C: n=13; 1V: n=15; 4V: n=15) were dissected and the pollen grains removed. The pollen grains were diluted in 15 ml of a saline solution for numerical enumeration using a particle counter (MultisizerT 3 COULTER COUNTER®). The total number of pollen grains in a sample was estimated from three subsamples of 0.5 ml each. The amount of pollen removed from anthers by bees was determined by subtracting the mean amount of pollen left inside anthers after the visits (treatments 1V and 4V) from the mean amount of pollen found in virgin flowers (treatment C).

Data analysis

Statistical analyses were performed using the software packages BioEstat, Statistica, and Sigma Plot. Because data were not normally distributed (Shapiro-Wilk test; $P<0.05$), we performed non-parametric statistical tests only. The respective tests are given in the results section. The α-level of significance was $P \leq 0.05$. Throughout the text, data are presented as mean values ± standard deviation.

RESULTS

Are fruit set and weight and seed number of tomatoes related to the number of bee visits?

Fruit set was similar in most visitation treatments. In the treatment groups C (n=18), 2V (n=16), 4V (n=18), and C2 (n=16), fruit set was 100%. In groups 1V (n=16) and 3V (n=17), one flower (1V) and two flowers (3V) were aborted after one day of lack of water caused by a failure in the irrigation system, resulting in reduced fruit sets of 93.8% (1V) and; 88.2% (3V). Fruit set in treatment group SV (n=20) was 90%. The only group with clearly reduced success was BP (anther pores blocked, n=16), where fruit set was 75%.

Fruit weight was significantly lower in the control group C compared to the treatment groups 1V, 2V, 3V, 4V, SV and C2 (Kruskall-Wallis test, $X^2=40.9$, $P<0.05$; Dunn's pairwise comparison, $P<0.05$). There were no statistically significant differences in fruit weight between C and BP, neither among treatment groups 1V, 2V, 3V, 4V, SV and C2 (Dunn's pairwise comparison, $P>0.05$) (Fig. 3A). Fruits of control group C produced significantly fewer seeds than fruits of the treatment groups 1V, 2V, 3V, 4V, SV and BP (Kruskall-Wallis test, $X^2=26.2$, $P<0.05$; Dunn's pairwise comparison, $P<0.05$). There were no statistically significant differences in seed number between C and C2, neither among treatment groups 1V, 2V, 3V, 4V, SV, C2, and BP (Dunn's pairwise comparison, $P>0.05$) (Fig. 3B).

Are bigger foragers more efficient in pollinating tomato flowers than smaller ones?

The mean thoracic weight of foragers was 48.3±11.4 mg (n=20; maximum: 68.8 mg; minimum: 26.1 mg). The investigated physical parameters of the thoracic vibrations, main frequency and velocity amplitude, did not correlate with the thoracic weight of the forager generating them (Tab. 1). Neither of these vibration parameters nor the thoracic weight correlated with the amount of pollen removed after one and after four flower visits (Tab. 1). Also concerning fruit quality, we found no significant correlations between the thoracic weight of the forager and fruit weight (Spearman Rank Correlation: r=0.00, $P>0.05$, n=88), or seed number (Spearman Rank Correlation: r=0.10, $P>0.05$, n=88) of the tomatoes that resulted from the respective bee's visit.

Are foragers capable of assessing the amount of available pollen during a flower visit?

The mean number of pollen grains in the anthers of virgin tomato flowers (control group C) was 96,561 ±

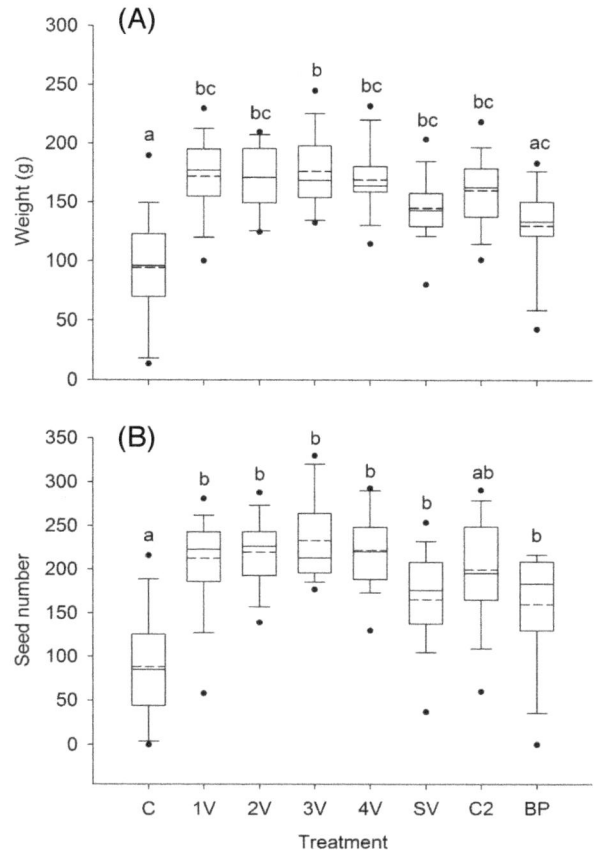

FIG. 3. (A) Mean weight and (B) mean seed number of tomato fruits produced by different treatments (C = control, 1V = one visit, 2V = two visits, 3V = three visits, 4V = four visits, SV = several visits: virgin flowers from plants kept for 8 hours in the bee compartment, C2 = control 2: a drop of silicon on the surface of the anther cone, BP = blocked pore: one visit flowers which had the pore of the anther cone blocked with silicon to stop pollen release. All visits were performed by *Bombus impatiens* (Apidae). Different letters (a, b) indicate statistical differences at $P<0.05$. *Box plots:* box indicates the distribution of 50% of the values, horizontal full line indicates median, horizontal dashed line indicates mean, whiskers indicate standard error (above 90% and below 10%) and spheres indicate outliers.

28,220 (n=13). After a single bumble bee visit, the number of pollen grains dropped to an average of 40,768 ± 32,701 (n=15) and after four visits further to 30,595 ± 36,794 (n=15). Thus, a forager removed, on average, 57.8% of a flower's pollen during the first visit, and 68.3% within four visits. The number of pollen grains in virgin flowers was significantly larger than the number of pollen grains after one and after four visits (Kruskall-Wallis test: $X^2=21.6$, $P<0.05$; Dunn's pairwise comparison: C vs 1V: $P<0.05$; C vs 4V: $P<0.05$; Fig. 5); however, there was no statistically significant difference concerning the number of pollen grains after the first visit or after four visits (Dunn's pairwise comparison: 1V vs 4V: $P>0.05$; Fig. 4).

During the first visit, foragers remained significantly longer on a flower than during all the following visits (Tab. 2), as expected. When the pores of the anthers were blocked (treatment BP, no access to pollen), the foragers' visits were

TABLE I. Spearman correlation coefficients among main frequency (Hz), velocity amplitude (mm/s), amount of pollen remaining on tomato flowers after one (1V) and four visits (4V) and thoracic weight of *Bombus impatiens* foragers (mg).

Treatment		Main frequency (Hz)	Velocity amplitude (mm/s)	Amount of pollen left
1V	Amount of pollen left	0.30[ns]	0.19[ns]	—
	Thoracic weight of forager (mg)	-0.23[ns]	-0.10[ns]	-0.45[ns]
4V	Amount of pollen left	-0.32[ns]	-0.51[ns]	—
	Thoracic weight of forager (mg)	0.36[ns]	0.23[ns]	-0.47[ns]

ns: not significant at $P<0.05$

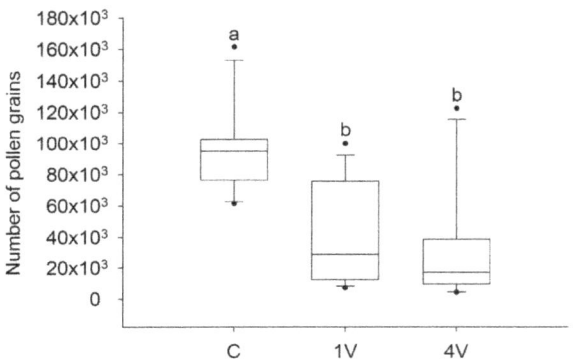

FIGURE 4. Mean number of pollen grains of tomato virgin flowers (C), of flowers visited once (1V) and four times (4V) by a forager of *Bombus impatiens*. Different letters (a, b) indicate statistical differences at $P<0.05$ (Kruskall-Wallis, pair comparison: Dunn's method). *Box plots:* box indicates the distribution of 50% of the values, horizontal full line indicates median, horizontal dashed line indicates mean, whiskers indicate standard error (above 90% and below 10%) and spheres indicate outliers.

significantly shorter compared to visits to flowers with pollen access (Tab. 2).

Although visit duration changed significantly, the physical parameters of thoracic vibrations generated during the visits did not. Both mean main frequency (1V = 334.3 ± 11.4 and 4V = 344.7 ± 20.4) and velocity amplitude of the foragers' vibrations (1V = 195.0 ± 16.4 and 4V = 223.6 ± 64.4) did not differ between the first and the fourth visit (Wilcoxon Signed Rank-Test: for frequency - T=24.0, Z=1.5, $P>0.05$; for velocity - T=19.0, Z=1.8, $P>0.05$). The vibration parameters did also not differ

between the first visit to untreated flowers (1V: mean main frequency = 332.8 ± 19.3, velocity amplitude= 194.7 ± 20.5), flowers of control treatment 2 (C2: mean main frequency = 337.9 ± 18.0, velocity amplitude=188.8 ± 27.5) and flowers with no access to pollen (BP: mean main frequency = 344.0 ± 18.5, velocity amplitude= 188.8 ± 27.5) (Kruskal-Wallis test: for frequency - X^2=2.5, $P>0.05$; for velocity - X^2=2.8, $P>0.05$).

DISCUSSION

Bumble bees are important pollinators for tomato crops. With the present study, we wanted to contribute to a more profound understanding of the biological background of the efficiency of *Bombus impatiens* in pollinating tomato flowers. Our results indicate that (1) foragers are capable of assessing the pollen reward of the flowers, (2) the first flower visit is the most effective concerning pollen removal and, consequently, pollination, and (3) small foragers are as efficient in pollinating tomato flowers as are big ones. From these results, we can draw important conclusions concerning both the foraging biology of this bee species and their use and management for tomato crop pollination.

Foraging biology of Bombus impatiens

Foragers of *B. impatiens* spent less time visiting flowers with low amounts of pollen than they spent on virgin flowers (Tab. 2). This points to the bees' capability of evaluating the resource value during their visit. The perceptual mechanism by which bees assess the amount of pollen available while removing it from a flower is still far from understood. Hodges & Miller (1981) proposed that bees are not capable of evaluating the amount of pollen obtained during a single visit. Also, assessing the quantity of pollen loaded may be difficult for bees because, in contrast to nectar, pollen is deposited on the body, not ingested (Hodges & Miller 1981).

Our results, however, in line with some earlier studies (Buchmann & Cane 1989, Harder 1990, Shelly et al. 2000) provide clear evidence that bees are indeed capable of perceiving the amount of pollen obtained while visiting a flower. In accordance with Buchmann & Cane (1989), we observed that bees groomed several times during a visit, thereby transferring pollen to their corbiculae. Thus, one possibility is that bees evaluate the amount of pollen during grooming. A second possibility is that bees directly register the pollen that falls on their body (head, thorax and abdomen) through mechano-sensitive hair, the sensilla trichodea, which are highly sensitive to tactile stimuli (McIver 1975). A third possibility is that bumble bee foragers make use of scent marks, deposited by previous flower visitors, to evaluate whether or not the flower still provides pollen (e.g. Stout et al. 1998; Goulson et al. 2000; Stout & Goulson 2001). Although these possibilities are not mutually exclusive, and bees may use more than one information for reward evaluation, our finding that foragers spent significantly less time on flowers without pollen reward (visitation treatment BP, pollen release blocked) that on virgin flowers (Tab. 2) corroborate the direct perception-mechanism. Just like the virgin flowers, flowers with blocked

TABLE 2. Mean, minimum (Min) and maximum (Max) visit duration of *Bombus impatiens* foragers to tomato flowers and respective standard deviations (SD) and sample sizes (N). Treatments: (1V) one visit; (2V) two visits; (3V) three visits; (4V) four visits; (C2) control 2: a drop of silicon on the surface of the anther cone; and (BP) blocked pore. Different letters (a, b) indicate statistical differences at P<0.05

| | | Visit duration (s) | | | | Statistics | |
		Mean ± SD	Min	Max	N	Test	P
1V		89 ± 71a	11	242	17		
C2		89 ± 70a	10	292	16	Kruskal-Wallis	X^2=17.4, P<0.05
BP		12 ± 8b	1	29	16		
2V	1st visit	78 ± 61a	24	269	16	Wilcoxon Matched Pairs	T=27.0, Z=2.1, P <0.05
	2nd visit	32 ± 42b	3	147			
3V	1st visit	107 ± 83a	1	323	17	Friedman ANOVA	X^2= 10.5, P<0.05
	2nd visit	24 ± 29ab	1	105			
	3rd visit	14 ± 16b	1	66			
4V	1st visit	112 ± 71a	2	276	17	Friedman ANOVA	X^2= 19.9, P<0.05
	2nd visit	15 ± 18b	2	66			
	3rd visit	30 ± 60b	1	247			
	4th visit	11 ± 9b	1	30			

pores had not been visited before, thus they carried no scent marks.

Although visit duration changed significantly with pollen reward (Tab. 2), the mechanical characteristics of the thoracic vibrations (main frequency and velocity amplitude) generated by the forager during the flower visits did not. Also, the thoracic vibrations did not differ between bees of different size (Tab. 1). This result seems surprising on first sight, given that the force of the thoracic vibrations is determined, in parts, by the mass of the flight muscles and, thus, depends on thorax size (Buchmann et al. 1977; Buchmann & Hurley 1978; Morse 1981; King & Buchmann 1995, 1996; Hrncir et al. 2008). Furthermore, in a recent study, De Luca et al. (2012) found a significant correlation between bumble bee (*B. terrestris*) forager mass and peak amplitude of the vibrations of *Solanum rostratum* flowers caused by the foragers. Probably, these differences between our findings and those by De Luca et al. (2012) stem from methodological differences. Whereas de Luca et al. (2012) measured vibrations on the petals of flowers, we picked up the vibrations directly from the thoraces of the bees. From this, we can assume that small bees, even when generating thoracic vibrations of similar amplitudes as big bees (Tab. 1), vibrate the flowers with reduced force due to reduced body mass compared to big bees (force = mass × acceleration, where acceleration is proportional to amplitude × frequency). Consequently, flower vibrations caused by small bees are of smaller amplitude than those caused by big bees. Here, future investigations on the vibration transfer between bees and flowers shall test our assumption.

The use and management of Bombus impatiens for tomato crop pollination

Bumble bee pollination increases tomato production, weight and seed number, which guarantees a better market price (Kevan et al. 1991; Velthuis & van Doorn 2006). In addition to weight as important factor for the value of a crop, a recent study indicates that the number of seeds is important for the sensory characteristics of tomatoes, resulting in the preference of bee-pollinated tomatoes over wand-pollinated ones by consumers (Hogendoorn et al, 2010).

Our results show that a single bee visit is enough to guarantee heavier fruits with elevated seed number (Fig. 3). Interestingly, both these fruit characteristics did not increase significantly when flowers were visited more than once by bumble bees. Our finding corroborates the results by Morandin et al. (2001a), who examined the relation between tomato weight and seed number and the bruising (caused by the bees biting the anther cones) on the anthers, which indicate the approximate number of bee visits. In compliance with our findings, these authors observed that tomato weight did not increase with bruising levels above one (one visit) and seed number did not increase with bruising levels above two (one or two visits). A possible explanation for this observation that a single bee visit is sufficient to promote high-quality tomatoes is that bees remove a significantly larger amount of pollen from flowers during the first visit than on subsequent visits (Fig. 4). This elevated pollen removal may result in the deposition of enough pollen grains to fertilize most ovules during the first visit.

Controlling the intensity of bumble bee visits is important for tomato production because a high level of visitation damages the reproductive organs of flowers, causing abortion (Morandin et al. 2001b; Morse 2009). In our study, the plants in the bee compartments were intensely visited, causing flower destruction and abortion (Fig. 5). Tomato growers need to consider this potential damage through bumble bees when planning the pollination management of their crop. Morandin et al. (2001b) suggest that 7 to 15 colonies of *B. impatiens* per hectare are

FIGURE 5.   Tomato flowers damaged from intense visitation by *Bombus impatiens* and showing signs of abortion (red arrows).

sufficient to guarantee adequate pollination (one visit). In fact, a single bumble bee visit considerably increases the economic value of tomato crops. From our results on fruit set and fruit quality (weight), the estimated base-value of 100 tomatoes (control group C, no visits) would be C$14.62 (estimated value = mean fruit weight: 94.34g × fruit set: 100% × price/kg tomato in 2010, 2011: C$1.55, Shalin Khosla, personal communication). For the visitation treatments 1V, 2V, 3V, 4V and SV the estimated values were C$25.00, C$26.48, C$24.11, C$26.23, and C$20.26, respectively. Thus the estimated yield due to bumble bee visits was 71.0% (1V), 81.1% (2V), 64.9% (3V), 79.4% (4V), and 38.5% (BP) higher than that for the control group C.

The second important result from our study concerning the commercial use and management of *B. impatiens* as tomato pollinators was that fruit quality (weight, number of seeds) was not related to forager size (thorax weight). This independence of bee size may be due to tomato floral morphology: tomato flowers have a cone of anthers surrounding the pistil, thus forming a chamber (Rick & Robinson 1951; McGregor 1976; Fig. 1B). This structure allows for self-pollination because the pollen, when released from the anthers, falls directly onto the stigma inside the chamber (Rick & Robinson 1951; Rick & Dempsey 1969). This self-pollination mechanism and the fact that the first bee visit causes the release of sufficient pollen to fertilize most ovules of a flower, independently of forager size (Tab. 1), could explain why fruit quality is not related to the size of the pollinator.

This finding may have an important impact for the management of bumble bees as pollinators for tomato crops. First, the size of *B. impatiens* workers decreases with increasing age of the colony (Couvillon et al. 2010). Hence, if forager size affected pollination efficiency, tomato growers would have to substitute their colonies frequently in order to assure the availability of big foragers. Our results now suggest that there is no real need for this practice. Second, the independence between bee size and pollination efficiency indicates that other bumble bee species may have potential for commercial tomato pollination. Recently, Torrez-Ruiz and Jones (2012) showed that *B. ephippiatus* is as efficient

in pollinating tomatoes as is *B. impatiens*, despite differences in foraging pattern. Colony foraging pattern, however, may have an important contribution to pollination efficiency. Whittington and Winston (2004) compared the behaviour of *B. occidentalis* and *B. impatiens* in tomato greenhouses. The observed differences in foraging pattern lead these authors to the conclusion that *B. impatiens* is a better pollinator for tomatoes than *B. occidentalis*. Therefore, to explore the possibility of using different bumble bee species for tomato pollination, more comparative studies are needed. We suggest that these studies explore not only colony foraging pattern and yield, but also the bees' efficiency in pollen removal. Although our results indicate that tomato pollination efficiency is not related to bee size, this relationship should be further investigated.

ACKNOWLEDGEMENTS

We thank CAPES (BEX: 5344/09-3) and Kinross Canada-Brazil Network for Advanced Education and Research in Land Resource Management for funding. We also thank Dr. Brian Husband and Sarah Baldwin for the use of the particle counter and help with the technique, Bruno Nunes Silva for image editing and NSERC-CANPOLIN for which this will be Contribution #84.

REFERENCES

Banda HJ, Paxton RJ (1991) Pollination of greenhouse tomatoes by bees. Acta Horticulturae 288:194-198.

Buchmann SL (1983) Buzz pollination in angiosperms. In: Jones CE, Little RJ (eds) Handbook of Experimental Pollination Biology. Scientific and Academic Editions, New York, Van Nostrand Reinhold, pp 73-113.

Buchmann SL, Cane JH (1989) Bees assess pollen returns while sonicating *Solanum* flowers. Oecologia 81:289-294.

Buchmann SL, Hurley JP (1978) A biophysical model for buzz pollination in angiosperms. Journal of Theoretical Biology 72:639-657.

Buchmann SL, Jones CE, Colin LJ (1977) Vibratile pollination of *Solanum douglasii* and *Solanumxantii* (Solanaceae) in Southern California. The Wasmann Journal Biology 35:1-25.

Choi YH, Kang NJ, Park KS, Chun H, Cho MW, Um, YC,You HY (2009) Influence of fruiting methods on fruit characteristics in cherry tomato. Korean Journal of Horticultural Science & Technology 27(1):62-66.

Couvillon MJ, Jandt JM, Duong N, Dornhaus A (2010) Ontogeny of worker body size distribution in bumble bee (*Bombus impatiens*) colonies. Ecological Entomology 35:424-435.

De Luca PA, Bussière LF, Souto-Vilaros D, Goulson D, Mason AC, Vallejo-Marin M (2012) Variability in bumblebee pollination buzzes affects the quantity of pollen released from flowers. Oecologia doi: 10.1007/s00442-012-2535-1

Dogterom MH, Matteoni JA, Plowright RC (1998) Pollination of greenhouse tomatoes by the North American *Bombus vosnesenski* (Hymenoptera: Apidae). Journal of Economic Entomology 91(1):71-75.

Goulson D, Stout JC, Langley J, Hughes WHO (2000) Identity and function of scent marks deposited by foraging bumblebees. Journal of Chemical Ecology 26(12):2897-2911.

Harder LD (1990) Behavioral responses by bumble bees to variation in pollen availability. Oecologia 94:244-246.

Hodges CM, Miller RB (1981) Pollinator flight directionality and the assessment of pollen returns. Oecologia 50:376-379.

Hogendoorn K, Bartholomaeus F, Keller MA (2010) Chemical and sensory comparison of tomatoes pollinated by bees and by a pollination wand. Journal of Economic Entomology 103(4):1286-1292.

Hrncir M, Jarau S, Zucchi R, Barth FG (2004) Thorax vibrations of a stingless bee (*Melipona seminigra*). I. No influence of visual flow. Journal of Comparative Physiology A 190:539-548.

Hrncir M, Gravel AI, Schorkopf DLP, Schmidt VM, Zucchi R, Barth FG (2008) Thoracic vibrations in stingless bees (*Melipona seminigra*): resonances of the thorax influence vibrations associated with flight but not those associated with sound production. Journal of Experimental Biology 211:678-685.

Kevan PG, Straver WA, Offer M, Laverty TM (1991) Pollination of greenhouse tomatoes by bumble bees in Ontario. Proceedings of the Entomological Society of Ontario 122:15-19.

King MJ, Buchmann SL (1995) Bumble bee-initiated vibration release mechanism of *Rhododendron* pollen. American Journal of Botany 82(11):1407-1411.

King MJ, Buchmann SL (1996) Sonication dispensing of pollen from *Solanum laciniatum* flowers. Functional Ecology 10:449-456.

Lefebvre D, Pierre J (2006) Spatial distribution of bumblebees foraging on two cultivars of tomato in a commercial greenhouse. Journal of Economic Entomology 99:1571–1578.

McIver SB (1975) Structure of cuticular mechanoreceptors of arthropods. Annual Review of Entomology 20:381-397.

McGregor SE (1976) Insect pollination of cultivated crop plants. USDA, Washington, DC.

Morandin LA, Laverty TM, Kevan PG (2001a) Effect of bumble bee (Hymenoptera: Apidae) pollination intensity on the quality of greenhouse tomatoes. Journal of Economic Entomology 94(1):172-179.

Morandin LA, Laverty TM, Kevan PG (2001b) Bumble bee (Hymenoptera: Apidae) activity and pollination levels in commercial tomato greenhouses. Journal of Economic Entomology 94(2):462-467.

Morse A (2009) Floral scent and pollination of greenhouse tomatoes. University of Guelph, Guelph.

Morse A, Kevan P, Shipp L, Khosla S, McGarvey B (2012) The impact of greenhouse tomato (Solanales: Solanaceae) floral volatiles on bumble bee (Hymenoptera: Apidae) pollination. Environmental Entomology 41(4):855-864.

Morse PM (1981) *Vibration and Sound.* American Institute of Physics of the Acoustical Society of America, New York.

Palma G, Quezada-Euán JJG, Reyes-Oregel V, Meléndez V, Moo-Valle H (2008) Production of greenhouse tomatoes (*Lycopersicon esculentum*) using *Nannotrigona perilampoides, Bombus impatiens* and mechanical vibration (Hym.: Apoidea). Journal of Applied Entomology 132:79–85.

Rick CM, Dempsey WH (1969) Position of the stigma in relation to fruit setting of the tomato. Botanical Gazette 130(3):180-186.

Rick CM, Robinson J (1951) Inherited defects of floral structure affecting fruitfulness in *Lycopersicon esculentum*. American Journal of Botany 38(8):639-652.

Shelly T.E., Villalobos E. & OTS-USAP (2000) Buzzing bees (Hymenoptera: Apidae, Halictidae) on *Solanum* (Solanaceae): floral choice and handling time track pollen availability. Florida Entomologist 83(2):180-187.

Stout JC, Goulson D, Allen JA (1998) Repellent scent-marking of owers by a guild of foraging bumblebees (*Bombus* spp.). Behavioral Ecology and Sociobiology 43:317-326.

Stout JC, Goulson D (2001) The use of conspecific and interspecific scent marks by foraging bumblebees and honeybees. Animal Behaviour 62:183-189.

Torres-Ruiz A, Jones RW (2000) Comparison of the efficiency of the bumble bees *Bombus impatiens* and *Bombus ephippiatus* (Hymenoptera: Apidae) as pollinators of tomato in greenhouses. Journal of Economic Entomology 105(6):1871-1877.

Vaissière BE, Freitas BM, Gemmill-Herren B (2011) Protocol to detect and assess pollination deficits in crops: a handbook for its use. FAO, Rome.

Velthuis HHW, Van Doorn A (2006) A century of advances in bumblebee domestication and the economic and environmental aspects of its commercialization for pollination. Apidologie, 37:421-451.

Whittington R, Winston ML (2004) Comparison and examination of *Bombus occidentalis* and *Bombus impatiens* (Hymenoptera: Apidae) in tomato greenhouses. Journal of Economic Entomology 97(4):1384-1389.

14

THE IMPORTANCE OF BEE POLLINATION OF THE SOUR CHERRY (*PRUNUS CERASUS*) CULTIVAR 'STEVNSBAER' IN DENMARK

Lise Hansted[1], Brian W. W. Grout*[1], Jørgen Eilenberg[2], Ivar B. Dencker[1] & Torben B. Toldam-Andersen[1]

[1]*Department of Agriculture and Ecology, University of Copenhagen, Højbakkegård Allé 13, 2630 Taastrup, Denmark*
[2]*Department of Agriculture and Ecology, University of Copenhagen, Thorvaldsensvej 40, 1871 Frederiksberg C, Denmark*

Abstract— Low fruit set, despite normally-developed flowers, is often a significant contributor to poor yield of the self-fertile sour cherry (*Prunus cerasus*) cultivar 'Stevnsbaer' in Denmark. The aim of this study was to investigate the effect of insect, and particularly, bee pollination on the fruit set of this cultivar, in order to provide orchard management information for both Danish 'Stevnsbaer' growers and beekeepers. Visits to cherry flowers by honey bees (*Apis mellifera*), *Bombus* species and solitary bees, were recorded during the flowering of 'Stevnsbaer' in five separate Danish orchards. The results indicate that there is a significantly higher fruit set on open pollinated branches when compared to caged branches, where bees and other pollinating insects where excluded. The results were qualitatively consistent over three different seasons (2007, 2009 and 2010). A period of prolonged cold, humid weather before and during early flowering probably reduced fruit set significantly in 2010 compared to 2009. Regarding the apparent benefits of bee pollination on fruit set and subsequent implications for yield, we recommend placing honeybees in 'Stevnsbaer' orchards during flowering to sustain commercially viable production. Another valuable management strategy would be to improve foraging and nesting conditions to support both honey and wild bees in and around the orchards.

Keywords: Fruit set, Prunus cerasus, bee pollination, Apis mellifera, Bombus

INTRODUCTION

The numbers of nearby wild bees (Hymenoptera: Apidoidea) and *Apis mellifera* hives are important, determining factors of the level of bee pollination available as a free service to cultivated crops. Thus, the growing evidence for the global loss of bee diversity and decline of numbers (Potts et al. 2010) is of concern. Many interacting factors are implicated in this decline, such as changes of land-use resulting in fragmentation of habitats, increasing use of pesticides, environmental pollution, decreased plant resource diversity, alien species, spread of pathogens and climate change (Potts et al. 2010; Krewenka et al. 2011). Solitary bees in particular, seem to be more sensitive to agricultural intensification than bumble bees (Féon et al. 2010). In Denmark, the number of *A. mellifera* colonies has declined considerably between 1985 and 2008, mainly because of diseases and halving of the number of beekeepers (Branner & Vejsnæs 2007; Vejsnæs et al. 2010). Consequently on a nation-wide level, *A. mellifera* is unlikely to compensate for reduced wild bee pollination, even on crops where they are valuable and highly effective pollinators. Therefore, it becomes increasingly necessary to document the likely effect of bee pollination on specific, cultivated crops to aid decision-making concerning changing management practices that will support yield, such as introducing bees during flowering.

Pollinating insects are generally accepted as important for seed and fruit production of self-sterile crops such as sweet cherry and apples (Free 1993: McGregor 1976; Delaplane & Mayer 2000) but for self-fertile crops it is less clear whether such pollinating insects improve yield significantly (Benedek et al. 2005). However, reports have shown a positive effect of bee pollination on specific crops such as blackcurrant (Hansted 1993), raspberry (Chagnon et al. 1991) and strawberry (Blasse & Haufe 1989). Solitary bees (e.g. *Osmia aglaia*) have also been shown to be valuable pollinators of raspberry and blackberry (Cane 2005).

The sour cherry cultivar 'Stevnsbaer' is among the major fruit crops in Denmark, grown on 1400 ha with a commercial yield of 12,700 tonnes in 2010 (Faostat, 2012). It is self-fertile and wind plays a role in the transfer of its pollen (Hansen 1981, Ren 2005). Low fruit set in normally-developed flowers is a major problem and often the main reason for poor yield (Dencker et al. 1999). Preliminary studies by Ren (2005) where yield was increased by using an air blast orchard sprayer during flowering to increase air movement in the tree crowns indicated the limiting role of pollen availability in fruit set of 'Stevnsbaer'. Further, hand-pollination on the first day of flowering increased yield more effectively than on the second day (Ren 2005), suggesting that 'Stevnsbaer' stigma receptivity may be shorter than for many other cherry types (Toyama 1980; Furukawa & Bukovac 1989). Limited pollen tube growth through the style has also been suggested as a reason for a poor fruit set in 'Stevnsbaer', with slow growth rates, reduced receptivity of stigma surfaces and lower temperatures e.g. <15°C cited as contributing factors (Kühn 1988; Cerovic & Ruzic 1992).

*Corresponding author; email: bwg@life.ku.dk

An earlier study on a small number of 'Stevnsbaer' trees documented a positive effect of bee pollination on fruit set (Hansen 1981), confirming work by Benedek et al. (2005) who recorded increased fruit set in four self-fertile sour cherry varieties when pollinated by bees. In preliminary studies with 'Stevnsbaer', supplementary hand-pollination did not increase fruit set on freely-pollinated branches in orchards where bees were present (the authors, unpublished data), suggesting that bee pollination is indeed highly effective. However, a wide variation in yield is common between trees and orchards (Dencker & Toldam-Andersen 2003), even if the trees are clonally produced and expected to have a high degree of uniformity. Consequently, the positive effect of bee pollination on fruit set of 'Stevnsbaer' reported by Hansen (1981) for a limited number of trees and in a single orchard has to be validated for other, larger orchards.

The aim of this work was to study the effect of bee pollination on fruit set of 'Stevnsbaer' on a large number of trees in a number of different locations. Fruit set was recorded on branches of the same tree that were either open pollinated or caged to exclude bees and other pollinating insects. Numbers of bees and other insects visiting the flowers was also recorded. This information is intended to help 'Stevnsbaer' growers and beekeepers when considering the future role of bees in orchard management.

MATERIALS AND METHODS

Study sites

The experiments were carried out in 2007, 2009 and 2010 in five 'Stevnsbaer' orchards in the eastern part of Denmark and 20-60 trees were included each year in each orchard. Apis mellifera colonies, Bombus terrrestris colonies and Osmia rufa cocoons were placed in, or nearby, the orchards either before or at the beginning of bloom. A detailed description of the sites and bee introductions is shown in Tab. I.

Fruit set

Fruit set was measured on spurs of two year old branches. On each tree, two spurs of closely similar form and shoot development were selected in the middle of the crown. When the flowers were at the balloon stage, a 30-45 cm long section was marked on the innermost part of the branches and the number of flowers per spur recorded. Just before flowering, one of the branches (randomly selected) was covered with a chicken-wire and tulle net cage, shaped as an oblong tube (16-25 cm diameter). This was carefully placed to completely avoid contact between the net and the flowers (Fig. I). The tulle net had 30 perforations cm^{-2}, each with a diameter of 2 mm. The second selected branch on each tree was left uncaged. The cages were removed as soon as the petals had fallen from the flowers. Five weeks later the number of fruits per spur was counted and percentage fruit set calculated. The period between full bloom (> 90% open flowers) and the onset of senescence as indicated by the first colour change of the flowers was approximately four days in all orchards in all years.

Number of pollinators

Numbers of bees and other insects were recorded in each orchard in each year. In 2007, recordings were made on one day in all orchards; in 2009, they were made on one day in Slagelse and on 3 days in Taastrup and Ringsted;in 2010, they were made on 2 days in Slagelse and on 3 days in Taastrup and Ringsted. On each day, records were taken at 07:30, 10:00, 13:00 & 16:00 hours using the same rows of trees for all recordings. The observer walked slowly (15 m/minute) for 10 minutes along each of 4 rows. Numbers of A. mellifera, Bombus sp. queens and workers (the latter in

Location	Year	Study area (ha)	No. of trees	Bee populations within 130 m of the orchard	Timing of bee placement
Guldborg	2007	0.8	24	A 19	Before bloom
Klippinge	2007	1.0	40	A 2	Before bloom
Taastrup	2007	1.0	40	A 1	Start of bloom
	2009			A 8	A 2 at start of bloom
		1	40	B 2	A 6 at full bloom
	2010			O 100	B and O before bloom
Ringsted	2009	0.5	60	A 10 B 4	Before bloom
	2010	0.5	20	O 180	
Slagelse	2009			A 42 B 2	Before bloom
		0.2	20		
	2010			O 100	

TABLE I. Location and size of study sites in Denmark and number of investigated trees. The species and numbers of introduced bee colonies are given, together with the time they were placed in the orchard.

Legend: A = A. mellifera colony, B = B. terrestris colony with 75 workers, O = O. rufa cocoon.

FIGURE 1. Sour cherry (*Prunus cerasus*) cultivar 'Stevnsbaer' flowers covered with a net cage made from chicken wire and tulle. The cage was placed around the spurs on the inner part of two year old branches.

2009 and 2010 only), other wild bees and insects visiting flowers were recorded, with those on leaves or flying within the canopy being ignored. It should be noted that *Bombus* sp. workers are not naturally present at the time of 'Stevnsbaer' flowering (late April - first weeks of May) and were only recorded in 2009 and 2010 when commercially-sourced *B. terrestris* colonies had been placed in the orchards.

Climate

In 2010, temperature and relative humidity (2 m above ground level) and precipitation (1.5 m above ground level) measurements were taken from weather stations at Roskilde (14 km southwest of Taastrup) and Flakkebjerg (9 km southeast of Slagelse) and temperature only at Taastrup and Ringsted. These data are not reported here in detail but gave valuable weather information to aid the discussion of results.

Data analysis

Fruit set data were analyzed using a two-sample t-test for correlated samples. Data were analyzed separately for all years and orchards.

RESULTS

In any one season and in all orchards a significantly higher fruit set (t = 4.96-8.49, $P < 0.05$) was detected on open pollinated compared to caged branches (Tab.2). This difference held despite the inevitably different seasons, management practices (including the introduction of bees), variation in other plant species present and in insect abundance in the different orchards. Fruit set was significantly higher (t = 1.79-4.28, $P < 0.05$) in 2009 compared to 2010 for both caged and non-caged flowers in

Ringsted and Slagelse, and for caged flowers in Taastrup (Tab. 2).

Bees and other insects were visiting the flowers in all orchards in all years and *A. mellifera*, *Bombus* sp., wild bees (*Andrena* sp.) and Dipterans were recorded (Tab. 3). *A. mellifera* was proportionally the most numerous of these potential pollinators in each of the orchards, always exceeding 40% or even more of all flying insects (Tab. 3). Close observations confirmed that each bee species was in contact with both anthers and stigma while visiting a flower and would thus be able to transfer pollen.

DISCUSSION

It has been suggested that the window pollen transfer to flowers in full bloom leading to successful fruit set is as short as 1-2 days for 'Stevnsbaer' (Ren 2005) and that numbers of flying insects are significant for eventual pollination levels (Hansen 1981). If there is little or no wind to assist pollination then bees and other insects will be responsible for transferring the sticky pollen (Hansen 1981). Therefore, bees kept within or close to an orchard will raise fruit set and yield, particularly if there are few wild pollinators in the orchard. In this study it was noted that flowers both on the caged and open pollinated branches were similarly agitated under windy conditions, indicating that the net was not a major obstacle to wind pollination.

The consistently higher fruit set on open pollinated branches compared to caged ones (Tab. 2) is likely to result from the pollination activity of insects, given that wind pollination was possible in both circumstances In our study, we observed that wind could enter the cages and move flowers thus, allowing for wind-pollination. The number of bees visiting the 'Stevnsbaer 'flowers (Tab. 3) could have been influenced by the introductions close to the orchards as part of the

| Location | Fruit set (%) | | | | | |
| | 2007 | | 2009 | | 2010 | |
	Caged	Exposed	Caged	Exposed	Caged	Exposed
Guldborg	9.34 ± 2.48	18.25 ± 3.3				
Klippinge	4.75 ± 1.68	18.49 ± 3.19				
Taastrup	3.79 ± 1.39	14.29 ± 3.04	6.99 ± 1.60	13.33 ± 2.05	1.27 ± 0.95	9.43 ± 3.14
Ringsted			16.56 ± 2.57	29.28 ± 4.26	2.23 ± 1.14	15.65 ± 5.47
Slagelse			28.93 ± 5.13	44.21 ± 6.29	0.76 ± 0.68	11.93 ± 2.91

TABLE 2. Fruit set (%, presented as mean ± SE) for exposed and caged flowers of sour cherry (*Prunus cerasus*) cultivar 'Stevnsbaer' five weeks after petal drop. Values for exposed branches were significantly higher than for caged ones (t test, t = 4.96-8.49, *P* < 0.05) for each year at each location.

study (noted in Tab. 1) and the study clearly indicates the value provided by bees with respect to fruit set, particularly if there are relatively few other, wild pollinators in the orchard.

The low fruit set on both caged and exposed branches in 2010 compared to 2009 may be explained by a prolonged period of high humidity and low temperatures in early May 2010, when flower development in the orchards ranged from bud burst to white bud. These weather conditions probably slowed flower development but did not halt it. Ren (2005) showed that the amount of airborne 'Stevnsbaer' pollen correlates negatively with humidity but positively with temperature. Therefore, it may be assumed that due to the weather conditions the amount of pollen in the air in early May 2010 would have been low and, consequently, wind pollination would have been limited. At the same time, only few insects (and especially bees) were flying and so transfer of pollen by insects was likely to be limited as well. These adverse conditions continued until 17 May when the majority of branches had > 50% open flowers. Once the weather conditions improved and became more favourable

for bee activity, many of those hitherto unpollinated flowers would have been too old for successful pollination.

In 2006, Dencker (2010) had investigated several of the same orchards as in this study, using the same trees. In Guldborg and Taastrup, where colonies of *A. mellifera* had been available, he reported levels of fruit set similar to those for exposed branches recorded in 2007 in this study. At Klippinge, where *A. mellifera* colonies were not present in 2006, he reported markedly lower fruit set than as observed in this investigation in 2007, after the introduction of *A. mellifera*.

The present study is the first systematic investigation of the value of bee pollination for fruit set in sour cherry in Denmark, and it clearly documents that bees are important pollinators of 'Stevnsbaer', irrespective of individual orchard management practice. Compared to options such as developing new sour cherry varieties, keeping *A. mellifera* colonies in an orchard is, a simple, low-cost measure that can be implemented easily and will ensure the presence of pollinators during bloom. To maximize the abundance of

Location	Year	No.of sample days	*Apis mellifera*	*Bombus* sp. inc. queens	*B. terrestris* worker bees	Solitary bee	Flies incl. Syrphids
Taastrup	2007	1	6	4	-	1	1
Guldborg		1	24	14	-	7	8
Klippinge		1	26	15	-	5	17
Taastrup	2009	3	23	1	1	1	0
Ringsted		3	30	1	0	2	0
Slagelse		1	32	0	0	0	1
Taastrup	2010	3	95	1	3	2	3
Ringsted		3	37	0	3	1	1
Slagelse		2	45	0	2	0	5
Total			318	36	9	19	36

TABLE 3. Total number of bees and pollinating dipterans visiting sour cherry (*Prunus cerasus*) cultivar 'Stevnsbaer' flowers, recorded in 4x10 min observations on 1-3 days per location.

bees during bloom, landscape and orchard management practices should be integrated to improve foraging conditions for *A. mellifera* and wild bees and to increase the nesting possibilities for wild bees (Wittman et al. 2005; Klein et al. 2007).

Detailed knowledge on the pollination requirements of different crops and possible pollinator species, including fruit crops such as the 'Stevnsbaer' sour cherry, is often lacking, and it is of the utmost importance that more research into these aspects of crop pollination is undertaken (Williams 1994; Klein et al. 2007).

Globally, 35% of the food supply comes from crops that depend on animal (mainly bee) pollination (Klein et al. 2007). As agriculture is vulnerable to the risks and impacts of climate change (Fischer et al. 2002), then the yield of pollinator-dependent crops might be threatened due to continued pollinator shortages (Aizen et al 2009). It is therefore important that future agricultural practices are appropriately sensitive to pollinators (Howden et al. 2007; Ortiz 2011).

The present study is the first systematic one in Denmark on the value of bee pollination for fruit set in sour cherry, and it clearly documents that bees are important pollinators of 'Stevnsbaer', irrespective of individual orchard management practice and that without them fruit set is significantly reduced. Placing *A. mellifera* colonies in the orchards during bloom can support yield and appropriate management to provide sustainable nesting habitats and foraging possibilities for wild bees can also contribute to a higher fruit set.

ACKNOWLEDGEMENTS

We thank Lars Skou Hansen, Thomas Jensen, Ole Pedersen and Preben Troels-Smith for allowing us to work in their 'Stevnsbaer' orchards and Henning B. Madsen for identifying bee specimens. We thank Danmarks Biavlerforening and Promilleafgiftsfonden for Landbrug for supporting the study in 2007, Foreningen Plan-Danmark for support in 2009 and 2010 and EWH Bioproduction for supply of *Bombus terrestris* colonies and *Osmia rufa* cocoons in 2009.

REFERENCES

Aizen MA, Garibaldi LA, Cunningham SA, Klein AM (2009) How much does agriculture depend on pollinators? Lessons from long-term trends in crop production. Annals of Botany 103: 1579–1588.

Benedek P, Nyeki J, Szabó Z, Szabó T (2005) Both self-sterile and self-fertile sour cherries need insect (bee) pollination. Acta Horticulturae 667: 399-402.

Blasse W, Haufe M (1989) Beeinflussung von Ertrag und Fruchtqualität durch Bieneneinsatz bei Erdbeere. Archiv für Gartenbau 37(4): 235-245.

Branner S, Vejsnæs F (2007) Biavlen i Danmark 2006. Temahæfte i Tidsskrift for Biavl nr. 10.

Cane JH (2005) Pollination potential of the bee *Osmia aglaia* for cultivated red raspberries and blackberries (*Rubus*: Rosaceae). HortScience 40(6):1705-1708.

Cerovic R, Ruzic D (1992) Pollen tube growth in sour cherry (*Prunus ceratus* L.) at different temperatures. Journal of Horticultural Science and Biotechnology 67 (3): 333-340.

Chagnon M, Gingras J, Oliveira D de (1991) Honeybee (Hymenoptera: Apidae) foraging behaviour and raspberry pollination. Journal of Economic Entomology 84(2):457-460.

DelaPlane KS, Mayer DF (2000) Crop pollination by bees. Cambridge, USA.

Dencker I (2010) Slutrapport for forsknings- og udviklingsprojekter med tilskud fra Innovationsloven. Ministeriet for Fødevarer Landbrug og Fiskeri. Direktoratet for FødevareErhverv. Udviklingskontoret.

Dencker I, Kühn BF, Toldam-Andersen TB, Callesen, O (1999) Blomsterkvalitet og udbytte hos kirsebær. Frugt og Bær 5: 122-123.

Dencker I, Toldam-Andersen TB (2003) Forfejlet dyrkningsteknik i Stevnsbær. Frugt og Grønt 10: 394-395.

Faostat (2012) Food and Agriculture Organization of the United Nations. http://faostat.fao.org/.

Féon VL, Schermann-Legionnet A, Delettre Y, Aviron S, Billeter R, Bugter R, Hendrickx F, Burel F (2010) Intensification of agriculture, landscape composition and wild bee communities: A large scale study in four European countries. Agriculture, Ecosystems and Environment 137: 143–150.

Fischer G, Shah M, Velthuizen HV (2002) Climate Change and Agricultural Vulnerability. International Institute for Applied Systems Analysis, Laxenburg, Austria.

Free JB (1993) Insect pollination of crops. Academic Press, New York.

Furukawa Y, Bukovac MJ (1989) Embryo sac development in sour cherry during the pollination period as related to fruit set. Hortscience 24(6):1005-1008.

Hansen P (1981) Bestøvning og frugtsætning hos surkirsebærsorten 'Stevnsbaer'. Tidsskrift for Planteavl 85: 411-19.

Hansted L (1993) Bibestøvning af solbær og production af nektar og aroma i blomsterne. Ph.D. Thesis. Den Kongelige. Veterinær- og landbohøjskole, Frederiksberg.

Howden SM, Soussana J-F, Tubiello FN, Chhetri N, Dunlop M, Meinke H (2007). Adapting agriculture to climate change. Proceedings of the National Academy of Sciences 104(50): 19691–19696

Klein A-M, Vaissière BE, Cane JH, Steffan-Dewenter I, Cunningham SA, Kremen C, Tscharntke T (2007) Importance of pollinators in changing landscapes for world crops. Biological Sciences 274 (1608): 303-313.

Krewenka KM, Holzschuh A, Tscharntke T, Dormann CF (2011)Landscape elements as potential barriers and corridors for bees, wasps and parasitoids.Biological Conservation 144: 1816–1825.

Kühn BF (1988) Examination of reasons for poor fruit set in the sour cherry cultivar 'Stevnsbaer' by means of fluorescence microscopy. Tidsskrift for Planteavl 92: 169-174.

McGregor SE 1976 Insect pollination of cultivated crop plants. Agricultural Handbook no. 496, USDA, Washington.

Ortiz R (2011) Agrobiodiversity Management for Climate Change. In: Agrobiodiversity Management for Food Security, A Critical Review. Editors:. J.M. Lenné and D. Wood. CABI, UK: 189-211

Potts SG, Biesmeijer JC, Kremen C, Neumann P, Schweiger O, Kunin WE (2010) Global pollinator declines: trends, impacts and drivers . Trends in Ecology and Evolution 25(6): 345-353.

Ren KC (2005) Bestøvning i surkirsebærsorten "Stevnsbaer". Speciale i Frugt og Bær. Den Kongelige Veterinær- og landbohøjskole, Frederiksberg.

Toyama TK (1980) The pollen receptivity period and its relation to fruit setting in the stone fruits. Fruit Varieties Journal 34: 2-4.

Vejsnæs F, Nielsen SL, Kryger P (2010) Factors involved in the recent increase in colony losses in Denmark. Journal of

Apicultural Research 49(1): 109-110.

Williams IH (1994) The dependences of crop production within the European Union on pollination by honey bees. Agricultural Science Reviews 6: 229-257.

Wittman D, Klein D, Schindler M, Sieg V, Blanke M (2005) Sind Obstanlagen geeignete Nahrungs- und Nisthabitate für Wieldbienen? Erwerbs-Obstbau 47: 27-36.

15

IDENTIFICATION OF PLANT SPECIES FOR CROP POLLINATOR HABITAT ENHANCEMENT IN THE NORTHERN PRAIRIES

Diana B. Robson*

The Manitoba Museum, 190 Rupert Avenue, Winnipeg, MB Canada R3B 0N2

Abstract—Wild pollinators have a positive impact on the productivity of insect-pollinated crops. Consequently, landowners are being encouraged to maintain and grow wildflower patches to provide habitat for important pollinators. Research on plant-pollinator interaction matrices indicates that a small number of "core" plants provide a disproportionately high amount of pollen and nectar to insects. This matrix data can be used to help design wildflower plantings that provide optimal resources for desirable pollinators. Existing interaction matrices from three tall grass prairie preserves in the northern prairies were used to identify core plant species that are visited by wild pollinators of a common insect-pollinated crop, namely canola (*Brassica napus* L.). The wildflower preferences of each insect taxon were determined using quantitative insect visitation and floral abundance data. Phenology data were used to calculate the degree of floral synchrony between the wildflowers and canola. Using this information I ranked the 41 wildflowers that share insect visitors with canola according to how useful they are for providing pollinators with forage before and after canola flowers. The top five species were smooth blue aster (*Symphyotrichum laeve* (L.) A. & D. Löve), stiff goldenrod (*Solidago rigida* L.), wild bergamot (*Monarda fistulosa* L.), purple prairie-clover (*Dalea purpurea* Vent.) and Lindley's aster (*Symphyotrichum ciliolatum* (Lindl.) A. & D. Löve). By identifying the most important wild insects for crop pollination, and determining when there will be "pollen and nectar gaps", appropriate plant species can be selected for companion plantings to increase pollinator populations and crop production.

Keywords: canola, core plants, insect visitors, pollination, restoration, wildflower plantings

INTRODUCTION

Many crop plants require or benefit greatly from pollination by insects, mainly bees and flies (Kevan et al. 1990; Klein et al. 2007). Worldwide wild pollinators are responsible for most crop pollination (Klein et al. 2007; Garibaldi et al. 2011; Garibaldi et al. 2013) although managed pollinators like honey bees can play an important role in areas where wild pollinator habitat is rare (Southwick & Southwick 1992). Unfortunately, wild pollinator populations are in decline due to numerous causes including land use intensification and pesticide use, climate change, the introduction of alien species, and the spread of pests and pathogens (Kevan 1999; Carvell et al. 2006; Vanbergen & the Insect Pollinators Initiative 2013). Researchers have found a link between landscape diversity, species richness and the abundance of wild pollinators in agroecosystems (Klein et al. 2009; Kennedy et al. 2013). This is likely because many crops, such as canola, do not provide adequate pollen and nectar resources for pollinators with life spans longer than the blooms (Morandin &Winston 2005). The general consensus is that patches of wild grassland and pasture, hedgerows, tree bluffs, windbreaks, grassy ditches (Lagerhöf 1992; Kells et al. 2001; Morandin et al. 2007;

Korpela et al. 2013; Kovács-Hostyánszki et al. 2013; Morandin & Kremen 2013) and even the presence of some flowering non-native 'weeds' (Carvalheiro et al. 2011, 2012) provides habitat for pollinators when crops are not in flower, which improves their survival and abundance. The process whereby floral species facilitate each other's persistence by supporting shared pollinators is called sequential mutualism (Waser & Real 1979). The more floral resources that are present in these natural and semi-natural areas, the better the habitat is for pollinators (Pywell et al. 2005) and the more stable the system will be (Winfree & Kremen 2009). The resulting increase in pollinator abundance can subsequently increase the productivity of insect-pollinated crops by reducing pollen limitation (Morandin & Winston 2006; Carvalheiro et al. 2011, 2012; Blaauw & Isaacs 2014).

As a result, some landowners are preserving natural and semi-natural habitats for pollinators. Active restoration of pollinator habitat, such as roadside wildflower plantings, can significantly increase wild pollinator abundance (Hopwood 2008; Haaland et al. 2011; Tarrant et al. 2012). Government programs to help support such initiatives are also forthcoming (Carvell et al. 2007; Tuell et al. 2008; Decourtye et al. 2010). The greatest improvements in pollinator richness will likely be in areas that have been the most intensively cultivated (Tscharntke et al. 2005; Kennedy et al. 2013; Scheper et al. 2013). The positive impact of wildflower plantings may be even greater when combined with organic farming methods (Morandin & Winston 2005; Winfree 2010; Kennedy et al. 2013) and/or methods to

---

*Corresponding author: drobson@manitobamuseum.ca;

preserve or increase nesting habitat for pollinators (Barron et al. 2000; Kremen et al. 2002; Williams & Kremen 2007).

In Canada, the native tall grass prairie in the Red River Valley of Manitoba has been almost completely cultivated with less than 1% remaining, leaving very little natural habitat for pollinators. Research suggests that growing some wildflowers adjacent to field margins in agricultural regions such as this is likely to increase pollinator populations and reduce the pollen deficit of crops (Tscharntke et al. 2005; Kennedy et al. 2013; Stanley et al. 2013; Blaauw & Isaacs 2014; Klatt et al. 2014). Native wildflowers are reported to be more attractive to wild pollinators than non-native ones (Williams & Kremen 2007; Menz et al. 2011; Morandin & Kremen 2013; Gill et al. 2014). Although appropriate wildflower mixtures have been identified for many regions in the United States (The Xerces Society 2014), many of the suggested species are not native to or particularly abundant in the northern prairies of Canada. Since native plants are adapted to grow in certain regions, the optimal wild plant species for attracting crop pollinators will vary depending on the local climate and soils. Several different methods to select the wild plants that are most attractive to pollinators have been suggested. Isaacs et al. (2009) recommend growing a selection of candidate plants together in 1-m² plots and monitoring the insect visitation using vacuum sampling when the plants are in flower to identify highly desirable species. Menz et al. (2011) suggest selecting plant species based on the number of insect taxa that they attract, as these "core" species appear to be more important for ecosystem maintenance than species that attract fewer pollinators (Memmott et al. 2004; Saavedra et al. 2011). However, Johnson (1980) cautioned that resource usage by animals is influenced by abundance; an animal may actually prefer a species that is less common. Kells et al. (2001) used Johnson's index to calculate the preference index (PI) of bees for different floral species found along uncropped field margins. They found that the plants that were most numerous were not necessarily the species that were preferred by honey bees (*Apis* spp.) and bumblebees (*Bombus* spp.) (Kells et al. 2001).

Plant-pollinator interaction data from three tall grass prairie preserves in south eastern Manitoba were obtained over a five year discontinuous period (Robson 2008, 2010, 2013). Using these previously collected data, I determined whether abundance of the plants in the plots was correlated with the number of insect visits. Then I ranked each plant species according to both the number of insect taxa visiting it and the mean insect PI to see if there were differences in the ranks using these methods. I also used data on floral synchrony and existing literature documenting the pollinators of canola in Canada to identify wild plant species that bloom either before or after the crop, and that attract its known pollinators. Using these three kinds of data (i.e. number of shared insect visitors, PI and floral synchrony) I was able to identify suitable wild plant species for crop pollinator habitat enhancement near canola fields.

MATERIALS AND METHODS

Study Sites

Data from previously published research in three tall grass prairie preserves were used for the calculations: Birds Hill Provincial Park (BHPP), Living Prairie Museum (LPM) and the Tall Grass Prairie Preserve (TPP) (Robson 2008, 2010, 2013). Birds Hill Provincial Park, located north of Winnipeg, MB (50.0167° N 96.8833° W), is a 35-km² protected area in the Lake Manitoba Plain ecoregion of the Prairie ecozone (Ecological Stratification Working Group 1995) containing some tall-grass prairie and oak savannah. The soils are well-drained, glaciofluvial deposits that consist of gravel, sand and silt. There are at least 492 vascular plant species that occur in the park (Manitoba Naturalists Society 1996).

Living Prairie Museum in is located within the city limits of Winnipeg, MB (49.8844° N 97.1463° W). It is a 0.12-km² remnant of tall-grass prairie set aside by the City of Winnipeg in 1971. The park is part of the Lake Manitoba Plain ecoregion (Ecological Stratification Working Group 1995). The soils are moderately well-drained, glaciofluvial in origin and consist of a mixture of clay, silt and sand. Approximately 160 vascular plant species occur in the park.

The Tall Grass Prairie Preserve is located near Gardenton, MB (49.1167° N 96.6667° W). The TPP is a 22-km² site located in the Lake Manitoba Plain ecoregion (Ecological Stratification Working Group 1995), about 100-km south of Winnipeg. The soils are highly calcareous glacial till deposits containing a mixture of gravel, sand, silt and clay. The preserve is hydric in many places, holding water well into summer during wet years. Approximately 475 native vascular plant species occur in the preserve.

Vegetation Surveys

I established 16 plots in BHPP, and 6 plots each in LPM and TPP. At BHPP each plot was 2.5 m² in size, and at LPM and TPP the plots were 5-m². The plots were at least 5-m apart. Sampling in BHPP was conducted for 37 non-consecutive days: 6 days in June (2011), 12 days in July (2010 and 2011), 11 days in August (2008 and 2010), and 8 days in September (2008). Sampling at LPM and TPP occurred from June to September on four non-consecutive days each month in both 2004 and 2005 for a total of 32 days at each site. The number of flowering stems in the plots was recorded each sampling day. The percentage of all flower stems contributed by each species was determined.

Floral Visitor Surveys

Flower-visiting insect sampling at BHPP occurred for 37 non-consecutive days from mid-June to mid-September (99 h total sampling time at each site), thus covering the main period of insect activity. At LPM and TGPP sampling occurred for 32 non-consecutive days from mid-June to mid-September (96 h total sampling time at each site). Thus the total number of sampling hours was similar at each site. As foraging activity is generally low in the early morning when temperatures are cooler (Kevan and Baker 1983), surveys were conducted between 10 am and 5 pm. The order in which the plots were visited was varied each day by using a random number table to determine the plot visitation sequence. Some of the flower-visiting insects may have also been predators of other flower-visiting insects. Regardless of whether insects were foraging for pollen, nectar, or other insects, all were considered potential pollinators with the exception of ambush bugs (*Phymata* spp.) and crab spiders (*Misumena* spp.) as they tend to remain stationary on one flower stem for a long period of time. More detailed vegetation and floral visitor survey methods and results are described in Robson (2008, 2010, 2013).

A direct observation technique was used to sample the insects. The first time an insect was observed on an inflorescence, the specimen was netted, placed in a killing jar and then transferred to a container with a unique reference number. When the same (or a very similar) species was observed later on, the reference number was used to link the insect visit to the plant. Although this technique does not allow for complete identification "on the wing" (resulting in an underestimate of insect taxa) it does enable evaluation of insect visitation frequency (Parachnowitsch and Elle 2005). All insect voucher specimens were identified by qualified zoologists using reference specimens at The Manitoba Museum (TMM) and the Wallis Roughley Museum, University of Manitoba in Winnipeg, Manitoba; the specimens were deposited in TMM's zoology collection.

The data from BHPP, LPM and TPP was used to create one large plant-insect visitor matrix consisting of 54 native plant species and 169 insect taxa in four orders: six Coleoptera, 85 Diptera, 62 Hymenoptera and 16 Lepidoptera. This matrix was used to determine how many insect taxa visited each plant species. To determine which wild plant species the various pollinators preferred, an index from Kells et al. (2001) was used to calculate the preferences of each insect taxon for each plant species present:

$$PI = \frac{(V_k/V_t)}{(A_k/A_t)}$$

Where V_k is the number of foraging visits of those insect taxa to plant species k, V_t is the total number of visits of those insect taxa to all plant species, A_k is the total number of flowers of species k, and A_t is the total number of flowers of all species. Thus the PI measures the relative attractiveness of plants to specific insect taxa. A plant with a high mean PI is visited frequently even when there are other flowering plants nearby. The PI ranges for *Bombus* and *Apis* species are reported to be between 0.1 and 13 (Kells et al. 2001). The PI was calculated for each plant species for each survey day. If a plant species was in bloom in a plot but not visited by any insect taxon that was observed, it received a PI of zero for that day. The mean PI and standard error (SE) of this mean for all days the plant species was in bloom to each insect taxon was then calculated. Additionally, the mean PI and SE of each plant species was also calculated for just the three insect genera noted to be most abundant on canola fields in Canada, as well as for all insect taxa. The sample sizes (N) varied depending on the number of days each plant was in flower and how many insect taxa were active on each sampling day. Both the number of visiting taxa and the PI

have been suggested as ways to identify highly attractive species to pollinators (Kells et al. 2001; Menz et al. 2011). To determine if these two methods selected similar plant species, I ranked them relative to each other and compared the rank differences using Spearman's rank correlation.

A literature search was conducted to determine which genera and species of insects typically pollinate canola in southern Canada (Turnock et al. 2006; Morandin et al. 2007; Gavloski et al. 2011; Zink 2013). Native plant species that are visited by the same insect taxa as canola were identified using the plant-insect visitor matrix. Supplemental observation data to determine the total number of plant-insect interactions that are likely for each plant species was obtained from existing literature (Robertson 1929; Reed 1993; Petersen 1996; Hilty 2002; Colla & Dumesh 2010).

In southern Canada, canola was noted to bloom from approximately June 20 to July 20, assuming late May seeding (Clay 2009). The exact flowering dates will of course vary slightly from year to year depending on the weather. To determine if the wild plant species were flowering during the same time as canola, the flowering synchrony was calculated from a method modified from Primack (1980). The index of synchrony (X) for a plant species (i) and canola (j) is given by:

$$Xi = (1/fi) \, ej \neq i$$

Where f is the total number of days individual i was in flower, and ej is the number of days individual i and j overlapped in their flowering. Any date when individual i had flowered in any year when it was studied was marked as being in flower. Thus the phenological data from other sites represents summed data across several years. A species with $X = 0$ does not overlap at all with canola while species with $X > 0$ overlaps to some degree.

Data Analysis

I used linear regression analysis to determine the relationship between number of flower stems and insect visits to each plant species. Spearman's rank correlation was used to determine if the insect taxa ranks were significantly different from the PI ranks. These statistical tests were performed using Analyze-It software.

RESULTS

The percentage of flowering stems each plant species contributed to the plots and the percentage of insect visits received by it were not significantly correlated ($y = 0.094x + 1.436$, $R^2 = 0.018$, $P = 0.283$) (Fig. 1). Thus plants with a large number of flowering stems in the plots did not necessarily receive the most insect visits and *vice versa*. This implies that factors other than abundance influence insect choice.

There were 54 native plant species that were visited by at least one insect taxon. An additional ten plants occurred in the plots but were not observed being visited by any insects: ground-plum (*Astragalus crassicarpus* Nutt.), lesser yellow lady's-slipper (*Cypripedium parviflorum* Salisb.), white prairie-clover (*Dalea candida* Michx. ex Willd.), silverberry

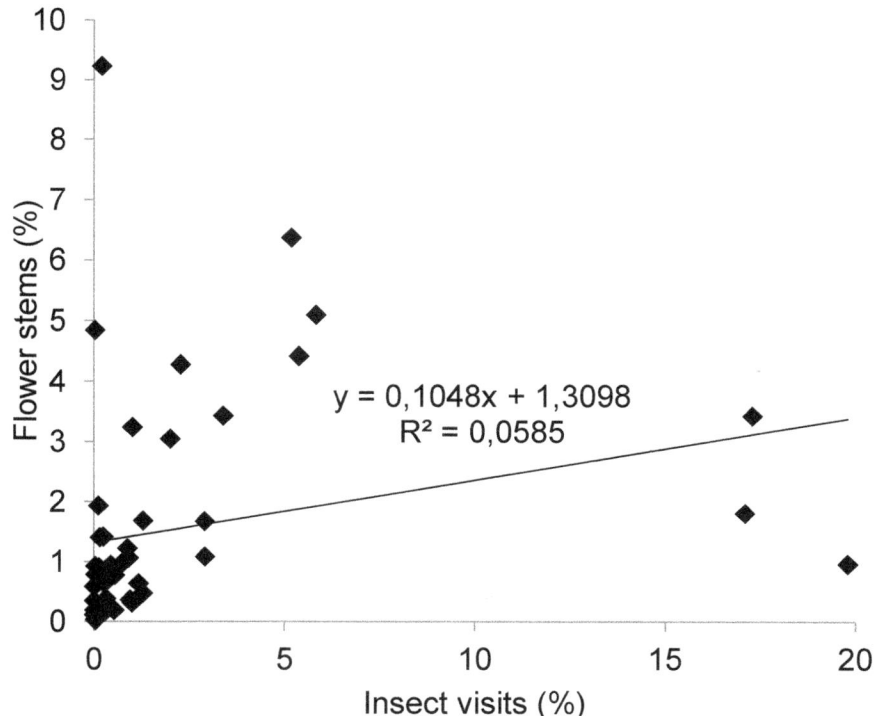

FIGURE 1. Relationship between the percentage of flowering stems in the plots and the percentage of insect visits observed to each of 54 wild plant species.

(*Elaeagnus commutata* Bernh. ex Rydb.), fringed gentian (*Gentianopsis crinita* (Froel.) Ma), prairie smoke (*Geum triflorum* Pursh), selfheal (*Prunella vulgaris* L.), veiny peavine (*Lathyrus venosus* Muhl. ex Willd.), whorled loosestrife (*Lysimachia quadrifolia* L.) and Indian breadroot (*Pediomelum esculentum* (Pursh) Rydb). Most of these species were observed for only a few days because they were less abundant in the plots than the other species. The total number of insect taxa that visited each plant species and the mean PI for all of them was calculated (Tab. 1). The ranks were significantly different from each other (Spearman's r_s = 0.36; P = 0.008). There were 19 species with a signed difference (i.e. = insect taxa rank – PI rank) greater than zero and 35 species with a signed difference lower than zero. Among the 19 species with a positive signed difference, 11 had SE's greater than two, indicating small sample sizes; therefore the PI values for these plant species are less reliable and additional data are needed to assess the species' attractiveness. Species with negative signed differences (i.e. a high insect taxa rank but low PI rank) may have been generalists attractive to many different generalist insects with low fidelities to a single species. The low SE (< 1) of most of the species with negative signed differences indicates that this data is generally more reliable.

Because the purpose of wildflower plantings adjacent to cropland is to provide wild and managed pollinators with floral resources when the crop is not in flower, plant species that are highly attractive to the most common pollinators of crop plants are particularly important to identify. There are 28 bee genera known to visit canola in Alberta; *Lasioglossum* (sweat bees) is the most abundant genus followed by *Bombus* and *Andrena* (Andrenid bees) (Morandin et al. 2007; Zink 2013). The only data from Manitoba was collected on *Bombus* abundance only. Turnock et al. (2006) indicates that there are at least 13 species of *Bombus* found in canola fields in Manitoba with the red-belted bumblebee (*B. rufocinctus* Cresson, 1863) being the most abundant one followed distantly by the northern amber bumblebee (*B. borealis* Kirby, 1837). Zink (2013) discovered two additional *Bombus* species in canola fields in Alberta but the most abundant ones observed were also *B. rufocinctus* and *B. borealis*. Wildflower visitation data on 11 of the bee genera and 12 of the species from these three studies (Turnock et al. 2006; Morandin et al. 2007; Zink 2013) are summarized in Tab. 2; however, only the taxa that were also observed in Manitoba's tall grass prairies during my research (Robson 2008, 2010, 2011) were listed. My data indicate that at least 41 species of wildflowers are visited by these bees in tallgrass prairie (Appendix 1). Seven species of plants shared ten or more insect visitor taxa with canola when data from other sources (i.e. Robertson 1929; Reed 1993; Petersen 1996; Hilty 2002; Colla & Dumesh 2010) were included: *D. purpurea* with 18 insect visitor taxa, gray goldenrod (*S. nemoralis* Ait.) and *S. rigida* each with 17, Canada goldenrod (*Solidago canadensis* L.) with 15, and *Monarda fistulosa*, *Symphyotrichum laeve* and western silvery aster (*S. sericeum* (Vent.) G.L. Nesom) each with 11. Five insect taxa visit more than ten wild plant species: *Lasioglossum pruinosum* (Halictidae) visited 20 species, orange-belted bumblebee (*Bombus ternarius* Say, 1863) visited 19, half-black bumblebee (*B. vagans* Say, 1837) visited 16, green metallic sweat bee (*Agapostemon texanus texanus* Cresson, 1872) visited 12 plant species and broad-handed leaf cutter bee (*Megachile latimanus* Say, 1823) visited 11 species. The only fly species that has been confirmed as a visitor to canola is the introduced drone fly (*Eristalis tenax* (L. 1758) (Jauker

TABLE 1.   Number of visiting insect taxa and mean preference indices (PI) of all insects observed for 54 Manitoba wildflowers ordered according to the signed difference[1]. The ranks were significantly different ($r_s = 0.36$, $P = 0.008$) using Spearman's rank correlation.

Plant Species	Insect taxa visiting (#)	Insect taxa rank	PI of all insects (mean±SE)[2]	PI rank	Signed difference	Days surveyed (#)
Symphoricarpos occidentalis	10	16	0.28±0.13	44	-28	12
Dalea purpurea	19	6	0.84±0.23	33	-27	38
Campanula rotundifolia	12	12	0.57±0.17	38	-26	29
Galium boreale	11	14	0.46±0.31	40	-26	17
Symphyotrichum sericeum	21	5	1.36±0.12	29	-24	13
Heterotheca villosa	15	10	0.83±0.22	34	-24	25
Symphyotrichum ericoides	11	14	0.75±0.35	35	-21	17
Symphyotrichum ciliolatum	15	10	1.18±0.64	30	-20	8
Solidago nemoralis	45	1	2.55±0.70	20	-19	34
Solidago rigida	42	2	2.41±0.41	21	-19	20
Solidago canadensis	34	3	2.57±0.97	19	-16	16
Erigeron glabellus	17	8	2.05±0.44	23	-15	25
Solidago ptarmicoides	16	9	1.86±0.85	24	-15	15
Lithospermum canescens	8	24	0.56±0.28	39	-15	20
Astragalus agrestis	9	20	0.92±0.34	32	-12	13
Lobelia spicata	2	40	0.02±0.01	51	-11	12
Vicia americana	9	20	1.18±0.29	30	-10	16
Erigeron strigosus	10	16	1.84±0.70	25	-9	24
Helianthus subrhomboideus	3	35	0.28±0.16	44	-9	6
Sisyrinchium montanum	3	35	0.37±0.25	43	-8	12
Parnassia palustris	1	46	0.01±0.01	54	-8	6
Houstonia longifolia	2	40	0.06±0.04	47	-7	9
Prunella vulgaris	1	46	0.01±0.01	53	-7	5
Rudbeckia hirta	19	6	4.21±1.63	12	-6	28
Symphyotrichum laeve	12	12	2.84±1.03	18	-6	13
Frageria virginiana	4	31	0.60±0.27	37	-6	8
Anemone cylindrica	2	40	0.12±0.08	46	-6	13
Zizia aptera	31	4	6.04±0.97	9	-5	8
Pediomelum argophyllum	1	46	0.02±0.02	51	-5	4
Viola nephrophylla	1	46	0.03±0.03	50	-4	3
Achillea millefolium	8	24	1.79±0.92	27	-3	35
Hypoxis hirsuta	1	46	0.04±0.03	49	-3	6
Monarda fistulosa	8	24	1.80±0.76	26	-2	15
Astragalus adsurgens	1	46	0.05±0.05	48	-2	8
Dasiphora fruticosa	10	16	3.00±2.26	17	-1	9
Amorpha nana	2	40	0.60±0.42	36	4	1
Euthamia graminifolia	9	20	3.62±2.14	15	5	7
Polygala senega	1	46	0.39±0.39	41	5	8
Penstemon gracilis	1	46	0.39±0.39	41	5	8
Helianthus maximilliani	6	29	2.29±1.71	22	7	7
Liatris ligulistylis	10	16	6.48±3.30	7	9	8
Zizia aurea	8	24	4.21±0.97	12	12	14
Allium stellatum	2	40	1.38±1.00	28	12	3
Rosa blanda	6	29	3.38±1.83	16	13	7
Zygadenus elegans	8	24	6.18±2.66	8	16	4
Cirsium flodmanii	4	31	3.72±1.86	14	17	6
Packera plattensis	9	20	11.09±6.30	2	18	7
Comandra umbellata	4	31	4.59±4.20	11	20	4
Gaillardia aristata	3	35	5.36±4.01	10	25	10
Asclepias ovalifolia	4	31	10.46±9.01	3	28	15
Crepis runcinata	3	35	7.83±7.39	6	29	4
Agoseris glauca	3	35	38.64±30.85	1	34	2
Potentilla arguta	2	40	9.84±6.56	4	36	2
Heuchera richardsonii	1	46	7.93±7.93	5	41	2

[1]Signed difference = Insect taxa rank − PI rank.
[2]Sample sizes (N) for each species ranged from 8 to 368 (mean $N = 111$).

TABLE 2. Bee taxa observed visiting canola in western Canada. Values are the percentage of all bees collected during the course of the study.

Bee taxa	Zink (2013) Bees collected (%)	Morandin et al. (2007) Bees collected (%)	Turnock et al. (2006)[1] Bees collected (%)
Agapostemon	2.39	0.30	-
Andrena	16.99	14.30	-
Bombus spp.	23.09	26.10	-
B. borealis	1.66	-	9.37
B. fervidus	0.46	-	0.16
B. perplexus	-	-	0.47
B. rufocinctus	9.69	-	80.80
B. ternarius	1.66	-	0.51
B. vagans	-	-	0.38
Other *Bombus* spp.	9.62	-	4.73
Coelioxys spp.	0.20	-	-
C. rufitarsis	0.20	-	-
Colletes	2.39	3.40	-
Epeolus	0.40	0.45	-
Hylaeus	0.86	0.15	-
Lasioglossum spp.	34.64	42.5	-
L. leucozonium	3.52	-	-
L. pruinosum	1.73	-	-
L. succipenne	0.13	-	-
Other *Lasioglossum* spp.	29.26	-	-
Megachile spp.	1.13	1.20	-
M. frigida	0.13	-	-
M. melanophae	0.13	-	-
Other *Megachile* spp.	0.87	-	-
Melissodes	2.19	0.61	-
Nomada	1.53	0.30	-
Osmia	0.27	0.45	-
Perdita	0.19	0.30	-
Pseudopanurgus	0.79	-	-
Sphecodes	0.66	2.60	-
Other bee genera	12.28	7.34	-

[1]Turnock et al. (2006) only collected data on *Bombus* spp.

& Wolters 2008; Jauker et al. 2012), which has also been observed visiting five wild plant species in Manitoba.

Most of the plant species that were visited by the pollinators of canola were in the Asteraceae (51%) followed by the Fabaceae (16%) (Tab. 3). Four plants were woody and the remainder perennial. Most of the species possessed yellow flowers (37%) followed by purple (24%), white (15%), pink (2%), orange (2%) and blue (2%); some of the Asteraceae species (17%) possessed yellow disk flowers with various colours of ray flowers (e.g. white, purple or pink). The inflorescences were mostly capitula with regularly symmetrical flowers owing to the abundance of Asteraceae species. The calyx/corolla tubes ranged from zero to 25 mm so a wide range of flower types were visited.

The mean PI of canola pollinator taxa to various wild plant species was calculated. Six of the *Bombus* species that visit canola were also observed visiting 25 wild plant species (Tab. 4). The high SE of some mean PI values indicates a relatively small sample size for that insect species. One plant species was visited by all six *Bombus* spp. (i.e. *Solidago rigida*) and two plants (i.e. purple milkvetch (*Astragalus agrestis* Dougl. and *S. canadensis*) were visited by four each.

The *Bombus* sp. most abundant in canola fields (Turnock et al. 2006) was the relatively short tongued *B. rufocinctus*; this insect was observed visiting six wildflowers in Manitoba but it preferred American vetch (*Vicia americana* Muhl. ex Willd.). The second most commonly seen species in canola fields, namely *B. borealis*, has a relatively longer tongue, visited eight plant species and also preferred *V. americana*. *Bombus ternarius*, which has a similar abundance in canola fields in Alberta as *B. borealis*, has a relatively short tongue, was observed visiting 18 plants, and preferred smooth rose (*Rosa blanda* Ait.) and *S. canadensis*.

The wildflower preferences of the three insect genera most commonly found in canola fields (Morandin et al. 2007; Zink 2013), namely *Andrena*, *Bombus* and *Lasioglossum*, were also determined (Fig. 2). Not included in Fig. 2 were four plants with mean PI's of 30 or greater for one of the three insect genera: prairie dandelion (*Agoseris glauca* (Pursh) Raf.), alumroot (*Heuchera richardsonii* R. Br.), dandelion hawksbeard (*Crepis runcinata* James T. & G.), and gaillardia (*Gaillardia aristata* Pursh). These species had high PI's of 243 ± 172, 143 ± 143, 111 ± 111, and 30 ± 30 respectively but the high SE's indicate low reliability;

TABLE 3. Floral characteristics of 41 wildflowers that share insect visitors with canola.

Plant species	Family	Rarity status[1]	Life habit	Flower colour	Inflorescence	Floral Symmetry	Calyx/corolla tube length (approx.)[2]
Achillea millefolium	Asteraceae	4	P	White	Corymbs of capitula	Regular	2-4.5 mm
Agoseris glauca	Asteraceae	4	P	Yellow	Capitula	Regular	4 mm
Amorpha nana	Fabaceae	4	W	Purple	Raceme	Irregular	2 mm
Astragalus adsurgens	Fabaceae	4	P	Purple	Raceme	Irregular	4 mm
Astragalus agrestis	Fabaceae	4	P	Purple	Raceme	Irregular	5-8 mm
Campanula rotundifolia	Campanulaceae	4	P	Blue	Solitary, raceme or panicle	Regular	5-7 mm
Cirsium flodmanii	Asteraceae	4	P	Purple	Capitula	Regular	12-15 mm
Crepis runcinata	Asteraceae	4	P	Yellow	Capitula	Regular	4-5.5 mm
Dalea purpurea	Fabaceae	4	P	Purple	Spike	Irregular	1.5-3 mm
Dasiphora fruticosa	Rosaceae	4	W	Yellow	Cluster	Regular	n/a
Erigeron glabellus	Asteraceae	4	P	Purple-yellow	Racemes of capitula	Regular	4-5.5 mm
Erigeron strigosus	Asteraceae	4	P	White-yellow	Racemes of capitula	Regular	1.5-2.5 mm
Gaillardia aristata	Asteraceae	4	P	Yellow-red	Capitula	Regular	0.5-1.5 mm
Galium boreale	Rubiaceae	4	P	White	Terminal and axillary cluster	Regular	n/a
Helianthus maximiliani	Asteraceae	4	P	Yellow	Racemes of capitula	Regular	5-7 mm
Helianthus pauciflorus spp. subrhomboides	Asteraceae	4	P	Yellow	Terminal capitula	Regular	6.5-7 mm
Heterotheca villosa	Asteraceae	4	P	Yellow	1-several terminal capitula	Regular	5-6 mm
Heuchera richardsonii	Saxifragaceae	4	P	Orange	Terminal raceme	Irregular	2-3.5 mm
Houstonia longifolia	Rubiaceae	4	P	White	Cyme	Regular	2-2.5 mm
Liatris ligulistylis	Asteraceae	4	P	Purple	Racemes of capitula	Regular	8-11 mm
Lithospermum canescens	Boraginaceae	4	P	Yellow	Cyme	Regular	7-18 mm
Lobelia spicata	Lobeliaceae	4	P	White	Raceme	Irregular	2 mm
Monarda fistulosa	Lamiaceae	4	P	Purple	Dense terminal and axillary cluster	Irregular	16-25 mm
Packera plattensis	Asteraceae	3	P	Yellow	Corymbs of capitula	Regular	2.5-3.5 mm
Pediomelum argophyllum	Fabaceae	4	P	Purple	Spike	Irregular	3-5 mm
Rosa blanda	Rosaceae	4	W	Pink	Solitary	Regular	n/a
Rudbeckia hirta	Asteraceae	4	P	Yellow	Capitula	Regular	2 mm
Sisyrinchium montanum	Iridaceae	4	P	Purple	Solitary	Regular	n/a
Solidago canadensis	Asteraceae	4	P	Yellow	Panicles of capitula	Regular	2.2-2.8 mm
Solidago nemoralis	Asteraceae	4	P	Yellow	Panicles of capitula	Regular	2.5-4.6 mm
Solidago ptarmicodes	Asteraceae	4	P	White	Corymbs of capitula	Regular	3.8-4.1 mm
Solidago rigida	Asteraceae	4	P	Yellow	Cymes of capitula	Regular	4.3-6.1 mm
Symphoricarpos occidentalis	Caprifoliaceae	4	W	White-pink	Terminal and axillary clusters	Regular	4-5 mm

TABLE 3. continued

Plant species	Family	Rarity status[1]	Life habit	Flower colour	Inflorescence	Floral Symmetry	Calyx/corolla tube length (approx.)[2]
Symphyotrichum ciliolatum	Asteraceae	4	P	Pink-yellow	Panicles of capitula	Regular	4.3-6.4 mm
Symphyotrichum ericoides	Asteraceae	4	P	White-yellow	Panicles of capitula	Regular	2.5-4 mm
Symphyotrichum laeve	Asteraceae	3	P	Purple-yellow	Panicles of capitula	Regular	3.5-6.1 mm
Symphyotrichum sericeum	Asteraceae	I	P	Pink-yellow	Panicles of capitula	Regular	5-7 mm
Vicia americana	Fabaceae	4	P	Purple	Raceme	Irregular	3.5-5.6 mm
Zizia aptera	Apiaceae	4	P	Yellow	Compound umbel	Regular	n/a
Zizia aurea	Apiaceae	4	P	Yellow	Compound umbel	Regular	n/a
Zygadenus elegans	Liliaceae	4	P	White	Raceme	Regular	n/a

[1] *I* At Risk in Canada, *3* Sensitive in Canada, *4* Secure in Canada, (Canadian Endangered Species Conservation Council 2011).
[2] Data obtained from Flora of North America Editorial Committee 1993+, and Reaume 2009.

additional data are needed to truly assess the PI of these species. Twelve species that had PI's of less than one were also not included on Fig. 2. In total, there were eight species of plants that insects in all three genera visited, almost all of which were in the Asteraceae: Flodman's thistle (*Cirsium flodmanii* (Rydb.) Arthur), smooth fleabane (*Erigeron glabellus* Nutt.), *S. canadensis, S. nemoralis, S. rigida, Symphyotrichum ciliolatum, S. laeve* and golden Alexanders (*Zizia aurea* (L.) Koch). Wildflower species with a high PI value for one genus did not necessarily have a high PI for the others; several of the favourite plants of *Bombus* and *Lasioglossum* were not observed being visited by the other genera, although this may be partly due to the small sample size for some of these plants. The only plant with a high (> 5) PI value for more than one genus was *C. flodmanii*, which was highly attractive to both *Andrena* and *Lasioglossum* spp, although the high SE suggests that additional data are needed to confirm its attractiveness.

To determine which wild plant species are most likely to provide the wild pollinators of canola with optimal resources when the crop is not in flower, three factors were taken into account: synchrony, number of shared insect visitor taxa and the PI of canola pollinators (Tab. 5). There were seven species of plants that had a synchrony with canola equal to one; that is complete flowering overlap (Fig. 3). Eight species had a synchrony of zero, indicating no overlap at all. The synchrony of 14 species was greater than zero but 0.5 or less, and that of 12 species greater than 0.5 but less than one. The number of shared insect visitor taxa with canola ranged from 16 to just one: this variation is partially affected by a lack of supplementary data on pollinator visitation for some species. Regarding the PI, eight species had SE's greater than two and so were not ranked due to the unreliability of the data. The top five plant species (excluding those with SE > 2) in decreasing order were: *Monarda fistulosa, Solidago canadensis,* heart-leaved Alexanders (*Zizia aptera* (Gray) Fern.), *Symphyotrichum laeve* and dwarf false indigo (*Amorpha nana* Nutt.). Each plant species was given a rank

according to how high its' value was for the three indicated factors. The three plants with a synchrony of one were not ranked as they are potential competitors for canola pollinators and likely provide floral resources at a time when it is not lacking. Plants with a synchrony of zero were given the highest rank. Plants that shared the most insect visitors with canola or had the highest PI were given the highest ranks. If two species had the same value, they were given the same rank. I decided to use the average rank for these three values to identify the plant species most likely to provide optimal forage for the crop pollinators of canola. Of the top 20 ranked plants, five species reach their flowering peak in June, four in July, eight in August and three in September.

DISCUSSION

The purpose of this research was to analyse data on plant-pollinator interactions in the northern prairies to identify wildflowers that support the pollinators of the popular insect-pollinated crop, canola. I found that insect visitation was not strongly correlated with plant abundance, an observation that was not unexpected as previous research indicates that there are many factors that influence foraging behaviour including sensory information, learning ability, and floral rewards (Heinrich 1976; Waddington 1983; Ibanez 2012). Within a species, insects may favour different plants depending on whether they are searching for pollen or nectar (Rasheed & Harder 1997; Elle et al. 2012). Even the presence of pollinator predators, like crab spiders, influences pollinator visitation (Jones & Dornhaus 2011). Thus data on actual pollinator visitations are more valuable for assessing wildflower suitability than data on plant abundance in a community.

Menz et al. (2010) suggest that when creating habitat for pollinators the core species, which are the plants that are visited by the most insect taxa, should be grown as they form the core of the plant-pollinator network. These core species are likely to be actinomorphic-flowered plants that do not restrict nectar access (Elle et al. 2012) rather than species

TABLE 4. Wildflower preference indices (PI) of *Bombus* spp. known to visit canola in Manitoba in order of decreasing importance to canola.[1] Relative tongue length indicated in brackets[2].

Plant species	Preference Index (mean±SE)[3]						Insect taxa visiting (#)	All *Bombus* (mean PI±SE)
	B. rufocinctus (short)	*B. borealis* (long)	*B. ternarius* (short)	*B. perplexus* (medium)	*B. vagans* (medium)	*B. fervidus* (long)		
Solidago rigida	2.21±1.1	0.04±0.04	1.19±0.68	4.91±3.14	1.75±1.75	20.80±0	5	2.54±0.96
Astragalus agrestis	2.78±2.78	5.09±4.56	0	23.8±0	5.01±3.12	-	4	5.32±2.22
Solidago canadensis	0.76±0.38	0	10.76±9.36	1.34±0.75	1.33±1.33	0	4	4.82±3.76
Monarda fistulosa	0	7.25±4.24	3.61±1.93	23.30±0	0	-	3	5.52±2.03
Vicia americana	6.55±6.55	8.72±3.38	0	0	2.63±1.92	-	3	4.26±1.51
Liatris ligulistylis	0	3.65±0	3.23±2.01	0	2.80±0	-	3	2.14±0.98
Dalea purpurea	0	1.04±0.35	1.59±0.71	0	4.94±2.14	-	3	1.48±0.47
Symphyotrichum laeve	1.20±1.20	0	1.49±1.49	-	-	0	2	1.04±0.76
Solidago nemoralis	0	0.96±0.96	0.36±0.16	0	0	0	2	0.39±0.22
Lithospermum canescens	0	0.30±0.30	-	0	0.07±0.07	-	2	0.13±0.10
Symphyotrichum ciliolatum	0	0	0.21±0.18	0.59±0.49	-	-	2	0.34±0.21
Zizia aurea	4.28±4.28	0	-	0	-	-	1	0.71±0.71
Rosa blanda	-	-	22.30±22.19	-	-	-	1	22.3±22.19
Amorpha nana	-	-	1.82±0	-	-	-	1	1.82±0
Helianthus pauciflorus ssp. *subrhomboides*	0	0	1.62±1.62	-	-	-	1	0.97±0.97
Symphoricarpos occidentalis	-	-	0.97±0.69	0	0	-	1	0.62±0.44
Erigeron glabellus	-	-	0.52±0.52	0	-	-	1	0.28±0.28
Rudbeckia hirta	0	0	0.51±0.51	-	-	-	1	0.17±0.17
Pediomelum agrophyllum	-	-	0.23±0.23	0	-	-	1	0.13±0.13
Helianthus maximiliani	0	-	0.04±0.04	0	-	0	1	0.02±0.02
Solidago ptarmicoides	0	0	0.03±0.03	0	0	-	1	0.02±0.02
Symphyotrichum ericoides	0	0	0.01±0.01	0	-	0	1	0.003±0.003
Campanula rotundifolia	0	0	0	-	3.04±2.10	-	1	0.66±0.49
Cirsium flodmanii	-	-	0	-	2.86±0.00	-	1	0.57±0.57
Houstonia longifolia	0	0	-	-	0.37±0.37	-	1	0.15±0.15

[1] Importance of *Bombus* spp. to canola as recorded in Turnock et al. (2006)

[2] Williams et al. (2014)

[3] Sample sizes (*N*) for each species ranged from 1 to 16 (mean *N* = 4).

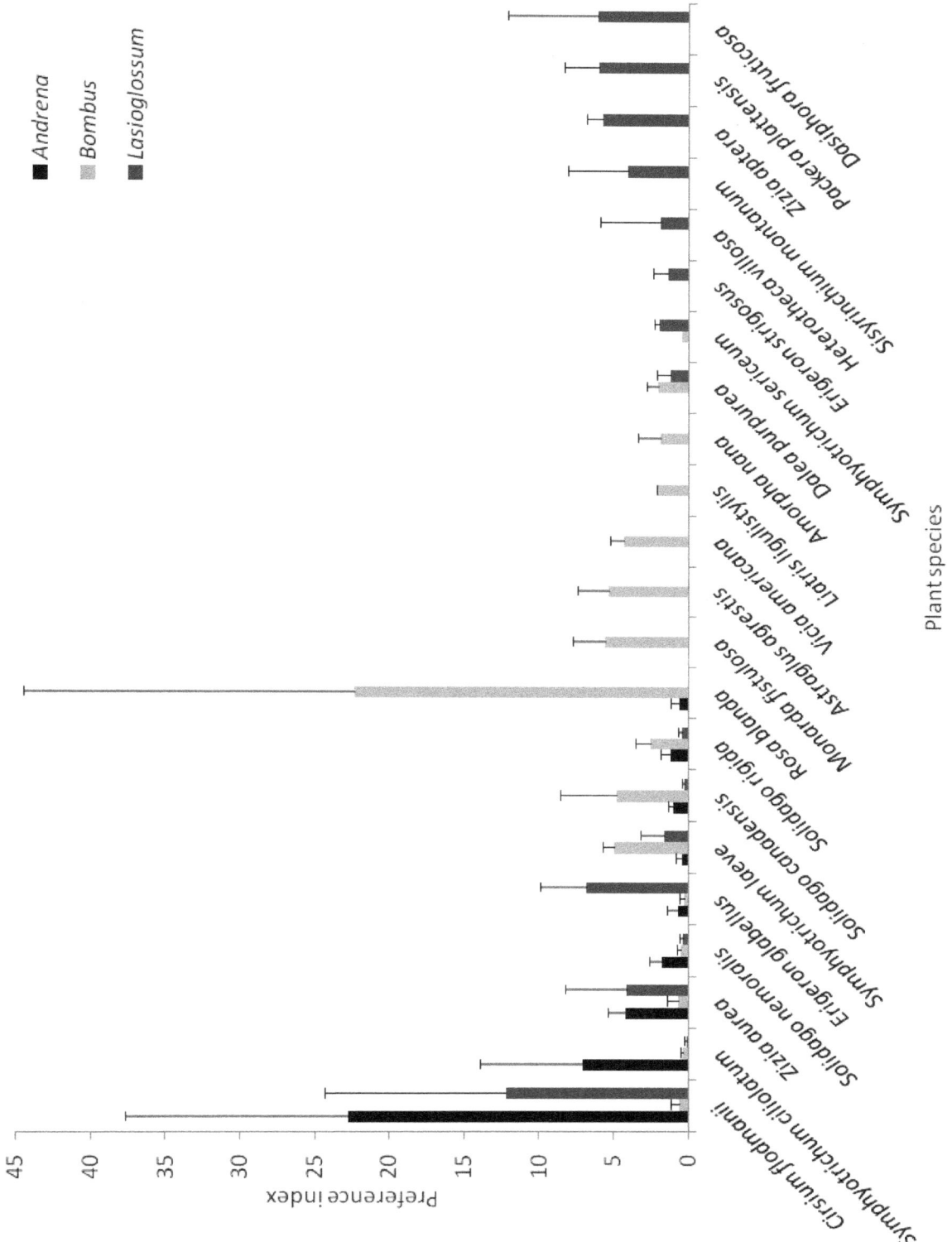

Figure 2.   Preference indices (PI) of three insect genera to 22 wildflowers. Error bars indicate +SE.

TABLE 5. Data on synchrony, insect visitors and preference index (PI) to 41 plant species that share insect visitors with canola in Manitoba.

Plant species	Suitability rank	Flowering peak (month)	Total synchrony	Shared insect visitor taxa[1] (#)	PI of shared visitors (mean±SE)[2]
Symphyotrichum laeve	1	September	0	11	2.71±0.85
Solidago rigida	2	August	0	17	1.60±0.47
Monarda fistulosa	3	July	0.31	11	4.00±1.94
Dalea purpurea	4	August	0.16	18	1.96±0.66
Symphyotrichum ciliolatum	5	August	0	5	2.18±1.98
Zizia aurea	6	June	0.14	11	1.86±0.76
Symphyotrichum sericeum[3]	6	September	0	11	0.38±0.19
Solidago canadensis	8	July	0.55	15	3.51±1.98
Erigeron glabellus	9	July	0.31	9	2.20±0.89
Liatris ligulistylis	9	August	0	4	1.68±0.65
Solidago nemoralis	9	August	0.04	17	0.53±0.13
Symphyotrichum ericoides	12	September	0	8	0.23±0.16
Vicia americana	13	June	0.46	7	2.40±0.72
Heterotheca villosa	14	August	0.04	5	0.75±0.36
Helianthus maximiliani	15	August	0.09	3	0.77±0.43
Helianthus pauciflorus ssp. subrhomboides	15	August	0	2	0.32±0.32
Astragalus agrestis	17	June	0.85	6	2.41±0.98
Amorpha nana	18	June	0.5	2	2.48±0.66
Campanula rotundifolia	19	July	0.52	6	0.86±0.41
Zizia aptera	20	June	0.83	2	2.86±1.54
Rudbeckia hirta	21	July	0.52	8	0.27±0.20
Dasiphora fruticosa	22	July	0.3	1	1.14±1.14
Erigeron strigosus	22	July	0.56	7	0.55±0.34
Pediomelum argophyllum	24	August	0	1	0.07±0.07
Solidago ptarmicodes	25	August	0.39	3	0.08±0.05
Sisyrinchium montanum	26	June	0.48	1	0.74±0.74
Lithospermum canescens	27	June	0.56	3	0.12±0.09
Houstonia longifolia	28	June	0.85	3	0.07±0.07
Lobelia spicata	28	August	0.37	1	0.04±0.04
Achillea millefolium	30	August	0.57	2	0.01±0.01
Symphoricarpos occidentalis	n/r[4]	July	1	6	0.87±0.53
Galium boreale	n/r[4]	July	1	2	0.07±0.06
Astragalus adsurgens	n/r[4]	July	1	1	0.15±0.15
Cirsium flodmanii	n/r[5]	July	0.42	3	9.84±5.63
Packera plattensis	n/r[5]	June	0.58	2	59.50±54.82
Crepis runcinata	n/r[5]	June	0.67	1	37.0±37.0
Rosa blanda	n/r[5]	June	0.7	4	8.80±8.52
Agoseris glauca	n/r[4,5]	July	1	2	145.60±111.30
Heuchera richardsonii	n/r[4,5]	July	1	1	20.41±20.41
Gaillardia aristata	n/r[4,5]	June	1	1	4.33±4.33
Zygadenus elegans	n/r[4,5]	July	1	1	3.05±3.05

[1] Information on the number of interactions obtained from Robertson (1929), Reed (1993), Petersen (1996), Hilty (2002), Colla & Dumesh (2010), Robson (2008, 2010, 2013).

[2] Sample sizes (*N*) for each species ranged from 2 to 126 (mean *N* = 36).

[3] This species is nationally rare in Canada.

[4] Suitability not ranked due to high synchrony with canola.

[5] Suitability not ranked due to high standard error of the PI (i.e. > 2).

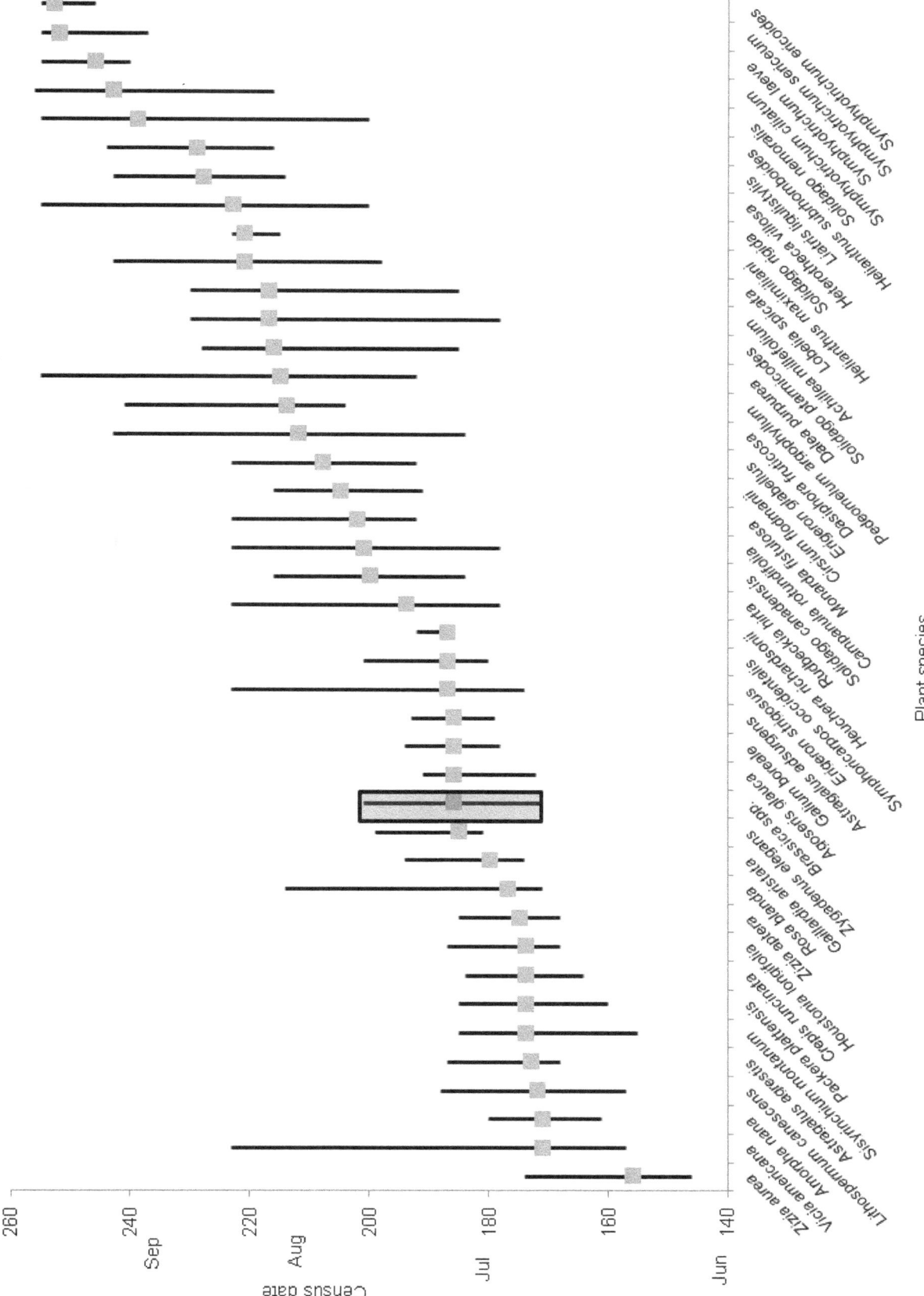

FIGURE 3. Flowering phenology of plant species that share insect visitors with canola over 110 calendar days. The line indicates the flowering duration and the square the day of the flowering peak in the research plots in south eastern Manitoba. The box in grey is the approximate flowering period of canola in Canada.

with deep, narrow nectar tubes (Stang et al. 2006) as most pollinators have relatively short mouthparts. Using plant-insect interaction matrices from the tall grass prairie in Manitoba I identified these core species. However, Winfree (2010) cautioned that basing flower selection decisions solely on use rather than preference may result in some desirable species being overlooked. By calculating the PI for each species (Johnson 1980; Kells et al. 2011), I was able to determine that some plants that were visited by a small number of insect taxa were actually highly preferred by them. As suggested by previous research (Stang et al. 2006; Elle et al. 2012), these species tended to have relatively deep nectar tubes. *Vicia americana*, for example, has a deep nectar tube and was only visited by nine insect taxa. Species such as this should not be discounted for inclusion in wildflower plantings, especially if they are favoured by a common pollinator of a crop. As both *Bombus rufocinctus* and *B. borealis*, the most common bumblebee pollinators of canola, preferred *V. americana* over all other plants, including it would likely be beneficial. Plants that were visited by many species (e.g. *Solidago nemoralis*) often had a low PI. This is because plants that are visited by many insects may not be highly important to any one species but rather moderately important to many. In summary, looking only at the number of insect visitor taxa may result in highly preferred flowers being overlooked and looking only at the PI may discriminate against species that are utilized by many taxa. Because both methods had shortcomings and both long and short-tongued insects visit canola, I concluded that ranking the plant species relative to each other and then averaging the ranks would likely identify the optimal mixture of species to use for pollinator habitat enhancement.

In western Canada, canola is visited most frequently by *Lasioglossum*, *Bombus* and *Andrena*. The relative efficacy of these three genera in pollinating canola has not been ascertained although *Bombus* tends to be more efficient than most other bee genera (Herrera 1987; Sahli & Connor 2007; Ali et al. 2011). Although these three genera tended to prefer slightly different wildflower species, there was still quite a bit of overlap: eight plants were visited by all three genera and ten species were visited by two of the three genera. The *Bombus* spp. preferred many of the purple, tubular flowers such as *Monarda fistulosa* and Rocky Mountain blazingstar (*Liatris ligulistylis* (A. Nelson) K. Schum.) while *Lasioglossum* preferred many of the asters with large, yellow capitula, such as hairy golden-aster (*Heterotheca villosa* (Pursh) Shin.) and fleabanes (*Erigeron* spp.). Carvell et al. (2007) noted that *Bombus* species were most attracted to a mixture of legumes, a finding also observed by Zink (2013) in Alberta. The legumes *Vicia americana* and *Dalea purpurea* were preferred by several of the *Bombus* spp. in this study as well. Legumes were more attractive to *Bombus* spp. than to *Lasioglossum* or *Andrena* likely because they often have longer mouthparts than the others, an observation also made by Lagerhof et al. (1992). However, because the shorter-tongued bees (i.e. *Andrena* and *Lasioglossum*) were more abundant than the longer tongued bumblebees in canola fields, a mixture of flowers attractive to all three taxa would likely be most effective in provisioning them. Supporting this is the observation that a diverse mixture of wildflowers

attracts more bee species and provides a better continuity of floral resources (Carvell et al. 2007). Plant mixtures with greater functional diversity would also likely provide more stable pollination service over time as insect populations fluctuate from year to year (Kremen et al. 2002; Klein et al. 2009; Albrecht et al. 2012).

Another factor to consider when selecting plant species for habitat enhancement is whether they supply nectar and pollen during resource limited times (Menz et al. 2010). By assessing the degree of flowering synchrony between canola and the wild plants, I was able to identify seven plant species that may compete with canola for pollinators and should either be grown sparingly or not at all. However, as canola provides mainly pollen, plants that provide primarily nectar may not necessarily compete with canola as much as they would complement it; the possibility of facilitation via resource complementarity (Ghazoul 2006) needs to be examined in more detail. There were 34 plant species that reached their flowering peak either before or after canola typically blooms. If the top 20 plants identified in this study were grown in a wildflower planting there would be a fairly even sequence of blooms throughout the year; five species in early June, four in late July, eight in August and three in September. Thus the objective of providing floral resources throughout the growing season would be achieved. Some of the plants identified were previously noted as being important pollen and nectar sources for pollinators in North America including: milkvetch (*Astragalus* spp.), bluebell (*Campanula rotundifolia* L.), prairie-clover (*Dalea* spp.), shrubby cinquefoil (*Dasiphora fruticosa* (L.) Rydb.), lobelia (*Lobelia* spp.), *Monarda fistulosa*, rose (*Rosa* spp.), ragwort (*Packera* spp.), goldenrod (*Solidago* spp.), western snowberry (*Symphoricarpos occidentalis* Hook.), aster (*Symphyotrichum* spp.) and *Zizia aurea* (Isaacs et al. 2009; Mader et al. 2011; Evans 2013). Additionally, plants in the genera *Cirsium*, *Fragaria*, *Helianthus*, *Penstemon*, *Prunella*, *Rudbeckia*, *Vicia* and *Viola* were noted to be attractive to one or more species of *Bombus* (Williams et al. 2014). Although *Symphyotrichum sericeum* was one of the top ten plants, this species is actually nationally rare in Canada. Thus even though *S. sericeum* provides good forage for the likely pollinators of canola, it may not be available for wildflower restoration in Canada due its legal protection. However, as this species is more common in the U.S., it may be useful to grow there as it supplies late summer forage. The remaining species are all common and many are already popular plants for wildflower plantings.

There were several assumptions and limitations that must be acknowledged. I assumed that canola would be in flower for about one month from approximately June 20 to July 20 but if spring seeding is impaired by cool or wet weather it may be in flower till the end of July. However, as cool, wet weather also hampers the bloom times of wildflowers by several weeks (Robson 2008), the synchrony between canola and wildflowers may not be significantly different. As well winter seeding of canola followed by an early spring may result in earlier bloom dates than I used. If canola will typically be seeded in August or September, fewer June-flowering plants and more July-flowering plants would be optimal for providing pollinators forage.

Good data on the abundance of wild pollinators in canola fields in Manitoba (aside from *Bombus* spp.) is lacking so I assumed that the genera found visiting canola in Alberta were likely to visit it in Manitoba (Morandin et al. 2006; Zink 2013). This study also assumed that the most important pollinators of canola were bees. This was simply due to a lack of data on the abundance of potentially pollinating fly species of canola in Canada. In Pakistan (Ali et al 2011), Germany (Jauker & Wolters 2008; Jauker et al. 2012) and New Zealand (Rader et al. 2009) some flies, particularly those in the Syrphidae were noted as effective pollinators of canola. However, the only species observed in those studies that is also found in North America is the introduced *Eristalis tenax* (Syrphidae). *Eristalis tenax* has been observed visiting five wildflowers in North America and may visit canola here as well. Gavloski et al. (2011) also noted that bee flies (Bombyliidae) are potential pollinators of canola although no data on their abundance or frequency in Canadian canola crops has been published. Further research is needed to truly understand the role that wild flies play in the pollination of canola and other Canadian crop plants.

Another limitation was the lack of data on pollinator visitation to wild plants. Although the data provided by other sources (Robertson 1929; Reed 1993; Petersen 1996; Hilty 2002; Colla & Dumesh 2010) was valuable and greatly increased the number of confirmed species interactions (49 additional links were added as a result), there were several plant species with no supplemental data (e.g. *Amorpha nana*, ascending purple milk-vetch (*Astragalus adsurgens* Pall.) and *Gaillardia aristata*). Thus some plant species are likely visited by more insect taxa than was reported, which resulted in them receiving a lower suitability rank than if better data were available. As well, many insects and plants were observed infrequently resulting in high SE's. Caution should be used when interpreting the PI of plant species with high SE's as the values may not reflect true attractiveness.

As the plots I selected did not contain any flowering plants in May, data on insect visitations to the earliest flowering plants are lacking. However, I did observe bumble and sweat bees visiting several shrubs growing near my plots in late May, including pin cherry (*Prunus pensylvanica* L.), chokecherry (*P. virginiana* L.), American plum (*P. americana* Marsh), and Saskatoon serviceberry (*Amelanchier alnifolia* Nutt.). Many of these flowering shrubs are popular in windbreaks on the prairies and provide the added benefit of supplying resources to pollinators early in the year. Thus planting later flowering wildflowers, such as *Solidago* and *Symphyotrichum*, alongside existing windbreaks of spring flowering shrubs would improve these habitats by providing a more stable supply of floral resources for pollinators.

Lastly, this study was restricted to just 56 species of wild plants common to the tall grass prairies. Visitation data to some of these plant species was inadequate and additional observations are needed to assess their relative importance. Some of these plants are common across the entire Canadian prairies (e.g. *Dalea purpurea*, *Monarda fistulosa*, *Symphoricarpos occidentalis*) while others are limited to Manitoba (e.g. *Liatris ligulistylis*) although species in the same genus may be present farther west (e.g. dotted blazingstar (*Liatris punctata* Hook.). Many plants that are common in the fescue prairies to the northwest or the drier mixed grass prairies to the southwest, where canola is also a common crop were not evaluated. However, the methodology presented here that integrates plant-insect visitor interaction networks, data on flower preference, and knowledge of crop pollinators and bloom times can be adapted for any crop or ecoregion.

Conclusions

Farmers are being encouraged to provide forage and breeding habitat for the wild pollinators of crop plants to improve crop productivity and resiliency of their agroecosystems. Existing data on crop pollinators, wild plant phenology, plant-insect interactions and quantitative data on insect visitation can be valuable for helping to identify the most appropriate plant species for wildflower plantings in agroecosystems. Using this approach I identified the plant species native to southern Canada that can provide insects with the resources they need for their survival. In particular, late-summer flowering plants like *Solidago* spp. and *Symphyotrichum* spp. were identified as excellent candidates for wildflower plantings to support wild pollination service to canola. Good early spring-flowering species include *Zizia* spp., *Vicia americana* and *Astragalus agrestis*. Exactly which of these species are selected for habitat enhancement will require the collection of more complete data on flower preferences by bees and possibly other pollinating insects, the relative importance of individual bee species to canola pollination and information on the degree to which canola competes with other concurrently flowering species for pollinators. As well, wild flower species selection will depend on the availability of seed as well as the local soil conditions as some plants cannot tolerate very dry or moist conditions. Finally, testing of a wildflower seed mixture using these species will be required to determine the impact on bee abundance and crop productivity. Protecting or providing nesting habitat as well as a steady supply of floral resources would likely be even more beneficial.

ACKNOWLEDGEMENTS

The Museum gratefully acknowledges the financial support of The Manitoba Museum Foundation Inc., WWF-Canada, the Government of Canada, and Manitoba Conservation and Water Stewardship. Thanks to the Government of Manitoba, the City of Winnipeg and Nature Manitoba for permitting me to conduct research their lands. Thanks to staff and volunteers at The Manitoba Museum for their help with specimen processing. Special thanks to Heather Flynn, Robert Wrigley and Sarah Semmler for preparation and identification of insect specimens. Karen Sereda, Mae Elsinger and Mark Wonneck graciously agreed to review this paper.

APPENDICES

Additional supporting information may be found in the online version of this article:

APPENDIX I. Plant-insect visitor matrix.

REFERENCES

Albrecht M, Schmid B, Hautier Y, Müller CB (2012) Diverse pollinator communities enhance plant reproductive success. Proceedings of the Royal Society B: Biological Sciences 279:4845-4852.

Ali M, Saeed S, Sajjad A, Whittington A (2011) In search of the best pollinators for canola (*Brassica napus* L.) production in Pakistan. Applied Entomology and Zoology 46:353-361.

Barron MC, Wratten SD, Donovan BJ (2000) A four-year investigation into the efficacy of domiciles for enhancement of bumble bee populations. Agricultural and Forest Entomology 2:141-146.

Blaauw BR, Isaacs R (2014) Flower plantings increase wild bee abundance and the pollination services provided to a pollination-dependent crop. Journal of Applied Ecology. doi: 10.1111/1365-2664.12257

Canadian Endangered Species Conservation Council (2011) Wild species 2010: the general status of species in Canada. National General Status Working Group, Ottawa, Ontario.

Carvalheiro LG, Veldtman R, Shenkute AG, Tesfay GB, Pirk CWW, Donaldson JS, Nicolson SW (2011) Natural and within-farmland biodiversity enhances crop productivity. Ecology Letters 14:251-259.

Carvalheiro LG, Seymour CL, Nicolson SW, Veldtman R (2012) Creating patches of native flowers facilitates crop pollination in large agricultural fields: mango as a case study. Journal of Applied Ecology 49:1373-1383.

Carvell C, Roy DB, Smart SM, Pywell RF, Preston CD, Goulson D (2006) Declines in forage availability for bumblebees at a national scale. Biological Conservation 132:481-489.

Carvell C, Meek WR, Pywell RF, Goulson D, Nowakowski M (2007) Comparing the efficacy of agri-environment schemes to enhance bumble bee abundance and diversity on arable field margins. Journal of Applied Ecology 44:29–40.

Clay H (2009) Pollinating hybrid canola-the southern Alberta experience. Hivelights 3:14-16.

Colla SR, Dumesh S (2010) The bumblebees of southern Ontario: notes on natural history and distribution. Journal of the Entomology Society of Ontario 141:39-68.

Decourtye A, Mader E, Desneux N (2010). Landscape enhancement of floral resources for honey bees in agro-ecosystems. Apidologie 41:264-277.

Ecological Stratification Working Group (1995) A national ecological framework for Canada. Agriculture and Agri-Food Canada, Research Branch, Centre for Land and Biological Resources Research and Environment Canada, State of the Environment Directorate, Ecozone Analysis Branch, Ottawa-Hull, Ontario.

Elle E, Elwell SL, Gielens GA (2012) The use of pollination networks in conservation 11This article is part of a Special Issue entitled "Pollination biology research in Canada: Perspectives on a mutualism at different scales". Botany 90:525-534.

Evans MM (2013) Influences of grazing and landscape on bee pollinators and their floral resources in rough fescue grassland. Dissertation. University of Calgary, Calgary, Alberta.

Flora of North America Editorial Committee (1993+) Flora of North America North of Mexico. 16+ vols. New York and Oxford.

Garibaldi LA, Aizen MA, Klein AM, Cunningham SA, Harder LD (2011) Global growth and stability of agricultural yield decrease with pollinator dependence. Proceedings of the National Academy of Sciences 108:5909-5914.

Garibaldi LA, Steffan-Dewenter I, Winfree R, Aizen MA, Bommarco R, Cunningham SA, ... Klein AM (2013) Wild pollinators enhance fruit set of crops regardless of honey bee abundance. Science 339:1608-1611.

Gavloski J, Cárcamo H, Dosall L (2011) Insects of canola, mustard, and flax in Canadian grasslands. In: Floate KD (ed) Arthropods of Canadian grasslands, volume 2: inhabitants of a changing landscape. Biological Survey of Canada Monograph Series No. 4, Biological Survey of Canada, pp 181-214.

Ghazoul J (2006) Floral diversity and the facilitation of pollination. Journal of Ecology 94:295-304.

Gill KA, Cox R, O'Neal ME (2014) Quality over quantity: buffer strips can be improved with select native plant species. Environmental Entomology 43:298-311.

Haaland C, Naisbit RE, Bersier LF (2011) Sown wildflower strips for insect conservation: a review. Insect Conservation and Diversity 4:60-80.

Heinrich B (1976) The foraging specializations of individual bumblebees. Ecological Monographs 46:105-128.

Herrera CM (1987) Components of pollinator" quality": comparative analysis of a diverse insect assemblage. Oikos 50:79-90.

Hilty J (2002) Insect visitors of Illinois Wildflowers. [online] URL: http://www.illinoiswildflowers.info/flower_insects/index.htm (accessed March 2014).

Hopwood JL (2008) The contribution of roadside grassland restorations to native bee conservation. Biological Conservation 141:2632-2640.

Ibanez S (2012) Optimizing size thresholds in a plant-pollinator interaction web: towards a mechanistic understanding of ecological networks. Oecologia 170:233-242.

Isaacs R, Tuell J, Fiedler A, Gardiner M, Landis D (2009) Maximizing arthropod-mediated ecosystem services in agricultural landscapes: the role of native plants. Frontiers in Ecology and the Environment 7:196-203.

Jauker F, Bondarenko B, Becker HC, Steffan-Dewenter I (2012) Pollination efficiency of wild bees and hoverflies provided to oilseed rape. Agricultural and Forest Entomology 14:81–87.

Jauker F, Wolters V (2008) Hover flies are efficient pollinators of oilseed rape. Oecologia 156:819-823.

Johnson DH (1980) The comparison of usage and availability measurements for evaluating resource preference. Ecology 61:65-71.

Jones EI, Dornhaus A (2011) Predation risk makes bees reject rewarding flowers and reduce foraging activity. Behavioral Ecology and Sociobiology 65:1505-1511.

Kells AR, Holland JM, Goulson D (2001) The value of uncropped field margins for foraging bumblebees. Journal of Insect Conservation 5:283-291.

Kennedy CM, Lonsdorf E, Neel MC, Williams NM, Ricketts TH, Winfree R ... Kremen C (2013) A global quantitative synthesis of local and landscape effects on wild bee pollinators in agroecosystems. Ecology Letters 16:584-599.

Kevan PG (1999) Pollinators as bioindicators of the state of the environment: species, activity and diversity. Agriculture, Ecosystems & Environment 74:373-393.

Kevan PG, Baker HG (1983) Insects as flower visitors and pollinators. Annual Review of Entomology 28:407-453.

Kevan PG, Clark EA, Thomas VG (1990) Insect pollinators and sustainable agriculture. American Journal of Alternative Agriculture 5:13-23.

Klatt BK, Holzschuh A, Westphal C, Clough Y, Smit I, Pawelzik E, Tscharntke T (2014) Bee pollination improves crop quality, shelf life and commercial value. Proceedings of the Royal Society B: Biological Sciences 281:2013-2440.

Klein AM, Müller C, Hoehn P, Kremen C (2009) Understanding the role of species richness for crop pollination services. In: Naeem S, Bunker DE, Hector A, Loreau M, Perrings C, (eds) Biodiversity, ecosystem functioning, and human wellbeing: an ecological and economic perspective. Oxford University Press, Don Mills, Ontario, pp 195-208.

Klein AM, Vaissiere BE, Cane JH, Steffan-Dewenter I, Cunningham SA, Kremen C, Tscharntke T (2007) Importance of pollinators in changing landscapes for world crops. Proceedings of the Royal Society B: Biological Sciences 27:303-313.

Korpela EL, Hyvönen T, Lindgren S, Kuussaari M (2013) Can pollination services, species diversity and conservation be simultaneously promoted by sown wildflower strips on farmland? Agriculture, Ecosystems and Environment 179:18-24.

Kovács-Hostyánszki A, Haenke S, Batáry P, Jauker B, Báldi A, Tscharntke T, Holzschuh A (2013) Contrasting effects of mass-flowering crops on bee pollination of hedge plants at different spatial and temporal scales. Ecological Applications 23:1938–1946.

Kremen C, Williams NM, Thorp RW (2002) Crop pollination from native bees at risk from agricultural intensification. Proceedings of the National Academy of Sciences 99:16812-16816.

Lagerhöf J, Stark J, Svensson B (1992) Margins of agricultural fields as habitats for pollinating insects. Agriculture, Ecosystems and Environment 40:117–124.

Mader E, Shepard M, Vaughan M, Hoffman Black S, LeBuhn G (2011) Attracting native pollinators: protecting North America's bees and butterflies. The Xerces Society guide.

Manitoba Naturalists Society (1996) Wild plants of Birds Hill Provincial Park, Manitoba, Canada, Eco Series 4. Manitoba Naturalists Society, Winnipeg, Manitoba.

Memmott J, Waser NM, Price MV (2004) Tolerance of pollination networks to species extinctions. Proceedings of the Royal Society of London Series B: Biological Sciences 271:2605-2611.

Menz MH, Phillips RD, Winfree R, Kreme C, Aizen MA, Johnson SD, Dixon KW (2011) Reconnecting plants and pollinators: challenges in the restoration of pollination mutualisms. Trends in Plant Science 16:4-12.

Morandin LA, Kremen C (2013) Hedgerow restoration promotes pollinator populations and exports native bees to adjacent fields. Ecological Applications 23:829-839.

Morandin LA, Winston ML (2005) Wild bee abundance and seed production in conventional, organic, and genetically modified canola. Ecological Applications 15:871-881.

Morandin LA, Winston ML (2006) Pollinators provide economic incentive to preserve natural land in agroecosystems. Agriculture, Ecosystems and Environment 116:289-292.

Morandin LA, Winston ML, Abbott VA, Franklin MT (2007) Can pastureland increase wild bee abundance in agriculturally intense areas? Basic and Applied Ecology 8:117-124.

Parachnowitsch AL, Elle E (2005) Insect visitation to wildflowers in the endangered Garry Oak, Quercus garryana, ecosystem of British Columbia. The Canadian Field Naturalist 119:245-253.

Petersen CE (1996) Bee visitors of four reconstructed tallgrass prairies in Northeastern Illinois. Proceedings of the 15th North American Prairie Conference, St. Charles, Illinois 23-26 October 1996. Natural Areas Association, Bend, Oregon.

Primack RB (1980) Variation in the phenology of natural populations of montane shrubs in New Zealand. Journal of Ecology 68:849-862.

Pywell RF, Warman EA, Carvell C, Sparks TH, Dicks LV, Bennett D, Wright A, Critchley CNR, Sherwood A (2005) Providing foraging resources for bumblebees in intensively farmed landscapes. Biological Conservation 121:479–494.

Rader R, Howlett BG, Cunningham SA, Westcott DA, Newstrom-Lloyd LE, Walker MK, Teulon DAJ, Edwards W (2009) Alternative pollinator taxa are equally efficient but not as effective as the honey bee in a mass flowering crop. Journal of Applied Ecology 46:1080-1087.

Rasheed S, Harder L (1997) Economic motivation for plant species preferences of pollen-collecting bumble bees. Ecological Entomology 22:209-219.

Reaume T (2009) 620 wild plants of North America. Canadian Plains Research Center, University of Regina, Regina, Saskatchewan.

Reed CC (1993) Reconstruction of pollinator communities on restored prairies in eastern Minnesota. Final Report to the Nongame Wildlife Program, Minnesota Department of Natural Resources, St. Paul, Minnesota.

Robertson C (1929) Flowers and insects. The Science Press, Lancaster, Pennsylvania.

Robson DB (2008) The structure of the flower-insect visitor system in tall-grass prairie. Botany 86:1266-1278.

Robson DB (2010) A comparison of flower visiting insects to rare Symphyotrichum sericeum and common Solidago nemoralis (Asteraceae). Botany 88:241-249.

Robson DB (2013) An assessment of the potential for pollination facilitation of a rare plant by common plants: Symphyotrichum sericeum (Asteraceae) as a case study. Botany 91:1-9.

Saavedra S, Stouffer DB, Uzzi B, Bascompte J (2011) Strong contributors to network persistence are the most vulnerable to extinction. Nature 478:233-235.

Sahli HF, Conner JK (2007) Visitation, effectiveness, and efficiency of 15 genera of visitors to wild radish, Raphanus raphanistrum (Brassicaceae). American Journal of Botany 94:203-209.

Scheper J, Holzschuh A, Kuussaari M, Potts SG, Rundlöf M, Smith HG, Kleijn D (2013) Environmental factors driving the effectiveness of European agri-environmental measures in mitigating pollinator loss—a meta-analysis. Ecology Letters 16:912-920.

Southwick EE, Southwick Jr L (1992) Estimating the economic value of honey bees (Hymenoptera: Apidae) as agricultural pollinators in the United States. Journal of Economic Entomology 85:621–633.

Stang M, Klinkhamer PG, Van Der Meijden E (2006) Size constraints and flower abundance determine the number of interactions in a plant–flower visitor web. Oikos 112:111-121.

Stanley DA, Gunning D, Stout JC (2013) Pollinators and pollination of oilseed rape crops (Brassica napus L.) in Ireland: ecological and economic incentives for pollinator conservation. Journal of Insect Conservation 17:1181-1189.

Tarrant S, Ollerton J, Rahman ML, Tarrant J, McCollin D (2012) Grassland restoration on landfill sites in the East Midlands,

United Kingdom: an evaluation of floral resources and pollinating insects. Restoration Ecology 21:560-568.

The Xerces Society (2014) Pollinator conservation seed mixes. [online] URL: http://www.xerces.org/pollinator-seed/ (accessed April 2014).

Tscharntke T, Klein AM, Kruess A, Steffan-Dewenter I, Thies C (2005) Landscape perspectives on agricultural intensification and biodiversity-ecosystem service management. Ecology Letters 8:857–874.

Tuell JK, Fiedler AK, Landis D, Isaacs R (2008) Visitation by wild and managed bees (Hymenoptera: Apoidea) to eastern US native plants for use in conservation programs. Environmental Entomology 37:707-718.

Turnock WJ, Kevan PG, Laverty TM, Dumouchel L (2006) Abundance and species of bumbles (Hymenoptera: Apoidea: Bombinae) in fields of canola, *Brassica rapa* L., in Manitoba: an 8-year record. Journal of the Entomology Society of Ontario 137:31-40.

Vanbergen AJ, Insect Pollinators Initiative (2013) Threats to an ecosystem service: pressures on pollinators. Frontiers in Ecology and the Environment 11:251-259.

Waddington KD (1983) Foraging behaviour of pollinators. Pages 213-241 in L. Real, editor. Pollination Biology. Academic Press, New York.

Waser NM, Real LA (1979) Effective mutualism between sequentially flowering plants species. Nature 281:670-672.

Williams NM, Kremen C (2007) Resource distributions among habitats determine solitary bee offspring production in a mosaic landscape. Ecological Applications 17:910-921.

Williams PH, Thorp RW, Richardson LL, Colla SR (2014) Bumblebees of North America. Princeton University Press, Princeton, New Jersey.

Winfree R (2010) The conservation and restoration of wild bees. Annals of the New York Academy of Science 1195:169-197.

Winfree R, Kremen C (2009) Are ecosystem services stabilized by differences among species? A test using crop pollination.Proceedings of the Royal Society B: Biological Sciences 276:229-237.

Zink L (2013) Concurrent effects of landscape context and managed pollinators on wild bee communities and canola (*Brassica napus* L.) pollen deposition. Dissertation. University of Calgary, Calgary, Alberta.

16

POLLINATION OF GREENHOUSE TOMATOES BY THE MEXICAN BUMBLEBEE *BOMBUS EPHIPPIATUS* (HYMENOPTERA: APIDAE)

Carlos H. Vergara[1]* and Paula Fonseca-Buendía[1]

Laboratorio de Entomología, Departamento de Ciencias Químico-Biológicas, Universidad de las Américas Puebla. Ex-Hacienda Santa Catarina Mártir, 72820 Cholula, Puebla, México;

Abstract—The Mexican native bumblebee *Bombus ephippiatus* Say was evaluated as a potential pollinator of greenhouse tomatoes (*Solanum lycopersicon* L.). The experiments were performed at San Andrés Cholula, Puebla, Mexico, from June to December 2004 in two 1 000 m² greenhouses planted with tomatoes of the cultivar Mallory (Hazera ®). For the experiments, we used two colonies of *Bombus ephippiatus*, reared in the laboratory from queens captured in the field. Four treatments were applied to 20 study plants: pollination by bumble bees, manual pollination, pollination by mechanical vibration and no pollination (bagged flowers, no vibration). We measured percentage of flowers visited by bumble bees, number of seeds per fruit, maturing time, sugar content, fruit weight and fruit shape. All available flowers were visited by bumblebees, as measured by the degree of anther cone bruising. The number of seeds per fruit was higher for bumble bee-pollinated plants as compared with plants pollinated mechanically or not pollinated and was not significantly different between hand-pollinated and bumble bee-pollinated plants. Maturation time was significantly longer and sugar content, fresh weight and seed count were significantly higher for bumblebee pollinated flowers than for flowers pollinated manually or with no supplemental pollination, but did not differ with flowers pollinated mechanically.

Keywords: *Bombus ephippiatus, Mexican bumble bees, tomato pollination, greenhouses*

INTRODUCTION

Greenhouse tomatoes, *Solanum lycopersicon* L., require supplemental pollination for fruit set (McGregor 1976, Review by Picken 1984, Free 1993) and were usually pollinated by mechanical vibration (manual pollination), which was labour intensive and thus expensive. In Europe, laboratory- or mass-reared colonies of *Bombus terrestris* L. have been in tomato greenhouses since 1987 and have subsequently replaced manual pollination (Ravenstijn and Nederpel 1988, Ravenstijn 1989, Heemert et al. 1990). Pollination by *B. terrestris* resulted in significantly heavier fruit (Banda and Paxton 1991, Ravestijn and Sande 1991) when compared with manual pollination, although fruit were of similar weight in another study (Kevan et al. 1991).

In North America, Agriculture Canada, the United States Department of Agriculture, and the Mexican SAGARPA (Secretaría de Agricultura, Ganadería, Desarrollo Rural, Pesca y Alimentación) restrict the importation of European bumble bee species. To provide for the greenhouse tomato market, Mexico has been importing *B. impatiens* from the United States and Canada since 1995. Between 2005 and 2009 more than 128 000 queens or small colonies of *B. impatiens* were imported into Mexico by two commercial companies that sell bumble bee colonies in the country (Campuzano-Hernández 2010). In an assessment report presented recently (Medina-Valdez 2010), importation of *B. impatiens* was regarded by the Mexican animal health authority as a potential risk to native Mexican bumble bees because, due to the possibility of accidental release of the species, it could become a competitor for pollen and nectar (Inari et al. 2005, Ishii et al. 2008, Ings et al. 2006) and transmit diseases to the native species, as has already happened in some other countries (Otterstatter and Thomson 2008).

One of the strongest recommendations of the same report is "to promote the study and use of native bumble bee species as greenhouse pollinators". Research and production of *Bombus ephippiatus* Say colonies at the laboratory level has been carried out since 2001 and, more recently, at the commercial rearing scale by greenhouse tomato producers in west Mexico (Cuadriello pers. com.). Since *B. impatiens* and *B. ephippiatus* are very closely related (Cameron et al. 2007), there is the additional risk of interbreeding, which could have a negative impact on the number of *B. ephippiatus* queens (Goka 1998).

Bombus ephippiatus occurs naturally from Northwest Mexico to West Panama, and is more abundant in association with pine-oak or cloud forests, above 800 masl (Ayala 2009). The abundance, wide distribution of the species and the fact that that we observed it buzz-pollinating tomato and potato flowers cultivated in the open, makes it a promising candidate to replace imports of the non-native *B. impatiens*.

Because no information exists on the effectiveness of *B. ephippiatus* as a greenhouse tomato pollinator, our objective was to evaluate this Mexican bumblebee as a pollinator of greenhouse tomatoes.

MATERIALS AND METHODS

The experiments were conducted in two 1 000 m² greenhouses located in San Andrés Cholula, Puebla, Mexico. The greenhouses were rectangular (78 x 12.8 m), covered with plastic, with liquid-feed systems and ambient lighting. Daily temperatures were maintained between 20 and 25°C. Density of "Mallory"® (Hazera) was three plants per square meter. Trusses were pruned to 6-7 flowers per truss and plants were maintained according to standard commercial practices. Plants were 2 months old in late June at the start of the experiment. Fruit was harvested during November and December.

Twenty test plants were selected at random in each greenhouse. To minimize the effect of inter plant variation, trusses 3, 4, and 5 were used on each of the test plants in the first greenhouse, to apply the first three treatments. The mechanical pollination treatment was applied to the plants in the second greenhouse.

Pollination treatments

Four pollination treatments were applied to the 20 study plants (n = 20):

- Pollination by *B. ephippiatus*. Two colonies of *B. ephippiatus* were reared under controlled conditions (Gretenkord 1996) at the Laboratory of Entomology, Universidad de las Américas Puebla. The colonies were started from queens captured in the field at the Volcán de Colima, Jalisco, Mexico in January 2004 and were introduced to the first greenhouse on 24 June 2004, when they had between 70 and 90 workers. Because tomato flowers do not produce nectar (Free 1993), the colonies were supplied with syrup (sugar mixture according to Kammerer 1994) until September 8, 2004, when the colonies were returned to the laboratory. Bumble bee foraging activity was assessed on 27 June and 7 July at 1300 hours by counting incoming and exiting bumble bees. Intensity of foraging activity was 12 bees per 5 min per colony (mean of 2 colonies) of which 0.75 bees were incoming pollen foragers. At midday on 30 June, the average bumble bee colony population estimate was 85 workers (range, 70-90). This colony population estimate excludes foragers working on the crop. The majority of bumble bee foraging occurred between 1000 and 1500 hours. Although this level of pollen foraging could be regarded as low, it is comparable to the level recorded in a similar study (Dogterom *et al.* 1998). Additionally, visitation by bumble bees was measured by the degree of anther cone bruising (Morandin et al. 2001), and indicated that a 100% of the flowers displayed bruising levels higher than 2, which can be considered as efficiently pollinated. One truss per plant was left uncovered and no manipulation was done on it.

- Manual Pollination. Upon anthesis of the flowers of the truss picked for this treatment, the anther cone of each flower was cut open and pollen was transferred to the stigma of the same flower, by using a fine brush. Once pollen transfer was performed, the truss was covered with a 500-μ white Nytex ® bag to prevent visitation by bumblebees. Once fruit set was confirmed, the fruit was uncovered to minimize the effect of bagging. Manual pollination was completed between 1000 and 1200 hours 3 times per week.

- No supplemental pollination. The truss chosen for this treatment was covered with a 500-μ white Nytex ® bag before anthesis and uncovered after fruit set was confirmed.

- Mechanical pollination was performed by vibrating the training wire associated with the test plants by hitting it lightly with a wooden rod. This system of mechanical vibration is traditionally used by Mexican tomato growers.

The effects of pollination by *B. ephippiatus* were determined by measuring its impact on fruit ripening time, fresh and dry weight, sugar contents, fruit roundness, and seed count of the fruits produced by the test plants.

All tomatoes were harvested at the same ripeness, based on visual assessment of colour. The tomatoes were considered ripe when they had a uniform orange-red colour. Ripening time was calculated by counting the days between fruit set and harvest date.

Tomatoes were weighed fresh and then dried for 48 h in an electric oven at 60°C. Both weights were measured to 0.01 g using an electronic scale (Ohaus CT200, Pine Brook, NJ). Sugar percentage was measured by extracting a sample of juice using a new Terumo ® disposable 5 ml syringe for each fruit sampled, The percentage of sugars and other dissolved contents in the juice was measured to one decimal place using a hand-held 0.0 ~ 90.0% refractometer (ATAGO, HSR500, Itabashi, Tokyo, Japan). Seed count was performed manually by rehydrating dry fruits and separating the seeds from the flesh. A roundness index was calculated by measuring the maximum and minimum diameters of the fruit (Morandin et al. 2001)

Statistical analyses

The data were first analyzed by MANOVA (multivariate ANOVA, Statistics, StatSoft, 1999), with roundness, weight, sugars, number of seeds, minimum diameter, the difference in diameter between the minimum and maximum diameter, and days until ripe as the response variables. MANOVA was followed by univariate ANOVA and Tukey's pairwise comparisons.

RESULTS

Multivariate ANOVA of the four pollination treatments (n = 20) showed a difference among pollination treatments with respect to roundness, weight, sugars, number of seeds, and days until ripe ($F = 0.239$; df = 18,11854; $P = 0.001$).

Univariate ANOVA showed that there was a difference in tomato fresh weight ($F = 168.29$; df = 3, 4196; $P = 0.0001$), dry weight ($F = 112.17$; df = 3, 4196; $P = 0.0001$), ripening time ($F = 160.67$; df = 3, 4196; $P = 0.0001$), percentage of sugars ($F = 472.80$; df = 3,4196; $P = 0.0001$), and number of seeds ($F = 1734.26$; df = 3, 4196; $P = 0.0001$), with respect to pollination treatments (Table 1). There was no difference among pollination

TABLE 1. Comparison of five measures of tomato quality recorded from four pollination treatment groups.

Treatment (n=20)	Ripening Time (Days)	Fresh weight (g)	% sugars	Number of seeds/fruit	Roundness
Pollination by bumble	55.45 ± 18.7a	62.60±23.7a	4.97 ± 1.8a	201.00 ± 80.5a	0.81 ± 0.06a
Mechanical Pollination	49.76 ± 18.6a	60.84 ± 22.9a	4.93 ± 1.9a	159.32 ± 31.1b	0.81 ± 0.03a
Manual Pollination	49.71 ± 19.6 a	59.36±22.2b	5.79 ± 3.1b	153.28 ± 25.3b	0.80 ± 0.05a
No supplemental Pollination	46.50 ± 17.3b	57.72±21.5b	4.70 ± 1.7c	139.03 ± 60.1c	0.88 ± 0.07a

Average ± S. E. of the response variables. Means followed by the same letter in any given column are not significantly different from one another (Tukey's HSD, $P < 0.05$)

treatments with respect to tomato roundness ($F = 2.16$; df = 3, 4196; $P = 0.09$).

Flowers pollinated by *B. ephippiatus* produced larger fruit than manually pollinated flowers (Table 1) as evidenced by significant increases ($P < 0.001$) in fruit weight and seed count. Mean fruit weight for bumble bee pollination was 5.46 % higher than for mechanical pollination.

DISCUSSION

Our results indicate that *B. ephippiatus* is a commercially and practical alternative to the use of imported bumble bees for pollination of greenhouse tomatoes in Mexico.

Pollination of tomato flowers by *B. ephippiatus* provides a greater yield of tomatoes (variety Mallory) than does manual-pollination or no pollination under greenhouse conditions. Fruit weight, percentage of sugars, and seed count were higher for bumble bee pollinated flowers than for non-bumble bee-pollinated flowers (manual and no pollination), although fruit quality, as indicated by roundness indices, did not significantly differ between treatments. Larger fruit size resulting from bumble bee pollination of flowers was reported in other studies (Sande 1989; Banda and Paxton 1991; Ravestijn and Sande 1991; Kevan et al. 1991; Dogterom et al. 1998). Dogterom et al. (1998), Banda and Paxton (1991), and Ravestijn and Sande (1991) reported that bumble bees produced heavier fruit than when manual pollination was used, although Kevan et al. (1991) found no significant difference in a similar study. Our results also suggest that perhaps pollination quality is different when pollination is completed by bumble bees versus manual pollination. Manual pollination is conducted according to a schedule of 3 times per week, whereas bumble bees may visit flowers at an optimal time (for fertilization) and perhaps visit flowers more than once. Thus, although differences between regression equations are significantly different, these differences did not appear to constitute practical differences between treatments.

The effect of bagging plants was not addressed in this study. It is possible that bagging decreased the amount of light that reached the developing fruit, thus inhibiting fruit development. However, the length of time that the top part of each tomato plant remained inside the bag was kept to a minimum by continually moving the bags when flowers were set.

The level of flower bruising can be used to monitor flower visitation by bumble bees (Morandin et al. 2001) and could be used to indicate when additional bumble bee colonies are required in the greenhouse. In this study we monitored flower visitation by bumble bees and found that 100% of the flowers showed a level of bruising (higher than 1) that guaranteed the transferral of enough pollen grains to set fruit of commercial value, according to Morandin et al. (2001).

Seed count may be the most accurate method for determining levels of pollination because fruit weight, but not seed count (Picken 1984), is influenced by environmental conditions such as plant resources. Manual pollination significantly increased seed count over the no-pollination treatment. A further increase in seeds resulted from bumble bee pollination, indicating that bumble bees are better pollinators than the manual pollination technique.

ACKNOWLEDGEMENTS

Antonio Aguirre provided contact with greenhouse owners and technical assistance during the study. Saioa Fernández helped with the work in the greenhouses. We thank also Granja Avícola Colorines, the owner of the greenhouses used. Fundación Produce Puebla provided funding for the study. David Inouye and an anonymous reviewer provided useful comments that improved the manuscript.

REFERENCES

Ayala R, Ortega-Huerta M (2009) El Abejorro *Bombus ephippiatus* Say, 1837, su Distribución Potencial y Estrategias para su Manejo. Memorias del VI Congreso Mesoamericano de Abejas Nativas. Antigua, Guatemala: 165-171.

Banda HJ, Paxton RJ (1991) Pollination of greenhouse tomatoes by bees. Sixth International Symposium on pollination. Acta Horticulturae 288: 194-198.

Cameron SA, Hines HM, Williams PH (2007) A comprehensive phylogeny of the bumble bees (*Bombus*). Biological Journal of the Linnean Society 91: 161-188.

Campuzano-Hernández R (2010) Situación actual de la importación de abejorros *Bombus impatiens*. http://www.conasamexico.org.mx/conasa/docs_17a_reunion/c

omite07/Rocio_Campuzano_Hernandez.pdf (accessed September 2011).

Dogterom MH, Matteoni JA, Plowright RC (1998) Pollination of greenhouse tomatoes by the North American *Bombus vosnesenskii* (Hymenoptera: Apidae). Journal of Economic Entomology 91: 71-75.

Free JB (1993) Insect Pollination of Crops. 2nd ed. Academic Press, Harcourt Brace Javanovich Publishers

Goka, K (1998) Influences of invasive species on native species: Will the European bumble bee, *Bombus terrestris*, bring genetic pollution into the Japanese native species? Bulletin of the Biogeographical Society of Japan 53(2): 91-101.

Gretenkord, C (1996) Laborzucht der dunklen Erdhummel *Bombus terrestris* L. (Hymenoptera: Apidae) und toxikologische Untersuchungen unter Labor- und Halbfreilandbedingungen. Ph. D. Thesis. Institut für Landwirtschaftliche Zoologie und Bienenkunde. Reinische Friedrich-Wilhelms-Universität, Bonn, Germany.

Heemert, C van, de Ruijter, A, van den Eijnde J and van der Steen J (1990) Year round production of bumble bee colonies for crop pollination. Bee World 71(2): 54-56.

Inari N, Nagamitsu T, Kenta T, Goka K, Hiura T (2005) Spatial and temporal pattern of introduced *Bombus terrestris* abundance in Hokkaido, Japan, and its potential impact on native bumblebees. Population Ecology, 47: 77-82.

Ings TC, Ward NL, Chittka, L (2006) Can commercially imported bumble bees out-compete their native conspecifics? *Journal of Applied Ecology*, 43, 940-948.

Ishii HS, Kadoya T, Kikuchi R, Suda SI, Washitani I (2008) Habitat and flower resource partitioning by an exotic and three native bumble bees in central Hokkaido, Japan. Biological Conservation 141(10): 2597-2607

Kammerer, FX (1994) Aktueller Stand der Erkenntisse über die Fütterung von Bienen mit Zucker. Deutsches Bienen Journal 1:18-20.

Kevan PG, Straver WA, Offer M, Laverty TW (1991) Pollination of greenhouse tomatoes by bumble bees in Ontario. Proceedings of the Entomological Society of Ontario 122: 15-17.

McGregor, SE (1976) Insect pollination of cultivated crop plants. Agricultural Research Services, United States Department of Agriculture.

Medina-Valdez R (2010) Conclusiones y recomendaciones del dictamen técnico sobre el riesgo que representa la introducción de especies exóticas de abejorros *Bombus impatiens*, para la polinización de vegetales y su impacto sanitario. http://www.conasamexico.org.mx/conasa/docs_18a_reunion/salon3miercoles900a1200/Rogelio_Medina_Valdez.pdf (accessed September 2011).

Morandin LA, Alberti TM, Kevan PG (2001) Effect of bumble bee (Hymenoptera: Apidae) pollination intensity on the quality of green house tomatoes. Journal of Economic Entomology 94 (1): 172-179.

Otterstatter MC, Thomson JD (2008) Does Pathogen Spillover from Commercially Reared Bumble Bees Threaten Wild Pollinators? PLoS ONE 3(7): e2771. doi:10.1371/journal.pone.0002771.

Picken AJF (1984) A review of pollination and fruit set in the tomato (*Lycopersicon esculentum* Mill.). J. Hortic. Sci. 59(1): 1-13.

Ravenstijn W van (1989) Hommels ook in ronde tomaat goede vervangers van trillen. Groenten en Fruit 45(20): 42-43.

Ravenstijn W van, Nederpel LSR (1988) Trostrillers in Belgie aan de kant: Hommels doen het werk. Groenten en Fruit 43(32): 38-41.

Ravenstijn W, Sande J van der (1991) Use of bumble bees for the pollination of glasshouse tomatoes. Sixth International Symposium on Pollination. Acta Horticulturae 288: 204-212.

Sande J van der (1989) Hommels goed alternatief voor trostrillen vleestomaat. Groenten Fruit 45(20): 40-41.

Statsoft (1999) Statistica 5.5 for Windows. StatSoft *Inc.* Oklahoma, USA.

17

Specialized and Facultative Nectar-Feeding Bats Have Different Effects on Pollination Networks in Mixed Fruit Orchards, in Southern Thailand

Tuanjit Sritongchuay[1,*], Sara Bumrungsri[1]

[1]*Department of Biology, Faculty of Science, Prince of Songkla University, Hat Yai, Thailand, 90122*

Abstract—Recent advances in the study of pollination networks have improved our ability to describe species interactions at the community level. In this study, we compared the abundance and network strength of facultative and obligate nectar-feeding bats to determine their roles in pollinating mixed fruit orchards. We were particularly interested in the effect of distance from forests and caves on the foraging activity of these two bat groups. For this study, we examined 10 pairs of orchards; each pair consisted of one orchard near to (< 1 km) and one orchard far from (> 7 km) the forest edge. We estimated the abundance of each bat group (nectarivorous vs. frugi-nectarivorous) using video observations to determine floral visitation rates. A pollination network was then created for each of the 20 study orchards and network strength was calculated for each bat group at each orchard. We found that nectarivorous bats showed higher abundance and network strength than frugi-nectarivorous bats. Both bat abundance and network strength were negatively correlated with distance to the nearest cave, however, only network strength was affected by distance to the forest. These results corroborate the importance of nectarivorous bats in pollinating crops within southern Thailand's mixed fruit orchards. Higher network strength of bats near forests and caves emphasizes the role of natural habitats as pollinator sources.

Keywords: bat, cave proximity, forest proximity, network strength, pollination network

Introduction

Pollination is a key mutualistic interaction. Although bat pollination is not as common as insect or bird pollination, approximately 250 genera of plants depend on bat pollinators (Sekercioglu 2006; Fleming et al. 2009). A recent study by Stewart et al. (2014) classified paleotropical, phytophagous bats into two feeding guilds, specialized nectarivores (which are obligate nectar feeders) and frugi-nectarivores (which visit flowers opportunistically). Since specialized nectarivores are dependent solely on floral resources, they may be more consistent visitors than frugi-nectarivores, and may provide greater pollination services. In Neotropical studies, nectar specialist and opportunistic bat species differ in their contributions toward plant reproductive success (Frick et al. 2013). However, knowledge about how these two bat groups contribute to plant-pollinator networks within an entire plant community is lacking.

Recent advances in the study of pollination networks have improved our ability to describe species interactions and the underlying structure, function, and stability of communities (Montoya et al. 2006). It has been demonstrated that some properties of pollination networks are influenced by spatial effects, such as habitat conversion and urbanisation (Geslin et al. 2013). Additionally, decreasing habitat availability at the landscape level can isolate patches of suitable habitat leading to altered pollinator diversity, frequency, and movement patterns (Holyoak et al. 2005; Greenleaf & Kremen 2006; Brosi et al. 2007; Zurbuchen et al. 2010). Proximity to natural habitats is important in enhancing ecosystem services provided by pollinators. However, effects of proximity to natural habitats may vary with the organism. In a previous study, we found that bat visitation and pollination success of durian are significantly negatively correlated with distance to the nearest cave (Sritongchuay et al. 2016). These results correspond to bat roosting behavior, as pteropodid species roost in foliage and limestone karst caves (Kunz & Fenton 2003; Bumrungsri et al. 2009). Previous studies have focused on only one species of plant. However, it is important to understand the effect of distance to natural habitats on the role of flower-visiting bats within the entire bat-pollinated plant community.

In this study, we aim to investigate the role of nectarivorous and frugi-nectarivorous bat species in southern Thailand's mixed fruit orchards (that vary in distance from forests and caves) by addressing these questions: 1) Do nectarivorous and frugi-nectarivorous bats have similar pollination roles? 2) Does the distance to forest patches and caves affect the abundance and/or network strength of either bat group (sum of dependencies across all plant species that

*Corresponding author: t.sritongchuay@gmail.com

a bat group interacted with), thus influencing the pollination services they provide to the plant community? We hypothesized that nectarivorous bats are more important pollinators, because they feed obligately on nectar and are therefore likely more frequent flower visitors. In addition, we predict that the abundance and network strength of all flower-visiting bats will be negatively correlated with distance to forest patches and caves due to pteropodid bat roosting habitats.

MATERIALS AND METHODS

Study sites

Mixed fruit orchards are commonly found around traditional villages in Southeast Asia. Each orchard consists of planted fruit crops and certain native tree species, as well as herb and shrub species. This multi-storied system thus resembles a forest in both structure and diversity. The main fruit trees are durian (*Durio zibethinus* L.), bitter beans (*Parkia speciosa* Hassk.), mangosteen (*Garcinia mangostana* L.), domestic jackfruit (*Artocarpus integer* (Thunb.) Merr.), longon (*Lansium parasiticum* (Osbeck) K.C.Sahni & Bennet), rambutan (*Nephelium lappaceum* L.), and mango (*Mangifera indica* L.). Durian, bitter bean (Bumrungsri et al. 2008, 2009), *Oroxylum indicum* (L.) Kurz (Srithongchuay et al. 2008), *Musa acuminate* Colla (Itino et al. 1991), and *Ceiba pentandra* (L.) Gaertn (Lobo et al. 2005; Nathan et al. 2005) are bat-pollinated and nectar-feeding bats commonly forage at the flowers of these species (Bumrungsri et al. 2013; Stewart et al. 2014).

Mixed fruit orchards in southern Thailand are distributed among forest patches. The study took place from September 2012 to June 2013, using 20 mixed fruit orchards situated at varying distances from 10 forest patches in southern Thailand (Nakhon Si Thammarat, Phattalung, Trang, Satun and Songkhla provinces; 6°20'to 8°20'S and 99°40' to 110°00'E). The actual size of the ten patches of tropical rain forest, excluding rubber and oil palm plantations, ranged in area between 3.6 to 650 km2 and occurred at altitudes between 230 to 1,090 m. We determined forest patch size, distance from each orchard to the nearest forest edge, and distance from each orchard to the nearest cave using 1:133 400 scale photographic imagery from Landsat Thematic Mapper data with a geographic information system (ARC GIS 10.2).

For each forest patch, we selected a pair of orchards (one near to and one far from the forest patch) that were managed without pesticide use. We used pollinator foraging distances to determine the cut-off distances for "near" and "far" orchards. Since previous work indicates that the mean foraging distance of local pollinator species ranges between 2-7 km (1,973 km for a stingless bee; 1.7-6.9 km for *Rousettus* bats (Wahala & Huang 2005; Bonaccorso et al. 2014); 4.4 km for *Eonycteris spelaea* bats, (Acharya et al. 2015), we classified orchards as 'near' if they were < 1 km away from the nearest rain forest patch and as 'far' if they were > 7 km away from rain forest. All pairs of orchards were at least 10 km apart. The distance from each study orchard to the nearest cave (potential roosts for nectarivorous bats and some frugi-nectarivorous bat species, such as *Rousettus* bats) ranged from 0.7 to 29 km (mean distance to caves ± SD: 9.42 ± 7.24 km). Bat roosting caves were identified from (Bumrungsri 1997) and the Shepton Mallet Caving Club (http://www.thailandcaves.shepton.org.uk).

Sampling the plant communities

In each study orchard, we marked a 50 × 150 m plot in which we set up 5 parallel 150-m transects at intervals of 10 m. We surveyed the plant communities from January 2012 to June 2013 by recording every individual of all flowering species in the study orchards every month. We counted the number of floral units (either individual flowers or capitula) for each plant. We determined the mean number of flowers in a capitulum from 20 capitula. We estimated the number of individuals of each plant species in each orchard by multiplying plant density (determined from the marked plot) by the total area of the orchard. Additionally, we calculated the total number of flowers by multiplying the number of individual plants by the mean number of open flowers for each plant.

Sampling the flower-visitors

To identify flower visitors and understand how the network of interaction changes with the proximity to forest, flower visitor observations were conducted monthly from April 2012 to June 2013. This was done in calm weather (i.e. sunny and without rain with the temperature ranging from 31°C to 38°C). In each orchard, we observed flower visitors while walking the five 150 m transects described above. Sampling took place 0800 – 1100 h and 1500 to 1830 h, recording both visitor frequency and visitor richness. We only collected data on insects coming into contact with the reproductive parts of the flower. For each plant species, pollinator observations were focally conducted from the four cardinal directions using 15 min observation sessions. Insects were collected with a long-handled net up to a height of 2 m and transferred to a euthanizing bottle containing ethyl acetate. Insects were identified from field guides or by professional taxonomists (see Acknowledgements). Insects that could not be identified to species were morphotyped (Memmott et al. 1993).

For nocturnal pollinators such as bats and moths, we used video cameras set to record for 15 min every hour from 1900 h to 0500 h. Because it is difficult to identify bats to species from camera traps, we also mist-netted at each site to identify the local species, allowing us to confirm our video identification. The mist nets were placed close to the flowering trees to avoid capturing the bats that visit to other fruit trees in the same orchard. Bats were identified to species following Francis (2008), mainly from external morphology and size. We categorized fruit bats into two groups, nectarivorous (*Eonypteris spelaea* (Dobson), *Macroglossus minimus* (Geoffroy), *M. sobrinus* (Andersen)) and frugi-nectarivorous (*Cynopterus brachyotis* (Muller), *C. horsfieldi* (Gray), *C. sphinx* (Vahl) and *Megaerops ecaudatus* (Temminck)), following criteria in Stewart et al. 2014.

Constructing the flower-visitation networks

The overall pollination network structures across all seasons were visualized using the bipartite package implemented in R (ver. 2.13.0, R. Development Core Team 2011 http://www.R-project.org). For each network, interactions were summarized as a bipartite matrix, with each cell containing the frequency of the pairwise interaction between a plant and animal species. To assess the abundance of each bat group (nectarivorous vs. frugi-nectarivorous), we summed all bat sightings captured by camera traps at each plant species. We then calculated a network strength value for each bat group by summing the dependencies across all plant species with which a bat group interacted. Dependency is calculated as the proportion of interactions performed by each animal species (Bascompte et al. 2006).

Statistical analyses

We used generalized linear mixed models (GLMM) to examine the effect of distance to forest edge (near vs. far), distance to the nearest cave, and bat group (nectarivorous vs. frugi-nectarivorous) on both bat abundance and network strength of each bat group in the pollination networks. We modelled the residuals with a normal distribution. Distance to forest, distance to nearest cave, and bat group were included as explanatory variables. Study orchard pairing was treated as a random factor. To determine the best predictive model, we selected the GLMM with the lowest AIC score.

RESULTS

Overall, we recorded 61 species of plant. The five bat-pollinated plant species were visited by 87 species of insect, 2 species of bird and 7 species of bat. Hymenoptera were common visitors to both orchard types; within this order, 32 species belonged to the family Apidae. Bats contributed to 0.2% to 0.4% of all visits at orchards near the forest and 0.3% to 0.8% at orchards far from the forest. We netted 553 individuals of six fruit bat species (193 *E. spelaea*, 81 *Macroglossus sobrinus*, 52 *Rousettus amplexicaudatus*, 30 *R. leschenaulti*, 29 *Cynopterus horsfieldi*, 126 *C. sphinx* and 46 *C. brachyotis*) during 480 hours of mist-netting. One pair of bipartite matrix, interaction between a plant and animal groups are shown in Fig. 1.

The model that best described bat abundance included negative effect of distance to cave ($P < 0.001$), bat group ($P < 0.001$), and the distance to cave x bat group interaction ($P = 0.024$) (Tab. 1). The overall abundance of nectarivorous bats (mean $\pm$ SD = 238 $\pm$ 81.41 visits) was greater than that of frugi-nectarivorous bats (155.2 $\pm$ 33.58 visits). Additionally, abundance was negatively correlated with distance from the nearest cave for both nectarivorous and frugi-nectarivorous bats. The negative correlation was much more pronounced in nectarivorous than frugi-nectarivorous bats, demonstrating the significant distance to cave x bat group interaction (Fig. 2).

The model that best described network strength included distance to the forest edge ($P = 0.01$), distance to the nearest cave ($P = 0.23$), and bat group ($P = 0.001$) (Tab. 1). The network strength of nectarivorous bats (0.83 $\pm$ 0.54) was greater than that of frugi-nectarivorous bats (0.41 $\pm$ 0.22). Since there is no bat group x forest interaction or bat group × cave interaction, we analyzed all bats together. Network strength of all bats was negatively correlated with both the distance to forest and distance to cave (Tab. 1).

DISCUSSION

Through the use of pollination networks, we demonstrate that nectar and frugi-nectarivorous bats differ in their impact on bat-pollinated plant species within southern Thailand's mixed fruit orchards. Additionally, these two groups are differentially affected by distance to caves and forests. Both bat groups are strongly integrated into pollination networks, playing important roles in the networks where they occur. However, nectar bats were more important than frugi-nectarivorous bats (higher network strength), and plants received more visits from nectar bats than frugi-nectarivorous bats. This pattern may result from foraging strategies; nectar bats only forage on floral resources, while frugi-nectarivorous bats predominantly forage on fruit resources (Stewart et al. 2014). Moreover, nectar-specialist *E. spelaea* has strong fidelity to its foraging area and visits the same area each night (Acharya et al, 2015), whereas frugi-nectarivorous bats visit up to six feeding areas each night (Bumrungsri, 2002). Additionally, species-specific morphological traits may constrain the opportunity for interactions between bats and flowers. Nectar-specialist *Eonycteris* and *Macroglossus* species have elongated rostrums and tongues. In contrast, the other (frugi-nectarivorous) bat species have relatively robust rostrums and short tongues (Bumrungsri et al. 2008, 2013; Francis 2008; Hodgkison et al. 2004; Marshall 1983, 1985). Consequently, tubular flowers (e.g. *Musa*, *Oroxylum*) are more likely to be visited by nectar-specialist bats (Srithongchuay et al. 2008; Fleming et al. 2009; Stewart et al. 2014), while frugi-nectarivorous bats may be more likely to visit flowers with "shaving brush" morphologies (e.g. *Parkia*) because tubular shaped corollas limit frugi-nectarivorous bat access to flowers.

Nectar bats showed higher network strength in orchards closer to the caves. Our study indicates that there is a higher abundance of bats visiting flowers near caves, and this emphasizes the role of caves as sources of pollinators for surrounding trees. We found that the most abundant bat species was the nectar bat *E. spelaea*, which roosts in caves (Bumrungsri et al. 2009), although most other pteropodid bat species roost in foliage (with the exception of cave-roosting Rousettus bats; Campbell et al. 2006, Kunz & Fenton 2003). Previous studies have shown that *E. spelaea* is the main pollinator of *Parkia* (Bumrungsri et al. 2008; Acharya et al. 2015), durian (Bumrungsri et al. 2009), and *Oroxylum indicum* (Srithongchuay et al. 2008). Furthermore, we previously found that bat visitation to durian flowers is significantly negatively correlated with distance to the nearest cave. Additionally, the number of durian fruits set per inflorescence was not significantly affected by distance to forest, but it was influenced by distance to the nearest cave (Sritongchuay et al 2016). In our current study, by examining the entire plant community, we

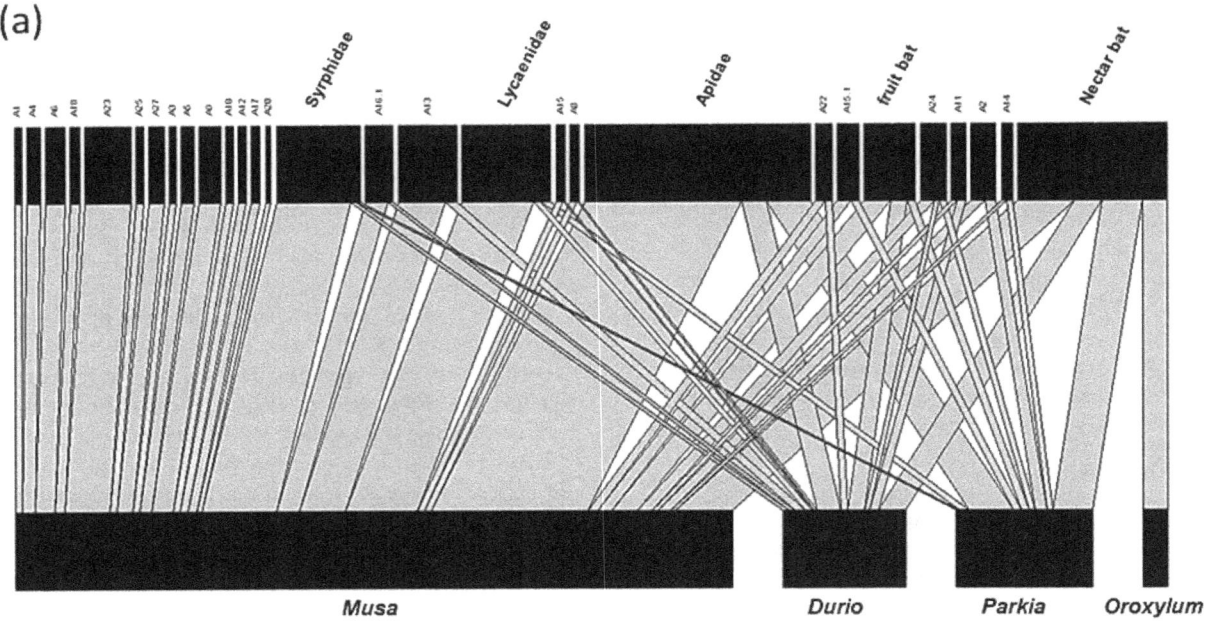

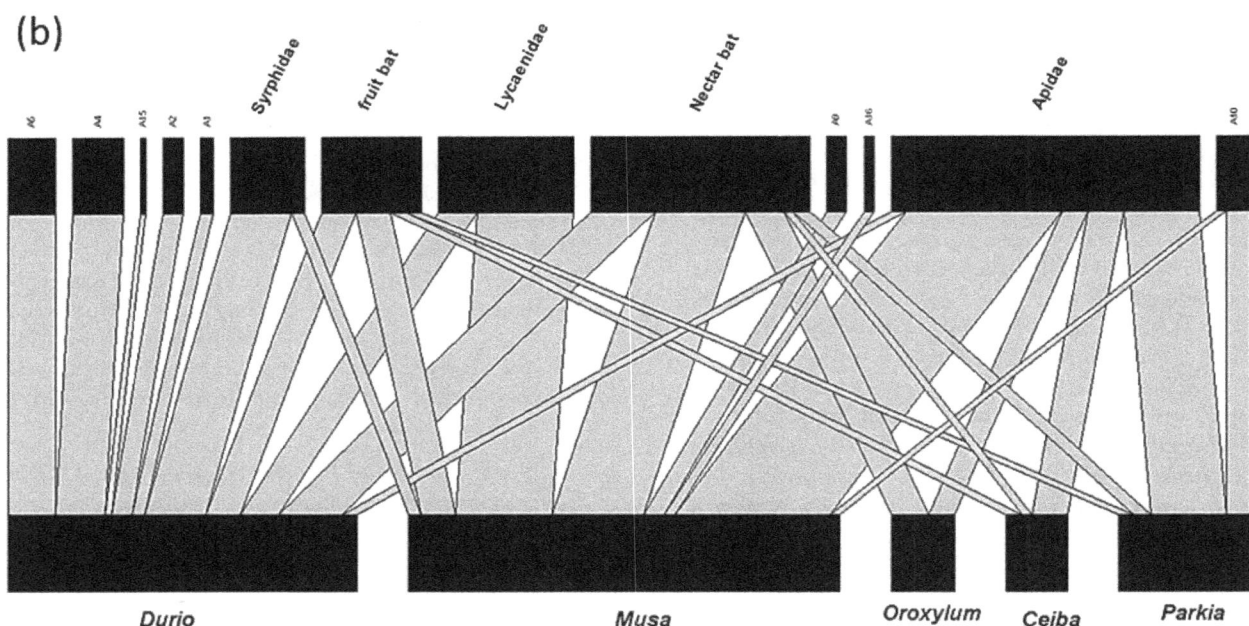

FIGURE I.   Quantitative pollination networks at mixed fruit orchards in southern Thailand (A) near the forest edge and (B) far from the forest edge. For each web, lower bars represent plant abundance and upper bars represent animal visitor abundance. Linkage width indicates frequency of each plant-animal interaction.

have also demonstrated that forests (not just caves), are important sources for bat pollinators. Similarly, previous authors have also found that the pollination success of chiropterophilous plants in the neotropics was affected by forest fragmentation (Stoner et al. 2002; Quesada et al. 2003, 2004).

Our findings emphasize how plant-bat interactions within mixed fruit orchards may undergo severe transformations due to isolation from pollinator sources. Moreover, our study provides solid evidence that increasing the distance to pollinator sources limits the abundance and network strength of pollinators. In quantifying the impact of forest and cave proximity on pollination networks, our results can provide potential conservation recommendations concerning both plants and animals. Conservation practices aiming to preserve plant-pollinator interactions should promote the maintenance of both groups of bats and

TABLE 1.   Results of generalized linear mixed models for (A) bat abundance and (B) network strength of bat. Fixed effects include distance to the forest edge, distance to cave, and bat group.

Explanatory fixed variable	Estimate	SE	t-value	P-value
A) Bat abundance (AIC = 84.21)				
Intercept	322.422	19.543	16.498	<0.001***
Distance to forest edge (Near)	-32.127	15.798	-2.034	0.0523
Distance to cave	-7.185	1.595	-4.505	<0.001***
Bat group	-134.079	26.229	-5.112	<0.001***
Distance to cave * Bat group	5.371	2.240	2.398	0.024*
B) Network strength of bat groups (AIC = 62.24)				
Intercept	1.251	0.146	8.555	<0.001***
Distance to forest edge (Near)	-0.294	0.105	-2.788	0.010*
Distance to cave	-0.029	0.012	-2.416	0.023*
Bat group	-0.676	0.174	-3.884	0.001**
Distance to cave * Bat group	-0.027	0.015	-1.861	0.074

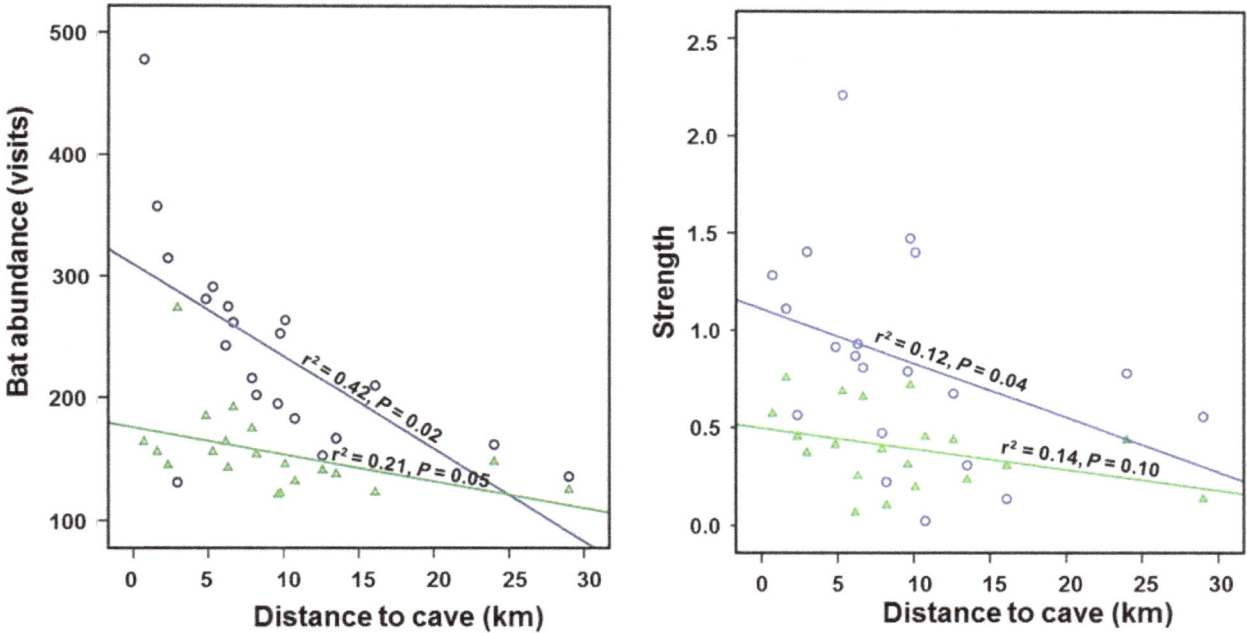

FIGURE 2.   The (A) abundance and (B) network strength of nectarivorous bats (blue circles) and frugi-nectarivorous bats (green triangles) plotted against distance from the nearest cave. Each point represents a single fruit orchard in southern Thailand.

specialist plant species (e.g., *Oroxylum indicum*). Moreover, to maintain pollinators in orchards, we recommend including plant species that flower year-round, such as *Musa*. Disseminating information about the ecological and economic significance of pollination services to farmers can help raise awareness about natural habitats and nectarivorous bats, which can promote local protection of forest and caves. This knowledge will further advance our understanding of how sustainable conservation policies and practices can be adopted.

ACKNOWLEDGEMENTS

Funding for this project was provided by the Thailand Research Fund (TRF) through the Royal Golden Jubilee Ph.D. Program jointly with Prince of Songkla University under Grant No. PHD/0225/2552, and Prince of Songkla University's graduate school. We are extremely grateful to S. Ith, S. Bilasoi, N. Chaichart, and D. Sanamxay for providing invaluable assistance with field sampling, C. Pankeaw for providing identification to bee species, and L. M'Gonigle for assistance with statistical analyses. We thank the members of the Small Mammal & Bird Research Unit, the Kremen's lab at UC Berkeley, and the community Ecology group at the University of Bristol who helped develop ideas. We also thank A. Stewart for comments on earlier versions of this manuscript.

REFERENCES

Acharya PR, Racey PA, Sotthibandhu S, Bumrungsri S (2015) Feeding behaviour of the dawn bat (*Eonycteris spelaea*) promotes cross pollination of economically important plants in Southeast Asia. Journal of Pollination Ecology 15:44–50.

Bascompte J, Jordano P, Olesen JM (2006) Asymmetric coevolutionary networks facilitate biodiversity maintenance. Science 312:431–433.

Bonaccorso FJ, Winkelmann JR, Todd CM, Miles AC (2014) Foraging movements of epauletted fruit bats (Pteropodidae) in relation to the distribution of sycamore figs (Moraceae) in Kruger National Park, South Africa. Acta Chiropterologica 16:41–52.

Brosi BJ, Daily GC, Ehrlich PR (2007) Bee community shifts with landscape context in a tropical countryside. Ecological Application 17:418–430.

Bumrungsri S (1997) Roost selection of cave dwelling bats in Songkla and Satun Provinces. M. Sc. Thesis, Kasetsart University, Bangkok.[in Thai with English abstract]

Bumrungsri, S (2002). The foraging ecology of the short-nosed fruit bat, *Cynopterus brachyotis* (Müller, 1838), in lowland dry evergreen rain forest, southeast Thailand. Doctoral dissertation, University of Aberdeen. Aberdeen.

Bumrungsri S, Harbit A, Benzie C, Carmouche K, Sridith K, Racey P (2008) The pollination ecology of two species of *Parkia* (Mimosaceae) in southern Thailand. Journal of Tropical Ecology 24:467–475.

Bumrungsri S, Sripaoraya E, Chongsiri T, Sridith K, Racey PA (2009) The Pollination ecology of Durian (*Durio zibethinus*, Bombacaceae) in southern thailand. Journal of Tropical Ecology 25:85–92.

Bumrungsri S, Lang D, Harrower C, Sripaoraya E, Kitpipit K, Racey PA (2013) The dawn bat, *Eonycteris spelaea* Dobson (Chiroptera: Pteropodidae) feeds mainly on pollen of economically important food plants in Thailand. Acta Chiropterologica, 15:95-104.

Campbell P, Reid NM, Zubaid A, Adnan AM. Kunz TH (2006) Comparative roosting ecology of *Cynopterus* (Chiroptera: Pteropodidae) fruit bats in peninsular Malaysia. Biotropica 38:725–734.

Francis CM, Barrett P (2008) A field guide to the mammals of Thailand and South-East Asia. Asia Books, London,

Fleming TH, Geiselman C, Kress WJ (2009) The evolution of bat pollination: a phylogenetic perspective. Annual of Botany 104:1017–1043.

Frick WF, Price RD, Heady III PA, Kay KM (2013) Insectivorous bat pollinates columnar cactus more effectively per visit than specialized nectar bat. American Naturalist 181:137–144.

Geslin B, Gauzens B, Thébault E, et al (2013) Plant pollinator networks along a gradient of urbanisation. PLoS ONE 8:e63421.

Greenleaf SS, Kremen C (2006) Wild bee species increase tomato production and respond differently to surrounding land use in Northern California. Biological Conservation 133:81–87.

Hodgkison R, Balding ST, Zubaid A, Kunz,TH (2004) Temporal ariation in the relative abundance of fruit bats (Megachiroptera: Pteropodidae) in relation to the availability of food in a lowland Malaysian rain forest. Biotropica 36:522–533.

Holyoak M, Leibold MA, Mouquet N, Holt RD, Hoopes M (2005) A Framework for Large-Scale Community Ecology. In: Holyoak M, Leibold MA, Holt RD (eds) Metacommunities: spatial dynamics and ecological communities. The University of Chicago Press, Chicago, 1-31.

Kunz TH, Fenton MB (2003) Bat ecology. University of Chicago Press, Chicago.

Itino T, Kato M, Hotta M (1991) Pollination ecology of the two wild bananas, *Musa acuminata* subsp. *halabanensis* and *M. salaccensis*: chiropterophily and ornithophily. Biotropica. 151–158.

Lobo JA, Quesada M, Stoner KE (2005) Effects of Pollination by Bats on the Mating System of *Ceiba pentandra* (Bombacaceae) Populations in Two Tropical Life Zones in Costa Rica. American Journal of Botany 370–376.

Marshall AG (1983) Bats, flowers and fruit: evolutionary relationships in the Old World. Biological Journal of the Linnean Society 20:115–135.

Marshall AG 1985 Old world phytophagous bats (Megachiroptera) and their food plants: A survey. Zoological Journal of the Linnean Society 83:351–369.

Memmott J, Godfray HCJ, LaSalle J (1993) Parasitoid webs. In: Lasalle J, Gauld, ID (ed) Hymenoptera and Biodiversity. CAB International, Wallingford. pp 217-234.

Montoya JM, Pimm SL, Solé RV (2006) Ecological networks and their fragility. Nature 442: 259-264.

Nathan PT, Raghuram H, Elangovan V, Karuppudurai T, Marimuthu G (2005) Bat pollination of kapok tree, *Ceiba pentandra*. Current Science 25:1679-81.

Quesada M, Stoner KE, Rosas-guerrero V, Palacios-Guevara C, Lobo JA (2003) Effects of habitat disruption on the activity of nectarivorous bats (Chiroptera: Phyllostomidae) in a dry tropical forest: implications for the reproductive success of the neotropical tree *Ceiba grandiflora*. Oecologia 135:400–406.

Quesada M, Stoner KE, Lobo JA, Herrerias-Diego Y, Palacios-Guevara C, Munguia-Rosas MA, Salazar KA, Rosas-Guerrero V (2004) Effects of forest fragmentation on pollinator activity and consequences for plant reproductive success and mating patterns in bat-pollinated Bombacaceous trees. Biotropica 36:131–138.

Sekercioglu CH (2006) Increasing awareness of avian ecological function. Trends in Ecology and Evolution 21:464–471.

Srithongchuay T, Bumrungsri S, Sripao-Raya E (2008) The Pollination ecology of the late-successional tree, *Oroxylum indicum* (Bignoniaceae) in Thailand. Journal of Tropical Ecology 24:477–484.

Sritongchuay T, Kremen C, Bumrungsri S (2016) Effects of forest and cave proximity on fruit set of tree crops in tropical orchards in Southern Thailand. Journal of Tropical Ecology 32: 269–279.

Stewart AB, Makowsky R, Dudash MR (2014) Differences in foraging times between two feeding guilds within Old World fruit bats (Pteropodidae) in southern Thailand. Journal of Tropical Ecology 30:249–257.

Stoner KE, Quesada M, Rosas-Guerrero V, Lobo JA (2002) Effects of forest fragmentation on the colima long-nosed bat (*Musonycteris harrisoni*) foraging in tropical dry forest of Jalisco, Mexico. Biotropica 34:462–467.

Wahala S, Huang P (2005) Foraging distance in the stingless bee *Trigona thoracica*. In: Harrison RD (ed) Proceedings of the CTFS-AA International Field, pp 71-74.

Zurbuchen A, Landert L, Klaiber J, Müller A, Hein S, Dorn S (2010) Maximum foraging ranges in solitary bees: only few individuals have the capability to cover long foraging distances. Biological Conservation 143:669–676.

18

Controlled pollinations reveal self-incompatibility and inbreeding depression in the nutritionally important parkland tree, *Parkia biglobosa*, in Burkina Faso

Kristin Marie Lassen[a], Erik Dahl Kjær[a], Moussa Ouédraogo[b], Yoko Luise Dupont[c], Lene Rostgaard Nielsen[a]

[a]*Department of Geosciences and Natural Resource Management, Faculty of Science, University of Copenhagen, Rolighedsvej 23, 1958 Frederiksberg C, Denmark*
[b]*Centre National de Semences Forestières, Route de Kaya, 01 BP 2682 Ouagadougou, Burkina Faso*
[c]*Department of Bioscience, Aarhus University, Vejlsøvej 25, 8600 Silkeborg, Denmark*

Abstract—The socioeconomically important fruit tree *Parkia biglobosa* is becoming less abundant in the West African savannah, possibly due to poor regeneration. This decline can be self-enforcing if lower densities of fertile trees result in increasing self-pollination followed by increased abortion rates or poor regeneration due to inbreeding depression. Hence, we have studied the reproductive success and seedling viability of *P. biglobosa* after controlled self- and cross-pollination based on a full diallel crossing design with eight trees. Controlled cross-pollination tripled the pod set compared to open-pollinated capitula, suggesting that fruiting of *P. biglobosa* trees in the study area is already seriously pollen limited. Self-pollination and specific pairs of trees resulted in very few pods, suggesting a high level of self-incompatibility. Cross-pollination resulted in larger pods with more and heavier seeds than self-pollinated pods. The total amount of sugar in the fruit pulp was correlated with both the number of healthy and total seeds per pod. Growth rate of self-pollinated seedlings was lower than the cross-pollinated ones, suggesting significant inbreeding depression. Because the wild fruit trees play an important role in human nutrition, these results give rise to serious concerns. We recommend that future studies investigate how the level of cross-pollination can be increased and how the regeneration of *P. biglobosa*, whether natural or planted, can be improved.

Keywords: *Controlled pollination; Fruit quality; Inbreeding depression; Pollen limitation; Self-incompatibility*

Introduction

The majority of tropical tree species rely on animals for pollination (Ollerton et al. 2011), and many possess a system of self-incompatibility (Bawa et al. 1985; Ward et al. 2005). The high dependency on animal pollinators in the tropics (Ollerton et al. 2011), combined with pollination deficit due to historical decline of wild pollinators (e.g. Goulson et al. 2015) and limited alternative options for cross-pollination, may decrease fruit production and quality of important crop trees. In addition, anthropogenic impacts, such as farming (monoculture, pesticides), landscape fragmentation, and climate change are likely to reduce abundance and species diversity of pollinators (Kennedy et al. 2013). Understanding mechanisms of pollination and optimising fruit production in tropical fruit trees are crucial, given that many livelihoods depend on crop yield from animal-pollinated tropical trees.

Local decline of pollinators and/or increased distances between conspecific trees can reduce seed set due to pollen limitation, if fruit and seed set are limited by the supply of compatible pollen, rather than resource availability (Ashman et al. 2004; Knight et al. 2005; Aizen & Harder 2007).

Another potential consequence is increased selfing, which may decrease fruit production, and reduce the fitness of seedlings due to inbreeding depression (Husband & Schemske 1996). Hence, we expect a decrease in reproductive success, combined with reduction of seedling fitness when conspecific trees are more widely spaced in a disturbed landscape.

Parkia biglobosa (Jacq.) R. Br. ex G. Don (Fabaceae: Mimosoideae), is a West African parkland tree with high nutritional importance for the rural people due to its sweet fruit pulp and seeds with high protein content (Uwaegbute 1996; Hall et al. 1997). In Burkina Faso, pods from *P. biglobosa* are an important food source (Lykke et al. 2002), especially during periods of food scarcity (Nyadanu et al. 2017). Due to over-exploitation (Gaisberger et al. 2017), low regeneration (Ræbild et al. 2012), and reduced annual rainfalls (Maranz 2009; Funk et al. 2012), there is a high risk of seriously decreasing tree densities (Gaisberger et al. 2017) as already witnessed by the local population (Lykke et al. 2002). This may increase selfing (Lassen et al. 2017), but the potential effects of increased self-pollination on seedling growth and nutritional content of the fruits of *P. biglobosa* are unknown. Related species possess a self-incompatibility system (Hinata et al. 1993), and hence increased selfing is expected to reduce seed set. Studies of other plant species have documented positive correlations between seed number and fruit traits including fruit weight, size, and oil content (Hopping 1976; Roldán Serrano & Guerra-Sanz 2006; Abrol

*Corresponding author: kristin_lassen@yahoo.dk

2012). Thus, increased self-pollination, by causing reduced seed number, may also reduce the nutritional value of *P. biglobosa* fruits.

To understand productivity of this important crop tree species, it is important to determine if *P. biglobosa* is pollen limited, and how increased levels of self-pollination will affect fruit and seed set, nutrition contents and seedling fitness of this species. We addressed these questions by performing controlled self- and cross-pollinations of *P. biglobosa* capitula in order to compare 1) the reproductive success of selfed, crossed, and open-pollinated capitula, 2) the carbon and nitrogen contents of seeds and sugar content of fruit pulp from selfed and outcrossed pods, and 3) the germination percentage and seedling vigour of selfed and outcrossed seeds. We discuss the probability of *P. biglobosa* experiencing pollen limitation, self-incompatibility and inbreeding depression.

MATERIALS AND METHODS

Plant species

Parkia biglobosa is pollinated mainly by bees and bats (Baker & Harris 1957; Hopkins 1983; Ouédraogo 1995; Lassen et al. 2012; Lassen et al. 2017). It is predominantly outcrossing (Ouédraogo 1995; Sina 2006), although Lassen et al. (2017) found that selfing can occur in areas with low tree density. *Parkia biglobosa* flowers for around four weeks in the dry season in Burkina Faso from January to April, depending on latitude, and with year to year variation (pers. obs.).

Flowers are grouped in ball-shaped capitula with around 2,200 tiny bright red flowers packed closely on a bulbous receptacle hanging on a long peduncle (Hopkins 1983). Nectar is produced by sterile flowers close to the peduncle and accumulates in a nectar ring (Hopkins 1983). At the study site, buds opened during the afternoon. The capitula started producing nectar around 18:45 h (local time, UTC + 0 h) and shedding pollen around 19:30 h (pers. obs.). Each tiny flower has ten anthers and one style. *Parkia biglobosa* is andromonoecious, and the functionally male capitula have styles, which fail to elongate (Hopkins 1981). Pollen is shed in polyads with 32 pollen grains (pers. obs.) clumped together. The cup-shaped stigma can hold only one polyad; therefore all seeds per pod are full siblings (Lassen et al. 2014). According to our observations, based on $N = 15$ ovaries distributed on three trees in Burkina Faso, the ovary of a hermaphroditic capitulum contained a mean of 23 ovules (SE = 0.63, min-max: 16-29). The hermaphroditic capitula are protandrous (Ouédraogo 1995) and the female phase begins around 23:00-24:00 h when the stigmas have extended to reach the same level or above that of the anthers (pers. obs.). Furthermore, flowering within a capitulum is highly synchronized. Each capitulum blooms one night and during the morning both hermaphroditic and functionally male capitula start to wilt (pers. obs.). In the present study, each capitulum is treated as one unit.

One week after pollination and fertilisation, tiny green pods are visible, and after around two months the indehiscent, brown pods are mature (pers. obs.). Even though each hermaphroditic flower has the potential of producing a pod, only a few pods per capitulum develop (Hopkins 1984).

Study site

The present study took place in the village Pinyiri (syn. Kacheli) (11°14'34.89"N, 1° 8'1.73"W), eight km north of Pô, Nahouri province. The site is within the Sudanian climatic zone with a unimodal rainy season and an average precipitation (1981-2010) of 900-1,000 mm (Sanfo 2012). In 2011, preceding the fruiting season of *P. biglobosa* in 2012, the annual precipitation in Pô was 927 mm (Météo 2015).

Controlled pollination experiment

Prior to the experiment, swollen buds (expected to open the following night) were covered with cheesecloth with an inner band of chicken wire to keep the nettings from touching the flowers. We used inflorescences from the lowest part of the tree crown, which could be reached from a ladder. In the experiment, only the most apical bud within the compound inflorescence was used. During 8-18 March 2012, the crossing experiment was carried out as a diallelic cross (i.e. all trees were crossed with each other) of eight trees (mean DBH = 1.4 m, SD = 0.81), although one of the trees was not used as a pollen donor. Each treatment was replicated nine times per tree.

Treatments:

- Open = open-pollinated capitula (control, $N = 72$)

- Self = self-pollinated capitula ($N = 72$)

- Cross = cross-pollinated capitula ($N = 423$)

- 'Both' = half of a capitulum was pollinated with self-pollen and the other half with cross-pollen (only on six trees due to lack of capitula, $N = 54$)

Prior to the pollination treatments, we checked the sex (hermaphroditic or functionally male) of the flowers on each receiving capitulum, since only hermaphroditic capitula can develop pods. When in doubt, we measured the distance between stigmas and anthers, as this difference was suspected to influence the functional sex of the capitulum, due to the probability of the many densely packed anthers acting as a carpet keeping pollen away from the shorter stigmas. The dividing line between hermaphroditic and functionally male capitula was unknown at the time we carried out the controlled pollinations. Hence, these were carried out regardless of the assigned sex, from midnight until early morning (03-04 h) using capitula directly as pollen brushes by dabbing the donor capitula on the receiving capitula, which were previously protected by bags. No flowers were emasculated due to the high number of anthers per capitulum. To standardise the dose of pollen, one donor capitulum was used on three receiving capitula (1:3) for the treatments of self- and cross-pollinations (approximately one third of the donor on each of the receiving capitula). To study the effect of self- and cross-pollination under identical conditions, we applied self and cross-pollen to flowers on the same capitulum (but not the same flowers). We refer to this treatment as 'both', and for this treatment we used two donor capitula (self and cross) for two receiving capitula (half self and half cross on each capitulum) (1:1). Half of the peduncle was marked with a black marker (for the cross-pollination) and we performed the two kinds of pollination by eye. After pollination, the capitula were re-bagged. The trees were visited daily for five weeks; dropped capitula were collected and examined for sex (hermaphroditic or functionally male) and signs of developing pods and/or predation.

After harvest, pods were counted and weighed. Their lengths were measured as the mean of each side of the pod, excluding the pedicel. We did not count the seeds of these pods (see below).

DNA extraction and genotyping

To assess the level of self-pollination (on purely self-pollinated capitula and on 'both' capitula), we extracted DNA, genotyped seeds ($N = 142$) and compared their genotypes with the genotype of the mother trees. Likewise, genotypes of the eight experimental trees were compared to genotypes of a sub-sample of pods, to test if the cross had been performed correctly. DNA was extracted directly from dehulled seed; endosperm is absent because the cotyledons provide resources to the embryo (Hopkins 1983). DNA extraction used the DNeasy 96 Plant Kit (QIAGEN, Hombrechtikon, Switzerland), following the manufacturer's protocol. The genotyping was based on ten microsatellite primer pairs developed for *P. biglobosa* (Lassen et al. 2014) with PCR reactions and fragment analysis following Lassen et al. (2017).

Comparison of self- and cross-pollinated pods

The self-pollinated capitula in the controlled pollination experiment yielded extremely few pods. Thus, in order to compare the quality of seeds in *P. biglobosa* after self- and cross-pollination we assessed the quality from 48 pods obtained from a previous experiment from the same study site (Lassen et al. 2017). These pods were collected from eleven trees and were randomly selected, although the selection was balanced with an equal number of pods per tree being self- and cross-pollinated. Seeds were categorised as 'healthy', 'eaten' (hole in the husk and/or seed remains due to predation by parrots and/or worms), 'aborted' (weight < 0.05 g and/or with a flat shape), 'missing' (empty cavity in the pulp), and from these four numbers a total number of seeds was established.

Moisture, carbon and nitrogen content of seeds

One seed from each of 48 pods (24 selfed and 24 outcrossed) was analysed for moisture, carbon and nitrogen content. The seeds were weighed with and without testa, placed in open 2 ml Eppendorf tubes and dried at 65°C for three days until stable weight. Moisture content of seeds without testa was calculated as fresh weight minus dry weight, divided by fresh weight and multiplied by 100 (%). Then the dehulled seeds were ground in a mortar and weighed into tin capsules with 5 mg in each sample. Total carbon (C) and nitrogen (N) content were measured using the Dumas principle: samples were combusted at 1,850°C on a FLASH 2000 NC Analyzer (Thermo Scientific) according to the manufacturer's manual. To calibrate the measurements, we used a standard ('spruce needles' Forest Foliar Coordinating Centre, FFCC), and two reference samples ('maple leaves' FFCC, run twice, and 'corn gluten organic' Sercon, run for every nine samples of *P. biglobosa*). The FLASH 2000 Analyzer is comparable with the Kjeldahl method (Krotz & Giazzi 2014).

We have converted the amount of nitrogen (N) to crude protein by multiplying N with the commonly used factor 6.25 (AOAC 1990), although Ezeagu et al. (2002) found a lower nitrogen-to-protein conversion factor on 4.97 for Fabaceae seeds (mean for ten species) and Yeoh & Wee (1994) found an even lower conversion factor on 4.23 for leaves of *Parkia timoriana* (DC.) Merr. (syn. *P. javanica*).

Pod pulp analysis

We analysed the sugar content in the pulp of the same pods as above, except for two self-pollinated pods, which had no pulp (i.e. 22 self- and 24 cross-pollinated pods were used in this analysis). We kept the pulp from each pod separately. The pulp was dried (103°C for 3 h), ground and sifted, and two sub-samples of 100 mg per pod were used. The soluble sugars were extracted and analysed by HPLC according to the method described by Liu et al. (2004).

Germination rate and seedling vigour of selfed and crossed seeds

In order to compare seed viability under optimal conditions, we tested the germination by using healthy-looking seeds from the 24 self-pollinated pods (90 seeds) and 24 cross-pollinated pods (163 seeds) and assessed the seedling vigour by growth and dry weight. Each seed was weighed and scarified because of the hard seed coat (testa), which must be broken before the seeds can germinate (Etejere et al. 1982). The seeds were germinated in plastic boxes in a growth chamber in a 25°C day and night, and 12 h light/darkness regime. We defined the seeds as germinated when the radicle protruded for 3 mm, and we monitored the seed germination daily.

On the 19th day after sowing, seedlings were weighed and planted individually in pots (Ø = 13 cm) with planting peat soil. The pots were placed randomly in a greenhouse at 28°C day and 20°C night and with a 12 h light/darkness regime from 08:00 h (local time, UTC + 1 h). The seedlings were watered daily with demineralised water without fertiliser for the first 2 months, and thereafter with fertiliser. Plant height (from soil level to top of main stem, or to the highest stem in case of more stems), stem diameter (measured with an electronic caliper) and number of pinnae (i.e. primary division of a bipinnate compound leaf) were measured on five occasions (43, 76, 104, 144, and 222 days after sowing). At the last measuring, we included fresh weight of the seedlings before drying them at 80°C (Osonubi & Fasehun 1987). After 24 hours (until stable weight), we recorded dry weight of the entire plant, shoots (stem plus leaves) only, and roots only, in order to calculate the shoot:root ratio.

Data analysis

In the controlled pollination experiment, reproductive success was evaluated as 1) numbers of immature pods (reflecting the success of pollination) and mature pods (additionally reflecting available resources by the mother-tree) per hermaphroditic capitulum and 2) weight and length of these pods. Analysis of variance was performed based on average values per treatment and tree applying the statistical SAS software v.9.4 (SAS Institute 2011). The mean number of pods (immature and mature) per hermaphroditic capitulum included capitula without any pods. When testing differences between treatments for the controlled pollination experiment, we used the general linear model as implemented in the GLM procedure:

$$1. Y_{gh} = \text{Treatment}_g + \text{Mother-tree}_h + \mathcal{E}_{gh}$$

where Y_{gh} is the response variable, treatment g = (open, self, cross, 'both'), and mother-tree h = 1...8. Treatment$_g$ was considered a fixed effect, whereas Mother-tree$_h$ was considered a random effect with residual $\mathcal{E}_{gh}$ assumed independent and $N(0,\sigma_{\varepsilon}^2)$. We assessed and accepted the model assumptions by visual inspection of the residuals.

For testing differences among pairs of pollen donors (i.e. male parent) and mother-trees (i.e. female parent), we analysed number of pods per single capitulum (not averaged per tree), using the general linear model as implemented in the GLM procedure:

$$2. Y_{ij} = \text{Pollen donor}_i + \text{Mother-tree}_j + \text{Pollen donor}_i\text{*Mother-tree}_j + \mathcal{E}_{ij}$$

where Y_{ij} is the response variable (log transformed), pollen donor i = (T10, T14, T22, T76, T90, T92, and T93), mother-tree j = 1...8, and pollen donor$_i$*mother-tree$_j$ = the interaction between the pollen donor and the mother-tree. All effects were considered fixed with residual $\mathcal{E}_{ij}$ assumed independent and $N(0,\sigma_{\varepsilon}^2)$. We assessed and accepted the model assumptions by visual inspection of the residuals.

When testing differences between self- and cross-pollination for several parameters related to plant fitness, the results were averaged per type of pollination and tree, and we used a similar general linear model as above:

$$3. Y_{kl} = \text{Pollination type}_k + \text{Mother-tree}_l + \mathcal{E}_{kl}$$

where Y_{kl} is the response variable, pollination type k = (self, cross), and mother-tree l = 1...11. Pollination type$_k$ was considered a fixed effect, whereas Mother-tree$_l$ was considered a random effect with residual $\mathcal{E}_{kl}$ assumed independent and $N(0,\sigma_{\varepsilon}^2)$. Again, we assessed and accepted the model assumptions by visual inspection of the residuals.

When testing the difference between carbon and nitrogen contents of selfed versus outcrossed seeds, we used the following general linear model:

$$4. Y_{klm} = \text{Pollination type}_k + \text{Mother-tree}_l + \text{Seed weight}_m + \mathcal{E}_{klm}$$

where Y_{klm} is the response variable, pollination type k = (self, cross), mother-tree l = 1...11, and seed weight included as covariate. Seed weight varied from 0.0821 – 0.2733 g per seed. Pollination type$_k$ was considered a fixed effect, whereas Mother-tree$_l$ was considered a random effect with residual $\mathcal{E}_{klm}$ assumed independent and $N(0,\sigma_{\varepsilon}^2)$. Again, we assessed and

accepted the model assumptions by visual inspection of the residuals.

We used Fisher's exact test (as implemented in SAS procedure FREQ) to test differences between self- and cross-pollination for number of germination seeds and surviving seedlings.

The relationship between number of seeds per pod and sugar content in fruit pulp was analysed by calculation of Pearson correlation coefficients (using SAS procedure CORR).

RESULTS

Assessment of the diallel crossing experiment

In the controlled pollination experiment, 451 capitula were hermaphroditic, 139 capitula were functionally male, and 45 capitula were 'mixed' (i.e. containing both hermaphroditic and functionally male flowers in different ratios). Of capitula with hermaphroditic flowers, 83% developed pods.

The measurement of distances between stigmas and anthers of the capitula (N = 245 capitula) coupled with the reproductive success (setting fruit or not) revealed that capitula having anthers protruding > 5 mm longer than stigmas were typically functionally male (91%). Hence, this measure could be used as a rule of thumb.

Of the four pollination treatments (open, self, cross, and 'both'), self-pollination led to a significantly lower proportion of capitula with at least one immature pod (100%, 19%, 96%, 97%, respectively). In addition, the mature pod set also differed, and the experiment yielded 2,643 pods for the purely cross-pollinated capitula (423 pollinated capitula) and only 2 pods for the purely self-pollinated capitula (72 pollinated capitula). Genotyping of a subset of pods confirmed that cross-pollinated pods indeed were results of cross-pollinations while a few matured 'self-pollinated' pods, turned out to be cross-pollinated, probably due to small amounts of 'carryover' pollen. For treatment 'both', in which 502 pods were matured, no pods developed in the self-pollinated halves of the capitula and the subset of genotyped pods showed no self-pollination (N = 133).

Effects of pairs of trees and pollen doses on fruit set

In the controlled cross-pollination treatments, we found highly significant effects of both mother-trees (female parent, $F_{(7,231)}$ = 9.1, P < 0.001) and pollen donors (male parent, $F_{(6,231)}$ = 4.7, P < 0.001) on the number of immature pods per hermaphroditic capitulum. However, the number of mature pods was only significantly affected by the mother-trees ($F_{(7,199)}$ = 13.5, P < 0.001). Two trees (P14 and P22) were equally good as mother-trees and pollen donors, whereas three trees had highest reproductive success as mother-trees (i.e. producing many pods, P76, P92, and P93) and two trees were best as pollen donors (i.e. fathering many pods, T10 and T90) (Tab. 1 and 2).

Interactions between mother-trees and pollen donors were highly significant for both immature ($F_{(34,231)}$ = 3.4, P < 0.001) and mature pods ($F_{(34,199)}$ = 3.4, P < 0.001), i.e. fruit set depended on the combination of mother-trees and pollen

TABLE 1. Mean number ($\pm$ SE) of immature pods per hermaphroditic capitulum (i.e. small pods before maturation) of *Parkia biglobosa* (incl. capitula without pods). Self-pollination (grey colour) is shown in the last row. Tree P33 did not give pollen to the other trees.

	Treatment	P10	P14	P22	P33	P76	P90	P92	P93	Across trees
	Open[a]	4.1 (4.80)	5.6 (4.52)	1.3 (4.52)	2.8 (4.52)	7.1 (4.52)	2.9 (4.52)	2.8 (4.52)	4.3 (4.52)	3.9 (0.41)
Male parent (i.e. pollen donor)	T10	-	42.5 (5.54)	41.7 (7.83)	5.0 (7.83)	50.0 (4.52)	10.3 (7.83)	23.9 (5.13)	61.6 (4.80)	39.4 (4.74)
	T14	15.5 (4.80)	-	15.8 (4.52)	9.6 (4.80)	31.1 (4.52)	1.0 (6.07)	23.5 (5.54)	22.6 (5.13)	17.8 (2.08)
	T22	26.8 (5.54)	17.3 (5.13)	-	18.8 (6.07)	26.1 (4.52)	7.4 (5.13)	9.5 (5.54)	26.5 (4.80)	19.4 (2.23)
	T33	na	na	na	-	na	Na	na	na	na
	T76	10.7 (5.13)	15.3 (5.54)	15.4 (4.80)	na[d]	-	9.5 (5.54)	1.0 (7.83)	27.5 (5.54)	14.3 (1.70)
	T90	12.4 (6.07)	1.1 (5.13)	22.0 (6.07)	12.0 (4.80)	39.0 (4.52)	-	42.5 (6.78)	20.3 (7.83)	20.9 (3.11)
	T92	18.2 (5.54)	23.7 (5.54)	14.2 (6.07)	18.0 (13.57)	11.6 (5.13)	8.5 (5.54)	-	25.8 (6.78)	16.4 (1.97)
	T93	26.5 (5.54)	25.8 (4.52)	9.0 (6.07)	7.7 (7.83)	23.1 (4.52)	14.6 (5.13)	45.5 (9.59)	-	21.0 (2.48)
	Cross[b]	18.2 (2.27)	20.7 (2.46)	17.6 (2.01)	11.5 (1.64)	30.9 (3.56)	8.8 (0.93)	22.5 (3.78)	33.1 (3.77)	21.2 (1.12)
	Self[c]	0.0 (0.00)	0.0 (0.00)	0.4 (0.30)	0.3 (0.25)	0.4 (0.18)	0.1 (0.12)	0.5 (0.50)	0.0 (0.00)	0.2 (0.07)

[a]Open is open-pollination (control). [b]Cross denotes the mean of cross-pollination (i.e. across T10-T93 except self-pollination). [c]Self signifies self-pollination. [d]All nine P33-capitula pollinated with T76 were functionally male.

donors (Tab. 1 and 2). For instance, using P10 as a pollen donor resulted in more than twice as many immature pods per hermaphroditic capitulum on P93 compared to P92 (61.6 versus 23.9), while for P90 the result was vice versa (20.3 versus 42.5, Tab. 1).

Effects of pollen doses on fruit set (excluding self-pollination) and results of the open-treatment are shown in Tab. 3. The number of capitula with at least one immature pod was not influenced by the different pollen doses while the number of pods per hermaphroditic capitulum was significantly higher in hand-pollinated capitula (cross and 'both') compared to the open-pollinated capitula. The hand-pollinated capitula differ in that the 'both' treatment had around 3 times more cross-pollen than the cross treatment, but only on half of the capitulum. Hence the figures can be made comparable by multiplying those for the 'both' treatment with $\frac{2}{3}$: (30.8 x $\frac{2}{3}$ =) 20.5 for immature pods and (13.5 x $\frac{2}{3}$ =) 9.0 for mature pods, which are close to the actual figures for the cross treatment on 20.4 and 9.8 for immature and mature pods, respectively (Tab. 3). However, with an increasing number of pods per capitulum the abortion rate (immature minus mature pods) also increased. For pod weight and length, the differences between pollen doses were non-significant (Tab. 3).

In one pair of trees (P14 and P90), we found that cross-pollination resulted in a similar low pod set as for self-pollination, independently of which tree was mother-tree and which was pollen donor, suggesting that these two trees were not compatible. Finally, we found another pair of trees (P76 and P92) with very few pods when P92 was the mother-tree, but only a reduced pod set when P76 was the mother-tree (Fig. 1, Tab. 1 and 2), which suggest an incompatibility system.

Comparison of self- and cross-pollinated pods

Because the diallelic crossing experiment resulted in only two matured self-pollinated pods, we used 24 self-pollinated and 24 cross-pollinated pods originating from another experiment at the same study site, as stated above. Self-pollinated pods were significantly shorter and weighed less than cross-pollinated pods (Tab. 4). Furthermore, although not significant, the self-pollinated pods had half as many healthy seeds, significantly more aborted seeds and significantly fewer total seeds compared to the cross-pollinated pods (Tab. 4). The amounts of eaten seeds and missing seeds were low, and showed no difference between selfed and outcrossed pods (Tab. 4).

Table 2. Mean number (± SE) of mature pods per hermaphroditic capitulum (i.e. harvested pods) of *Parkia biglobosa* (incl. capitula without pods). Self-pollination (grey colour) is shown in the last row. Tree P33 was not used as a pollen donor.

	Treatment	Female parent (i.e. mother-tree)								
		P10	P14	P22	P33	P76	P90	P92	P93	Across trees
	Open[a]	4.1 (1.93)	4.7 (1.93)	1.3 (1.93)	2.7 (1.93)	6.8 (1.93)	2.9 (1.93)	2.4 (1.93)	3.7 (1.93)	3.6 (0.39)
Male parent (i.e. pollen donor)	T10	-	7.3 (2.36)	16.0 (5.79)	2.0 (4.09)	13.4 (3.34)	2.3 (2.19)	15.7 (2.05)	23.8	13.7 (1.84)
	T14	8.3 (2.05)	-	12.0 (4.09)	4.8 (2.36)	15.6 (2.05)	0.8 (2.59)	17.5 (2.36)	10.2 (2.36)	10.1 (1.33)
	T22	9.2 (2.36)	5.2 (2.59)	-	7.0 (2.59)	7.3 (2.19)	4.3 (2.19)	6.8 (2.59)	17.1 (2.19)	8.4 (1.04)
	T33	na	na	na	-	na	na	na	na	na
	T76	7.9 (2.05)	9.0 (2.36)	8.0 (2.36)	na[d]	-	4.8 (2.36)	1.0 (3.34)	18.8 (2.36)	8.9 (1.09)
	T90	8.0 (2.36)	0.9 (2.19)	7.0 (2.59)	5.3 (2.36)	20.0 (2.05)	-	25.3 (3.34)	6.7 (3.34)	9.9 (1.54)
	T92	13.5 (2.36)	11.8 (2.36)	11.0 (2.59)	1.0 (5.79)	8.0 (2.19)	3.4 (2.59)	-	15.5 (2.89)	10.1 (1.20)
	T93	13.3 (2.36)	11.0 (1.93)	4.2 (2.59)	1.7 (3.34)	9.3 (1.93)	6.9 (2.19)	18.5 (4.09)	-	9.1 (1.00)
	Cross[b]	9.8 (0.85)	7.7 (0.86)	8.3 (1.27)	4.6 (0.76)	12.4 (1.26)	4.1 (0.60)	14.0 (2.22)	16.6 (1.65)	10.0 (0.50)
	Self[c]	0.0 (0.00)	0.0 (0.00)	0.3 (0.19)	0.0 (0.00)	0.3 (0.16)	0.0 (0.00)	0.0 (0.00)	0.0 (0.00)	0.1 (0.05)

[a]Open is open-pollination (control). [b]Cross denotes the mean of cross-pollination (i.e. across T10-T93 except self-pollination). [c]Self signifies self-pollination. [d]All nine P33-capitula pollinated with T76 were functionally male.

TABLE 3. Effect of open-pollination (unknown pollen doses) and two known pollen doses on the percentage of hermaphroditic capitula with at least one immature pod, number of immature and mature pods per hermaphroditic capitulum (incl. capitula without pods), and the pod weight and length of *Parkia biglobosa*, including *F*-tests and significance levels. Except in the open-pollinated treatment, capitula were bagged until hand-pollination and re-bagged following pollination treatment.

Type of pollination	Pollen doses	No. of capitula[a], N	Capitula with ≥ 1 immature pod[b], %	No. of immature pods/capitulum[b]	No. of mature pods/capitulum[b]	No. of pods[c], N	Pod weight[c], g	Pod length[c], cm
Open[d]	unknown	71	100.0 (1.15)	3.9 (2.87)	3.6 (1.02)	258	14.0 (0.62)	22.6 (0.74)
Cross[e]	⅓	292	95.7 (1.15)	20.4 (2.87)	9.8 (1.02)	2,643	11.9 (0.62)	20.7 (0.74)
'Both'[f]	1	40	98.0 (1.41)	30.8 (3.52)	13.5 (1.25)	502	12.8 (0.77)	21.2 (0.90)
$F(P)$ Pollination			$F_{(2,12)}=3.5$ (ns)	$F_{(2,12)}=18.8$ (***)	$F_{(2,12)}=20.7$ (***)		$F_{(2,12)}=2.9$ (ns)	$F_{(2,12)}=1.8$ (ns)
$F(P)$ Tree			$F_{(7,12)}=0.7$ (ns)	$F_{(7,12)}=2.3$ (ns)	$F_{(7,12)}=3.0$ (*)		$F_{(7,12)}=5.6$ (**)	$F_{(7,12)}=5.2$ (**)

[a]Number of capitula used per type of pollination. [b]Values are least squares (LS) means with the standard error (SE) in brackets of the LS estimate. Significance level: ***=$P<0.001$, **= $P<0.01$, *= $P<0.05$, and ns= $P>0.05$. [c]Number of pods harvested. [d]Open was open-pollinated capitula, which were neither bagged nor hand-pollinated (control). [e]Cross was cross-pollinated capitula with one capitulum giving pollen to three capitula (1:3 ratio). [f]'Both' was capitula, which were self-pollinated on one half and cross-pollinated on the other half, with one capitulum giving pollen to two half capitula (1:1 ratio), but only cross-pollinated pods developed.

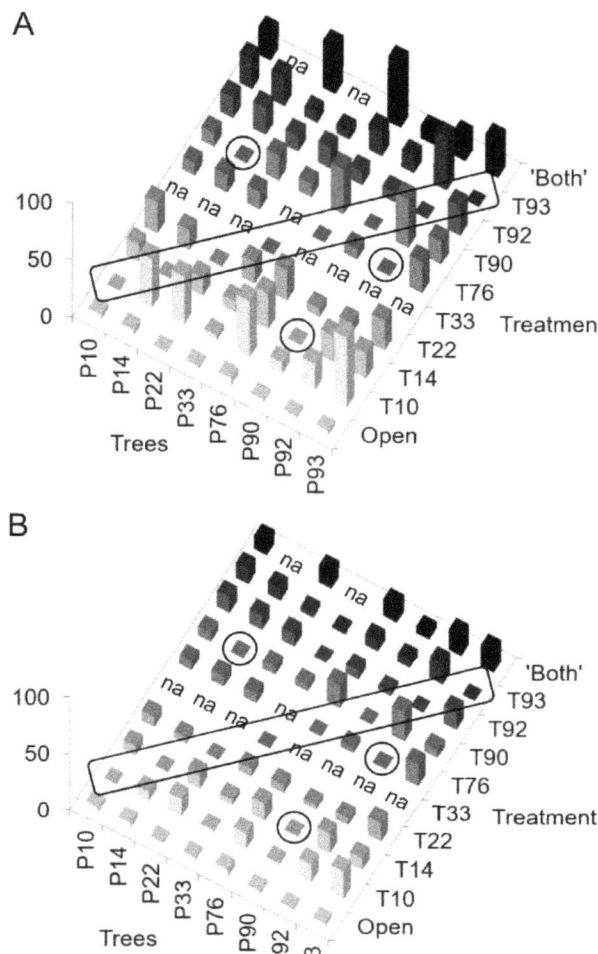

FIGURE 1. Overview of the number of immature (A) and mature (B) pods per hermaphroditic capitula for *Parkia biglobosa*, distributed on treatments (i.e. pollen donors) and trees. The box encompasses the events of self-pollination, which all resulted in very few pods, and the circles indicate low numbers of pods resulting from cross-pollinations. 'Open' indicate open-pollination (control) and 'Both' is a treatment in which capitula were pollinated with self-pollen on one half and cross-pollen on the other half. However, 'Both' yielded exclusively outcrossed pods. Pollen donor T33 was only applied to tree P33, pollen donor T76 was applied to only functionally male capitula on tree P33 and neither tree P14 or tree P33 had not enough capitula for the 'Both'-treatment.

Carbon & protein contents in seeds, and sugar contents in fruit pulp

Raw, dehulled seeds from cross-pollinated seeds had slightly more carbon compared to self-pollinated seeds (54.3% versus 53.4%), while percentage of protein (dry weight) was slightly lower (43.3% versus 45.2%, Tab. 4). The carbon content was highly dependent on seed weight ($F_{1,35}$ = 12.4, $P < 0.001$) with larger seeds containing more carbon. This was not the case for amount of protein ($F_{1,35}$ = 1.0, P = 0.3), hence larger seeds contained more carbon whereas protein content was constant, regardless of seed size.

Sugar content was higher in fruit pulp for the cross-pollinated pods, but only significantly so for glucose and fructose (Tab. 4). Numbers of healthy seeds and amounts of glucose, fructose and total sugars, respectively, were positively and significantly correlated (Tab. 5). No correlation was found between the amount of sucrose and number of seeds per pod (Tab. 5).

Germination and seedling growth of self- and cross-pollinated seeds

Seed weight and seedling growth was highly variable in both types of pollination (self and cross). The mean weight of cross-pollinated seeds was significantly higher than self-pollinated seeds, while germination percentage and germination speed did not differ significantly between pollination types (Tab. 4). Nine selfed and nine cross-pollinated seeds germinated but died before the first measurement of seedlings, and most seedlings, which died during the trial, perished before the second measurement. From the initial 90 self-pollinated and 163 cross-pollinated seeds, significantly fewer self-pollinated seedlings (69) than cross-pollinated (147) seedlings survived until the trial was terminated (Tab. 4). Initial fresh weights (19 days after sowing) and fresh and dry weights at harvest were significantly higher for the cross-pollinated seedlings compared to the self-pollinated ones. The shoot:root ratio was independent of the type of pollination (Tab. 4).

The growth of the seedlings is shown in Fig. 2 (A, B, and C) and the final height, stem diameter and number of pinnae 222 days after sowing in Tab. 4. Means of height, diameter, and numbers of pinnae were always lower for self-pollinated seedlings compared to cross-pollinated ones, and these differences in growth increased with time (Fig. 2). Seven months after sowing (222 days), the effect of type of pollination was significant for plant height and stem diameter, but not for number of pinnae ($F_{1,7}$ = 4.3, P = 0.08) (Tab. 4).

DISCUSSION

Parkia biglobosa is known to be mainly outcrossing (Ouédraogo 1995; Sina 2006; Lassen et al. 2017), and this study showed that self-pollination reduced the number of pods produced, pod size, number of seeds, sugar content in pulp, seed weight, and weight of seedlings (Tab. 1, 2, and 5). In addition, the diallel cross revealed that some combinations of mother-trees and pollen donors were more productive than others. The findings of variation in the success of male and female reproductive organs, i.e., an individual plant being good at either setting pods or at fathering pods on other conspecifics, has also been reported in other plant species such as the self-incompatible Trumpet creeper *Campsis radicans* (L.) Seem. (Bignoniaceae) (Bertin 1982) and the self-incompatible Crested dogstail grass *Cynosurus cristatus* L. (Poaceae) (Ennos & Dodson 1987).

Pollen limitation in Parkia biglobosa

Controlled pollinations with different doses of cross-pollen conducted in the present study yielded significantly more immature and mature fruits than in open-pollinated

TABLE 4. Influence of self- versus cross-pollination on the fitness of various parameters of pods and seeds of *Parkia biglobosa* including *F*-tests, Fisher's exact tests, and their significance levels.

	Type of pollination					Inbreeding depression[b], %
	Self, *N*	Cross, *N*	Self[a]	Cross[a]	*F*(*P*)Pollination	
Pod length, cm	24	24	15.8 (1.16)	20.0 (1.16)	$F_{(1,10)}$ = 6.4 (*)	21.0
Pod weight, g	24	24	7.9 (1.03)	12.5 (1.03)	$F_{(1,10)}$ = 10.1 (**)	36.8
Husk weight, g	24	24	4.0 (0.41)	5.5 (0.41)	$F_{(1,10)}$ = 7.5 (*)	27.3
Pulp[c] weight, g	22	24	2.5 (0.40)	4.2 (0.37)	$F_{(1,9)}$ = 8.7 (*)	40.5
Seed weight, g	24	24	1.6 (0.29)	2.6 (0.26)	$F_{(1,9)}$ = 6.9 (*)	38.5
Seeds per pod:						
Healthy seeds, *n*	24	24	6.6 (1.87)	12.4 (1.87)	$F_{(1,10)}$ = 4.8 (ns)	46.8
Eaten seeds, *n*	24	24	1.4 (0.70)	0.7 (0.70)	$F_{(1,10)}$ = 0.6 (ns)	-100.0
Aborted seeds[d], *n*	24	24	2.8 (0.46)	1.3 (0.46)	$F_{(1,10)}$ = 5.2 (*)	-115.4
Missing seeds[e], *n*	24	24	1.1 (0.29)	1.5 (0.29)	$F_{(1,10)}$ = 0.9 (ns)	26.7
Total seeds, *n*	24	24	11.9 (0.99)	15.9 (0.99)	$F_{(1,10)}$ = 8.1 (*)	25.2
Moisture in seeds, %	24	24	3.5 (0.14)	3.1 (0.14)	$F_{(1,10)}$ = 3.3 (ns)	-12.9
C and N in seeds, dry weight:						
Carbon in raw seeds, %	24	24	53.4 (0.29)	54.3 (0.29)	$F_{(1,10)}$ = 4.8 (ns)[f]	1.6
Nitrogen in raw seeds, %	24	24	7.2 (0.10)	6.9 (0.10)	$F_{(1,10)}$ = 5.0 (*)	-4.4
Protein in raw seeds[g], %	24	24	45.2 (0.60)	43.3 (0.60)	$F_{(1,10)}$ = 5.0 (*)	-4.4
Sugars in fruit pulp, dry weight:						
Sucrose[c], %	22	24	28.5 (0.81)	29.5 (0.74)	$F_{(1,9)}$ = 0.9 (ns)	3.4
Glucose[c], %	22	24	3.5 (0.68)	5.6 (0.62)	$F_{(1,9)}$ = 5.6 (*)	37.5
Fructose[c], %	22	24	4.2 (0.51)	5.9 (0.47)	$F_{(1,9)}$ = 5.6 (*)	23.8
Total sugar[c], %	22	24	36.2 (1.81)	41.3 (1.66)	$F_{(1,9)}$ = 4.3 (ns)	12.3
Seed germination:						
Seed weight, g	117	196	0.16 (0.006)	0.20 (0.005)	$F_{(1,8)}$ = 32.9 (***)	20.0
Seed germination[h], %	117	196	76.9	83.2	P < 0.2 (ns)	7.6
Days to germination	90	163	4.4 (0.28)	4.3 (0.23)	$F_{(1,8)}$ = 0.1 (ns)	-2.3
Seedlings:						
Fresh weight 19 days[i], g	88	161	0.85 (0.04)	1.02 (0.03)	$F_{(1,8)}$ = 10.2 (*)	16.7
Survival 222 days[h,i], %	90	163	76.7	90.2	P < 0.005 (**)	15.0
Height 222 days[i], cm	69	147	20.3 (0.97)	23.2 (0.74)	$F_{(1,7)}$ = 5.8 (*)	12.5
Stem diameter 222 days[i], mm	69	147	3.9 (0.20)	4.6 (0.15)	$F_{(1,7)}$ = 6.2 (*)	15.2
Number of pinnae 222 days[i]	69	147	35.0 (2.13)	40.5 (1.61)	$F_{(1,7)}$ = 4.3 (ns)	13.5
Fresh weight 222 days[i], g	69	147	17.0 (2.03)	23.4 (1.53)	$F_{(1,7)}$ = 6.3 (*)	27.4
Dry weight 222 days[i], g	69	147	6.1 (0.72)	8.4 (0.54)	$F_{(1,7)}$ = 6.5 (*)	27.4
Shoot, dry weight g	69	147	2.8 (0.41)	4.0 (0.31)	$F_{(1,7)}$ = 4.9 (ns)	30.0
Root, dry weight g	69	147	3.3 (0.48)	4.5 (0.37)	$F_{(1,7)}$ = 3.6 (ns)	26.7
Shoot:root ratio (dry weight)	69	147	1.0 (0.15)	0.9 (0.11)	$F_{(1,7)}$ = 0.1 (ns)	-11.1

[a]Values are least squares (LS) means with the standard error (SE) in brackets of the LS estimate. Significance level: ***= P<0.001, **= P<0.01, *= P<0.05, and ns= P>0.05. [b]Inbreeding depression is calculated as: (Cross-self) /cross*100. [c]Missing data for two self-pollinated pods (no pulp). [d]Aborted seeds: weight<0.05 g and/or with a flat shape. [e]Missing seeds: empty cavity in the pulp. [f]P = 0.054. [g]Nitrogen-to-protein conversion factor = 6.25 (AOAC, 1990). [h]Fisher's exact test. [i]Days after sowing.

capitula (Tab. 3). These results document pollen limitation of fruit set in *P. biglobosa* during the study year, indicating that fruit set could be increased by increasing pollen load above the natural level of pollination (but see discussion below). Freely exposed capitula attracted up to 50 honey bees foraging simultaneously per capitulum. As honey bees have been found to be good pollinators of *P. biglobosa* (Lassen et al. 2017), pollen limitation may be due to insufficient deposition of compatible cross-pollen compared to deposition of self- and incompatible cross-pollen. The density of *P. biglobosa* was relatively high (1.2 trees/ha), and seven of the eight mother-trees were separated by less than 60 m to the nearest *P.*

biglobosa tree. However, the mother-trees were large (mean crown area = 472 m², SD = 170) with many capitula, and this profuse blooming may have increased geitonogamy of the open-pollinated capitula. Higher fruit production of controlled cross-pollination compared to controlled self-pollination and open-pollination has also been reported for other tropical tree species e.g. the Coligallo palm, *Calyptrogyne ghiesbreghtiana* H. Wendl. (Cunningham 1996) (Arecaceae) and five species of neotropical *Inga* trees (Koptur 1984) (Fabaceae), where it has been linked to limitation of compatible pollen of self- and open-pollinations.

TABLE 5.   Correlation between number of seeds per pod and amount of sugar (sucrose, glucose, and fructose) in pulp for 46 pods (22 self-pollinated and 24 cross-pollinated) of *Parkia biglobosa* illustrated by Pearson correlation coefficients and significance levels.

	Sucrose	Glucose	Fructose	Total sugars
No. of healthy seeds	0.19 (ns)	0.48 (***)	0.38 (**)	0.48 (***)
No. of aborted seeds	0.08 (ns)	-0.48 (***)	-0.42 (**)	-0.28 (ns)
No. of total seeds	0.15 (ns)	0.26 (ns)	0.19 (ns)	0.31 (*)

Significance level: ***=$P<0.001$, **= $P<0.01$, *= $P<0.05$, and ns= $P>0.05$.

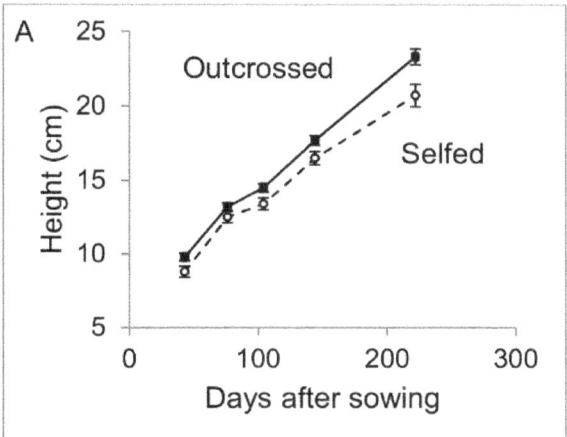

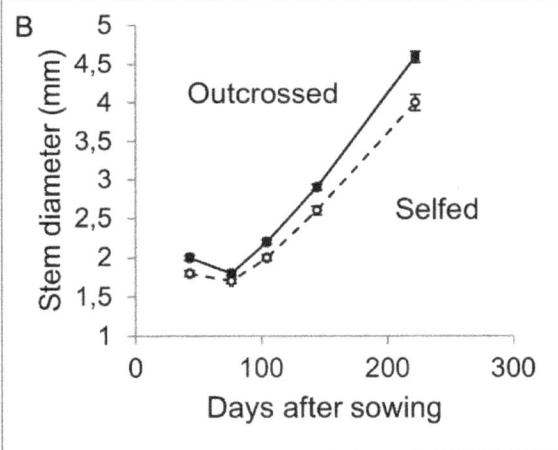

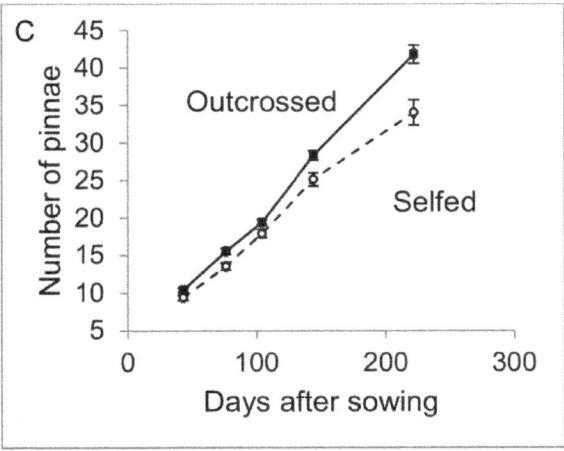

FIGURE 2. Growth of seedlings of *Parkia biglobosa* originating from 69 self-pollinated and 147 cross-pollinated seeds, showing the temporal development (days after sowing) of A) mean height, B) mean stem diameter, and C) mean number of pinnae. Error bars indicate ± 1 standard error of the mean. Only seedlings, which survived until harvest at 222 days after sowing are included.

Although half of the immature pods were aborted in controlled cross-pollinations, pollen dose was highly important for number of mature pods in *Parkia biglobosa*. The converting calculations between treatments with different pollen doses ('cross' and 'both') proposed a linear relationship between pollen dose and fruit set, suggesting that even more pods could have been initiated and matured, if the high dose of the cross-pollen in the 'both'-treatment (1:1) had been applied to the whole capitulum, and not only to half of it, or if the pollen dose had been even higher (e.g., 3:1). Most likely, the increased abortion of pods (i.e., immature pods minus mature pods) with higher pollen doses were due to lack of resources, suggesting that there is an upper limit for the number of produced pods per capitulum. However, we found no differences in pod weight or pod length between pods from open-, cross- or 'both'-treatments. The experiment had some limitations, and hence we cannot conclude whether the trees were truly pollen limited. First, the experiment lasted only one night at each tree; second, only a part of the blooming capitula per tree was included in the experiment; third, the experiment was only performed in one season. It is possible that the increased fruit set of hand-pollinated capitula came at the cost of open-pollinated capitula within the same tree (Obeso 2002). Nevertheless, some plant species show pollen limitation both on a whole plant level and in subsequent years (Ashman et al. 2004). More detailed hand-pollination experiments are needed to test whether a higher fruit production can be obtained in *P. biglobosa* by supplying cross-pollen night after night and year after year.

In the current study, a cross-pollinated capitulum was pollinated with only one cross-pollen donor, and hence the pod abortion was not due to selection between pollen donors, as has sometimes been suggested when explaining high rates of abortion (Bookman 1984). These patterns suggest a general lack of maternal resources to mature all or most of the initiated cross-pollinated pods.

Self-incompatibility in Parkia biglobosa

Self-incompatibility of *P. biglobosa,* as suggested by our study, is supported by other studies: Using controlled self- and cross-pollination ($N = 15$ trees), Ouédraogo (1995) concluded that *P. biglobosa* is largely a self-incompatible species: self-pollination was possible, but outcrossing was more successful. Likewise, Sina (2006) found high values of multi-locus outcrossing ($N = 238$ trees), also consistent with partial self-incompatibility. *Parkia* is a pantropical genus of around 35 species, of which most are believed to be bat-pollinated and the rest insect-pollinated (Hopkins 1998). The breeding system of most *Parkia* species has not been investigated (Bumrungsri et al. 2008), but their high pollen:ovule ratios are in the range characteristic of outcrossing (Cruden 1977; Hopkins 1984). Piechowski (2007) has tested *Parkia pendula* in Brazil for selfing, and as no pods were produced in spontaneous or controlled self-pollinations, he concluded that this species was self-incompatible. Furthermore, a study of two Asian species of *Parkia*, *P. speciosa* and *P. timoriana*, involving spontaneous (i.e. bagged capitula) and controlled self-pollination treatments, suggested that both species were self-incompatible (Bumrungsri et al. 2008). Finally, our finding of two cross-incompatible tree pairs (P14 & P90, and P76 & P92) fit with the self-incompatibility being controlled by a few specific loci (de Nettancourt 1977; Seavey and Bawa 1986).

In spite of the high pollen dose in the 'both'-treatment, no self-pollinated flowers developed into pods in this treatment. Since only one polyad pollinates one flower in *P. biglobosa* (Lassen et al. 2014), competition and selection between pollen donors is likely to take place between flowers rather than within flowers (Bawa & Buckley 1989). Pod set after self-pollination of the entire capitulum was rare. The 24 selfed pods contained significantly more aborted seeds and fewer total seeds, perhaps due to late acting self-incompatibility and/or early acting inbreeding depression. The 'missing' seeds were thought to be seeds that aborted very early, leaving only the empty cavity in the pulp, but we found few cavities per pod, and no differences between types of pollination.

Inbreeding depression in Parkia biglobosa

Initial seedling growth has been shown to depend on seed size, possibly due to the size of the cotyledons (Blackman 1919; Howe & Richter 1982; Boot 1996). Considering the correlation between carbon content and seed weight, and the higher seed weight of cross-pollinated seeds, we hypothesise that the nutritional differences between selfed and outcrossed seeds were due to relatively larger cotyledons in the cross-pollinated seeds. Because the differences in growth between self- and cross-pollinated seedlings increased with time in this study, we expect that the self-pollinated seedlings suffered from inbreeding depression. Few cases of inbreeding depression in early germination stages have been documented in other plant species while inbreeding depression at later life stages ('seed production of parent' and 'growth and reproduction') has more often been reported (Husband & Schemske 1996; Hardner & Potts 1995). A study by Mašková and Herben (2018) showed that larger-seeded species consistently had lower root:shoot ratios, explained by an advantage of faster development of shoots in asymmetric above-ground competition. We found no difference between shoot:root ratios of selfed and outcrossed seedlings, perhaps

due to lack of competition in the greenhouse. In the present study, the test of germination and growth took place under presumably optimal growth conditions (available water, light, nutrition and no competition), but the survival of seedlings was significantly lower for self-pollinated seeds. Walters & Reich (2000) observed that for ten tree species survival of seedlings in low light and/or low levels of N increased with seed weight. Therefore, it is likely that under natural conditions, inbreeding depression may have been more evident.

We expect that increased selfing in natural populations of *P. biglobosa* will negatively affect propagation by seeds, resulting in decreasing densities of adult trees in the future.

Quality of seeds and pod pulp from self- and cross-pollinated capitula

Pollination has been shown to impact the quality of fruits in different species (IPBES 2016). Because seeds and pulp from *P. biglobosa* are important food resources consumed by people and animals, it is highly relevant to understand the impact of self- versus cross-pollination on fruit and seed quality.

The content of protein in raw *P. biglobosa* seeds without testa (dry weight) of 43%- 45% (for outcrossed and selfed seeds, respectively) is similar to 43% found by one study (Ekpenyong et al. 1977), but much higher than reported by other studies: 27% (Esenwah & Ikenebomeh 2008), 30% (syn. *Parkia filicoidea* Welw.) (Fetuga et al. 1974) and 34% (Ijarotimi & Keshinro 2012). We found that self-pollinated seeds were more protein-rich but weighed less than cross-pollinated seeds. Hence, the increase in protein content can probably be explained by a simple concentration effect as found for various species grown under stress (Wang & Frei 2011). Inverse relationships of protein content and starch content of grains and yield, respectively, have been documented in maize hybrids (*Zea mays* subsp. *mays* L.) (Poaceae) (Idikut et al. 2009). Likewise, other studies of maize have found significantly higher protein content in self-pollinated kernels and significantly higher starch content in cross-pollinated kernels (Letchworth & Lambert 1998; Sulewska et al. 2014). Total seed protein produced by self-pollinated pods was much lower than for crossed pods, as self-pollinated pods contained much fewer healthy seeds (Tab. 4).

The positive correlation between seed number per pod and sugar content in pulp observed in the current study was also found in a similar study of *P. biglobosa* in The Gambia (Lassen et al. 2012). However, the percentages of sugars (dry weight) were much higher in the fruit pulp from The Gambia compared with the pulp from Burkina Faso (total sugar: 60% versus 36-41%) (Lassen et al. 2012). In the literature, carbohydrate content (dry weight) in fruit pulp of *P. biglobosa* is reported to be from around 40% in Nigeria (Nadro & Umaru 2004) to around 85% in Mali (Nordeide et al. 1996). The positive correlation between number of healthy seeds and total amount of sugar in the pulp in the current study may be due to seeds acting as sinks during pod development, attracting nutrients to their own growth and to that of the surrounding pod (Stephenson 1981; Lee 1988; Marcelis & Hofman-Eijer, 1997). Similarly, Valantin-

Morison et al. (2006) found a less-pronounced sweetness of the flesh in fruits with few filled seeds compared to fruits with a normal number of filled seeds in cantaloupe melon (*Cucumis melo* L.). In self-compatible, sweet orange (*Citrus sinensis* var. Red Junar) (Rutaceae), Partap (2000) found a higher number of seeds and more juice with a higher sugar content after honey bee pollination compared to wind-pollination.

Overall, our results suggest that fruit production, nutritional value of *P. biglobosa* pulp and seeds, and the fitness of seedlings, will decrease with increased levels of self-pollinations and affect the rural human populations negatively. We propose more research into how to increase the regeneration (natural or planted) of *P. biglobosa*. Our results could also be expanded by testing combinations of some of the acknowledged plus-trees (i.e. the superior trees) of *P. biglobosa* and by grafting the best combinations together, making it easier for the pollinators to bring about more cross-pollination with highly compatible pollen.

ACKNOWLEDGEMENTS

We thank the staff of the Centre National de Semences Forestières (CNSF) for their cooperative spirit. Thanks to Alassane Ouédraogo, Madi Tiemtoré, and Philbert Zoungrana for field assistance and to the farmers in Pinyiri for allowing us to use their trees. We also thank Sofie Fiona Hansen, Ruth Bruus Jakobsen, and Annalise Metz for measuring the pods; head laboratory technician Lene Korsholm Jørgensen from the Department of Plant and Environmental Sciences (UCPH) for analysing the content of carbohydrates in the fruit pulp of *P. biglobosa*; laboratory coordinator Preben Frederiksen from the Department of Geosciences and Natural Resources (UCPH) for analysing the content of carbon and nitrogen in the seeds of *P. biglobosa*; gardener Kurt Dahl and greenhouse supervisor Theodor Emil Bolsterli for taking good care of the *P. biglobosa* seedlings in the greenhouse (UCPH). The present paper is part of a PhD study financed by the Danish International Development Agency (Danida (FFU), research project no. 10-106-LIFE).

REFERENCES

Abrol DP (2012) Pollination biology. Biodiversity conservation and agricultural production. Springer, New York, USA.

Aizen MA, Harder LD (2007) Expanding the limits of the pollen-limitation concept: Effects of pollen quantity and quality. Ecology 88:271-281.

AOAC (Association of Official Analytical Chemists) (1990) Official methods of Analysis of the AOAC. (No. 954.01). Volume I. Association of Official Analytical Chemists Inc., Arlington, USA.

Ashman T-L, Knight TM, Steets JA, Amarasekare P, Burd M, Campbell DR, Dudash MR, Johnston MO, Mazer SJ, Mitchell RJ, Morgan MT, Wilson WG (2004) Pollen limitation of plant reproduction: Ecological and evolutionary causes and consequences. Ecology 85:2408-2421.

Baker HG, Harris BJ (1957) The pollination of *Parkia* by bats and its attendant evolutionary problems. Evolution 11:449-460.

Bawa KS, Buckley DP (1989) Seed: ovule ratios, selective seed abortion and mating systems in Leguminosae. In: Stirton CH, Zarucchi JL (eds.) Advances in legume biology. Monographs of systematic botany of the Missouri Botanical Garden No. 29. Missouri Botanical Garden, St. Louis, USA, pp 243-262.

Bawa KS, Perry DR, Beach JH (1985) Reproductive biology of tropical lowland rain forest trees. I. Sexual systems and incompatibility mechanisms. American Journal of Botany 72:331-345.

Bertin RI (1982) Paternity and fruit production in trumpet creeper (Campsis radicans). American Naturalist 119:694-709.

Blackman VH (1919) The compound interest law and plant growth. Annals of Botany 33:353-360.

Bookman SS (1984) Evidence for selective fruit production in *Asclepias*. Evolution 38:72-86.

Boot RGA (1996) The significance of seedling size and growth rate of tropical rain forest tree seedlings for regeneration in canopy openings. In: Swaine MD (ed.) The ecology of tropical forest tree seedlings. UNESCO, Paris, France, pp 267-283.

Bumrungsri S, Harbit A, Benzie C, Carmouche K, Sridith K, Racey P (2008) The pollination ecology of two species of *Parkia* (Mimosaceae) in southern Thailand. Journal of Tropical Ecology 24:467-475.

Cruden RW (1977) Pollen-Ovule ratios: A conservative indicator of breeding systems in flowering plants. Evolution 31:32-46.

Cunningham SA (1996) Pollen supply limits fruit initiation by a rain forest understorey palm. Journal of Ecology 84:185-194.

de Nettancourt D (1977) Incompatibility in angiosperms. Springer-Verlag, New York, USA.

Ekpenyong TE, Fetuga BL, Oyenuga VA (1977) Fortification of maize flour-based diets with blends of cashewnut meal, African locust bean meal and sesame oil meal. Journal of the Science of Food and Agriculture 28:710-716.

Ennos RA, Dodson RK (1987) Pollen success, functional gender and assortative mating in an experimental plant population. Heredity 58:119-126.

Esenwah CN, Ikenebomeh MJ (2008) Processing effects on the nutritional and anti-nutritional contents of African locust bean (*Parkia biglobosa* Benth.) seed. Pakistan Journal of Nutrition 7:214-217.

Etejere EO, Fawole MO, Sani A (1982) Studies on the seed germination of *Parkia clapertoniana*. Turrialba 32:181-185.

Ezeagu IE, Petzke JK, Metges CC, Akinsoyinu AO, Ologhobo AD (2002) Seed protein contents and nitrogen-to-protein conversion factors for some uncultivated tropical plant seeds. Food Chemistry 78:105-109.

Fetuga BL, Babatunde GM, Oyenuga VA (1974) Protein quality of some unusual protein foodstuffs. Studies on the African locust-bean seed (*Parkia filicoidea* Welw.). British Journal of Nutrition 32:27-36.

Funk C, Rowland J, Adoum A, Eilerts G, White L (2012). A climate trend analysis of Burkina Faso, U.S. Geological Survey, South Dakota, USA.

Gaisberger H, Kindt R, Loo J, Schmidt M, Bognounou F, Da SS, Diallo OB, Ganaba S, Gnoumou A, Lompo D, Lykke AM, Mbayngone E, Nacoulma BMI, Ouedraogo M, Ouédraogo O, Parkouda C, Porembski S, Savadogo P, Thiombiano A, Zerbo G, Vinceti B (2017) Spatially explicit multi-threat assessment of food tree species in Burkina Faso: A fine-scale approach. PLoS ONE 12:e0184457.

Goulson D, Nicholls B, Botias C, Rotheray E (2015) Bee declines driven by combined stress from parasites, pesticides and lack of flowers. Science 347:1255957.

Hall JB, Tomlinson HF, Oni PI, Buchy M, Aebischer DP (1997) *Parkia biglobosa*: a monograph. School of Agricultural and Forest Sciences Publications Number 9, University of Wales, Bangor, UK.

Hardner CM, Potts BM (1995). Inbreeding depression and changes in variation after selfing in *Eucalyptus globulus* ssp. *globulus*. Silvae Genetica 44:46-54.

Hinata K, Watanabe M, Toriyama K, Isogai A (1993) A review of recent studies on homomorphic self-incompatibility. International Review of Cytology 143:257-296.

Hopkins HC (1981) Taxonomy and reproductive biology of, and evolution in the bat-pollinated genus *Parkia*. PhD thesis, Oxford University, St. Hilda's College, UK.

Hopkins HC (1983) The taxonomy, reproductive biology and economic potential of *Parkia* (Leguminosae: Mimosoideae) in Africa and Madagascar. Botanical Journal of the Linnean Society 87:135-167.

Hopkins HC (1984) Floral biology and pollination ecology of the neotropical species of *Parkia*. Journal of Ecology 72:1-23.

Hopkins HCF (1998) Bat pollination and taxonomy in *Parkia* (Leguminosae: Mimosoideae). In: Hopkins HCF, Huxley CR, Pannell CM, White F (eds.) The biological monograph. The importance of field studies and functional syndromes for taxonomy and evolution of tropical plants. The Royal Botanical Gardens, Kew, UK, pp 31-55.

Hopping ME (1976) Effect of exogenous auxins, gibberellins, and cytokinins on fruit development in Chinese gooseberry (*Actinidia chinensis* Planch.). New Zealand Journal of Botany 14:69-75.

Howe HF, Richter WM (1982) Effects of seed size on seedling size in *Virola surinamensis*; a within and between tree analysis. Oecologia 53:347-351.

Husband BC, Schemske DW (1996) Evolution of the magnitude and timing of inbreeding depression in plants. Evolution 50:54-70.

Idikut L, Atalay AI, Kara SN, Kamalak A (2009) Effect of hybrid on starch, protein and yields of maize grain. Journal of Animal and Veterinary Advances 8:1945-1947.

Ijarotimi OS, Keshinro OO (2012) Comparison between the amino acid, fatty acid, mineral and nutritional quality of raw, germinated and fermented African locust bean (*Parkia biglobosa*) flour. Acta Scientiarum Polonorum Technologia Alimentaria 11:151-165.

IPBES (2016) The assessment report on pollinators, pollination and food production of the Intergovernmental Science-Policy Platform on Biodivversity and Ecosystem Services. Potts SG, Imperatriz-Fonseca VL, Ngo HT (eds). Secretary of the Intergovernmental Science-Policy Platform on Biodivversity and Ecosystem Services, Bonn, Germany.

Kennedy CM, Lonsdorf E, Neel MC, Williams NM, Ricketts TH, Winfree R, Bommarco R, Brittain C, Burley AL, Cariveau D, Carvalheiro LG, Chacoff NP, Cunningham SA, Danforth BN, Dudenhöffer J-H, Elle E, Gaines HR, Garibaldi LA, Gratton C, Holzschuh A, Isaacs R, Javorek SK, Jha S, Klein AM, Krewenka K, Mandelik Y, Mayfield MM, Morandin L, Neame LA, Otieno M, Park M, Potts SG, Rundlöf M, Saez A, Steffan-Dewenter I, Taki H, Felipe Viana B, Westphal C, Wilson JK, Greenleaf SS, Kremen C (2013) A global quantitative synthesis of local and landscape effects on wild bee pollinators in agroecosystems. Ecology Letters 16:584-599.

Knight TM, Steets JA, Vamosi JC, Mazer SJ, Burd M, Campbell DR, Dudash MR, Johnston MO, Mitchell RJ, Ashman T-L (2005) Pollen limitation of plant reproduction: Pattern and process. Annual Review of Ecology, Evolution and Systematics 36:467-497.

Koptur S (1984) Outcrossing and pollinator limitation of fruit-set - breeding systems of neotropical *Inga* trees (Fabaceae, Mimosoideae). Evolution 38:1130-1143.

Krotz L, Giazzi G (2014) Technical comparison of the Thermo Scientific FLASH 2000 Nitrogen/Protein Analyzer with the traditional Kjeldahl method Technical Note 42215. Thermo Fisher Scientific, Milan, Italy.

Lassen KM, Kjær ED, Ouédraogo M, Nielsen LR (2014) Microsatellite primers for *Parkia biglobosa* (Fabaceae: Mimosoideae) reveal that a single plant sires all seeds per pod. Applications in Plant Sciences 2:1400024.

Lassen KM, Ouédraogo M, Dupont YL, Kjær ED, Nielsen LR (2017) Honey bees ensure the pollination of *Parkia biglobosa* in absence of bats. Journal of Pollination Ecology 20:22-34.

Lassen KM, Ræbild A, Hansen H, Brødsgaard CJ, Eriksen EN (2012) Bats and bees are pollinating *Parkia biglobosa* in The Gambia. Agroforestry Systems 85:465-475.

Lee TD (1988) Patterns of fruit and seed production. In: Lovett Doust J, Lovett Doust L (eds.) Plant reproductive ecology. Patterns and strategies. Oxford University Press, New York, USA, pp 179-202.

Letchworth MB, Lambert RJ (1998) Pollen parent effects on oil, protein, and starch concentration in maize kernels. Crop Science 38:363-367.

Liu FL, Jensen CR, Andersen MN (2004) Drought stress effect on carbohydrate concentration in soybean leaves and pods during early reproductive development: its implication in altering pod set. Field Crops Research 86:1-13.

Lykke AM, Mertz O, Ganaba S (2002) Food consumption in rural Burkina Faso. Ecology of Food and Nutrition 41:119-153

Maranz S (2009) Tree mortality in the African Sahel indicates an anthropogenic ecosystem displaced by climate change. Journal of Biogeography 36:1181-1193.

Marcelis LFM, Hofman-Eijer LRB (1997) Effects of seed number on competition and dominance among fruits in *Capsicum annuum* L. Annals of Botany 79:687-693.

Mašková T, Herben T (2018) Root:shoot ratio in developing seedlings: How seedlings change their allocation in response to seed mass and ambient nutrient supply. Ecology and Evolution 2018;00:1-8. https//doi.org/10.1002/ece3.4238.

Météo (2015) Pluviometrie mensuelle (mm) dans Guilongou (Ziniare) et Pô 2010-2012 et temperature mensuelle (°C) dans Ouagadougou Aéroport et Pô 2011-2012, Direction Générale de la Météorologie du Burkina, Ouagadougou, Burkina Faso.

Nadro M, Umaru H (2004) Comparative chemical evaluation of locust bean (*Parkia biglobosa*) fruit pulp harvested during the dry and wet season. Nigerian Journal of Biotechnology 15:42-47.

Nordeide MB, Hatloy A, Folling M, Lied E, Oshaug A (1996) Nutrient composition and nutritional importance of green leaves and wild food resources in an agricultural district, Koutiala, in Southern Mali. International Journal of Food Sciences and Nutrition 47:455-468.

Nyadanu D, Adu Amoah R, Obeng B, Kwarteng AO, Akromah R, Aboagye LM, Adu-Dapaah H (2017) Ethnobotany and analysis of food components of African locust bean (*Parkia biglobosa* (Jacq.) Benth.) in the transitional zone of Ghana: implications for domestication, conservation and breeding of improved varieties. Genetic Resources and Crop Evolution 64:1231-1240.

Obeso JR (2002) The costs of reproduction in plants. New Phytologist 155:321-348.

Ollerton J, Winfree R, Tarrant S (2011) How many flowering plants are pollinated by animals? Oikos 120:321-326.

Osonubi O, Fasehun FE (1987) Adaptations to soil drying in woody seedlings of African locust bean, (*Parkia biglobosa* (Jacq.) Benth.). Tree Physiology 3:321-329.

Ouédraogo AS (1995) *Parkia biglobosa* (Leguminosae) in West Africa; biosystematics and improvement, Landbouwuniversiteit

Wageningen (Wageningen Agricultural University), Wageningen, The Netherlands.

Partap U (2000) Foraging behaviour of *Apis cerana* on sweet orange (*Citrus sinensis* var Red Junar) and its impact on fruit production. In: Asian bees and beekeeping. Progress of research and development. Proceedings of the 4th Asian Apicultural Association, Kathmandu, Nepal, pp 174-177.

Piechowski D (2007) Reproductive ecology, seedling performance, and population structure of *Parkia pendula* in an Atlantic forest fragment in Northeastern Brazil, Universität Ulm, Köln, Germany.

Ræbild A, Hansen UB, Kambou S (2012) Regeneration of *Vitellaria paradoxa* and *Parkia biglobosa* in a parkland in Southern Burkina Faso. Agroforestry Systems 85:443-453.

Roldán Serrano A, Guerra-Sanz JM (2006) Quality fruit improvement in sweet pepper culture by bumblebee pollination. Scientia Horticulturae 110:160-166.

Sanfo JB (2012) Apport de la Direction Générale de la Météorologie dans le processus d'information sur l'eau 13ème sommet de l'information sur l'eau, 11-13 April 2012, Ouagadougou, Burkina Faso.

SAS Institute (2011) The SAS system for Windows. Release 9.4, Cary, North Carolina, USA.

Seavey SR, Bawa KS (1986) Late-acting self-incompatibility in angiosperms. Botanical Review 52:195-219.

Sina S (2006) Reproduction et diversité génétique chez *Parkia biglobosa* (Jacq.) G.Don, Wageningen University, Wageningen, the Netherlands.

Stephenson AG (1981) Flower and fruit abortion: Proximate causes and ultimate functions. Annual Review of Ecology and Systematics 12:253-279.

Sulewska H, Adamczyk J, Cygert H, Rogacki J, Szymanska G, Smiatacz K, Panasiewicz K, Tomaszyk K (2014) A comparison of controlled self-pollination and open pollination results based on maize grain quality. Spanish Journal of Agricultural Research 12:492-500.

Uwaegbute AC (1996) African locust bean (*Parkia filicoidea* Welw.). In: Nwokolo E, Smartt J (eds.) Food and feed from legumes and oilseeds. Chapman & Hall, London, United Kingdom, pp 124-129.

Valantin-Morison M, Vaissiere BE, Gary C, Robin P (2006) Source-sink balance affects reproductive development and fruit quality in cantaloupe melon (*Cucumis melo* L.). Journal of Horticultural Science and Biotechnology 81:105-117.

Walters MB, Reich PB (2000) Seed size, nitrogen supply, and growth rate affect tree seedling survival in deep shade. Ecology 81:1887-1901.

Wang Y, Frei M (2011) Stressed food – The impact of abiotic environmental stresses on crop quality. Agriculture, Ecosystems & Environment 141:271-286.

Ward M, Dick CW, Gribel R, Lowe AJ (2005) To self, or not to self... a review of outcrossing and pollen-mediated gene flow in neotropical trees. Heredity 95:246-254.

Yeoh HH, Wee YC (1994) Leaf protein contents and nitrogen-to-protein conversion factors for 90 plant species. Food Chemistry 49:245-250.

19

Assessing the Risk of Stigma Clogging in Strawberry Flowers Due to Pollinator Sharing with Oilseed Rape

Lina Herbertsson[1]*, Ida Gåvertsson[1], Björn Klatt[1,2], Henrik G. Smith[1,2]

[1]*Lund University, Centre for Environmental and Climate Research, SE-223 62 Lund, Sweden*
[2]*Lund University, Department of Biology, SE-223 62 Lund, Sweden*

Abstract—Strawberry and oilseed rape are economically important and co-flowering insect-pollinated crops that may affect each other via shared pollinators. One potential negative effect of pollinator sharing is stigma clogging, i.e. that pollen from one plant species covers the stigma and prevents pollination in the other. We tested if application of oilseed rape pollen on strawberry receptacles reduces pollination with subsequent effects on strawberry weight, number of malformations and ripening time. We simulated real pollination situations by using dead bees mounted on toothpicks to mimic flower-visitation of foraging bees. Six strawberry flowers, usually on different plant individuals, were hand-pollinated sequentially per simulated foraging bout. In half of these foraging bouts, we started with an oilseed rape flower, and in those foraging bouts the proportion oilseed rape pollen was expected to decline with increasing number of visited strawberry flowers. Oilseed rape pollen had no effect on any of the tested variables. Increasing number of previously visited strawberry flowers in the simulated foraging bout enhanced the number of developed achenes, but this was marginally non-significant when accounting for the total number of achenes. Strawberry weight increased and ripening time decreased with increasing number of pollinated achenes, whereas none of the tested factors had any effect on the number of malformations. Our results have implications for strawberry farmers, because shortened ripening time could reduce the risk of yield loss from pests, diseases and unfavourable weather conditions. In addition, we show that oilseed rape pollen is unlikely to disturb pollination success of strawberry flowers.

Keywords: *Rapeseed, canola, mass-flowering crops, entomophilous crops, Fragaria, Brassica*

Introduction

Insect-pollinated crops constitute a third of the global agricultural yield (Klein et al. 2007), a share that is steadily increasing (Aizen et al. 2009). Because many of these crops can produce seeds and fruits without insect-pollination, less than 10% of the yield is directly attributable to insect pollination (Aizen et al. 2009). However, pollinators contribute to a large fraction of the non-starch nutrients that humans consume (Eilers et al. 2011) – with vital effects on human health (Smith et al. 2015). Thus, safeguarding pollination services may be crucial in optimizing crop yields (Bommarco et al. 2013) and healthy diets (Eilers et al. 2011; Smith et al. 2015). Crop pollination generally increases with wild pollinator abundance and species richness (Garibaldi et al. 2013), which in turn increase with reduced land management intensity and proximity to natural habitats (Ricketts et al. 2008; Tuck et al. 2014). While densities of managed pollinators can be elevated by placing hives near the flowering crops, management options available to improve crop yields by wild pollinators are associated with a higher uncertainty (e.g. Ricketts et al. 2008).

Increased availability of flowers can benefit crop pollination by wild insects (Blaauw & Isaacs 2014; Carvalheiro et al. 2012), in particular in the absence of natural habitats (Carvalheiro et al. 2011). However, a major obstacle for improving crop pollination by wild pollinators is related to the potential indirect interactions among flowering plants, since co-flowering plants may either compete for pollinators (Brown & Mitchell 2001), or facilitate the pollination of one another by attracting common pollinators (Ghazoul 2006). In addition, pollinator sharing increases the risk of interspecific pollen transfer, i.e. the movement of pollen between plant species. Interspecific pollen transfer can reduce pollination success, by affecting both male and female fitness (Morales & Traveset 2008). When pollen is removed from a flower and deposited on a stigma of another species, the pollen donor may suffer from pollen loss (Morales & Traveset 2008) and the pollen receiver from stigma clogging, i.e. that heterospecific pollen covers the stigma, preventing conspecific pollen from fertilizing the ovules (Brown & Mitchell 2001). In extreme cases, foreign pollen may inhibit pollen germination by producing toxic substances (Kanchan & Chandra 1980). Thus, co-flowering crops that share pollinators can potentially affect each other, either by competition, facilitation or interspecific pollen transfer, but to our knowledge such interactions among co-flowering crops have never been assessed.

Oilseed rape is an economically important insect-pollinated crop with an increasing global production since the 1960s (FAOSTAT 2017). Flowering oilseed rape attracts foraging pollinators (Kovács-Hostyánszki et al.

2013), thereby reducing pollination of co-flowering plants in nearby grasslands (Holzschuh et al. 2011), and facilitating or having no effect on pollination of adjacent plants (Cussans et al. 2010). Information on how interspecific pollen transfer from oilseed rape to other plants affects pollination is scant. In one study, the proportion of oilseed rape pollen on flowers of wild plants adjacent to the fields constituted around 3% of all pollen grains on the stigma, indicating that concentrations are too low to result in stigma clogging (Stanley & Stout 2014). On the contrary, Klatt (2013) showed that increasing proportion of honeybees correlated negatively with strawberry weight in field borders, where honeybees (but not other bees) carried pollen from both oilseed rape and strawberry crops. In field centers, where bees carried pure strawberry pollen loads, a positive correlation between honeybee densities and strawberry weight was found. This suggests that the reason for the observed pattern is interspecific pollen transfer, but the underlying mechanism remains to be examined.

Strawberry is an entomophilous crop that attracts insects by providing both pollen and nectar (Chagnon et al. 1993). All commercially used strawberry varieties are self-fertilising (Free 1993), but insects promote pollination by dispersing pollen between (Free 1993) and within flowers (Chagnon et al. 1993). The true fruits are the achenes on the strawberry surface (Free 1993), which produce growth hormones and induce strawberry growth and ripening (Nitsch 1950). Consequently, strawberry weight increases with the number of fertilized achenes (Abbott & Webb 1970), and a lack of pollinators can reduce the market value of strawberries by reducing the number of developed achenes or increasing the number of malformations (Klatt et al. 2014).

The aim of this study was to reveal if pollinator sharing with oilseed rape reduces pollination success in co-flowering strawberry crops as a result of stigma clogging. Based on the assumptions that strawberries deriving from flowers with a high number of developed (pollinated) achenes weigh more, have fewer malformations and shorter ripening time, we made several predictions. We predicted that oilseed rape pollen hinders strawberry pollination resulting in a lower number of developed achenes, lower strawberry weight, more malformations, and longer ripening time. On the contrary, we also predicted that increasing amounts of strawberry pollen on the receptacle result in higher numbers of developed achenes, higher strawberry weight, less malformations and shorter ripening time.

MATERIALS AND METHODS

Study design

We sowed spring-sown oilseed rape in a climate chamber, simulating Swedish spring, in February 2014. In April 2014, we acquired 96 strawberry plants with flower buds that had not yet opened, from a grocery store. We used the strawberry variety 'Salsa', which has a moderate insect-dependency compared to other varieties (Klatt et al. 2014). We potted strawberry plants in 1.5 L pots, kept them in a greenhouse and watered them daily. There was no possibility for insects to enter the greenhouse. When the first strawberry flowers started to open, we brought two flowering oilseed rape plants from the climate chamber to the greenhouse, to use as pollen donors.

Each strawberry branch produces several flowers that open sequentially, with the primary flower being the largest, containing most ovules (Free 1993) which results in the largest strawberry (Abbott & Webb 1970) and also the one most dependent on insect-pollination (Chagnon et al. 1989). Plants can produce more than one flower bearing branch, and can thus produce more than one primary strawberry per season (pers. obs.). In this experiment, we only regarded the primary flower/strawberry from the first branch as primary.

Open strawberry flowers were hand pollinated three times each, once per day for the first, second and third open day. We used dead red mason bees (*Osmia bicornis*) for the pollen application. The bees had never visited any flowers because they were frozen directly after they had emerged from the cocoons. The dead bees were attached to toothpicks and hand-pollination was performed by gently rotating the bee around the centre of the flowers, trying to mimic the pivoting flower-visitation behaviour of this and other similarly sized bee species (Albano et al. 2009; Chagnon et al. 1993; pers. obs.). After hand pollination, we cleaned the bees carefully with ethanol, and dried them at ambient temperature during at least 24 h.

We simulated two types of foraging bouts (oilseed rape or control), to mimic real pollination situations in which oilseed rape pollen was either involved or not. Each bee was only used in one type of simulated foraging bout to ensure that no oilseed rape pollen would contaminate the flowers in the control simulated foraging bouts. The oilseed rape foraging bout started with bees "visiting" an oilseed rape flower, after which it continued to six different strawberry flowers. We did not measure the amount of pollen on the bees' body nor how much pollen was transferred, but observed pollen on the bee bodies after the first flower visit (to oilseed rape). In the control simulated foraging bouts, bees visited six strawberry flowers, without first visiting the oilseed rape. The order of which the strawberry flowers were visited was the same during the first, second and third day of hand-pollination treatment. All flowers in the same foraging bout were usually from different plant individuals, but in a few cases we used more than one flower per plant. However, in the analyses we only used primary strawberries, which are the ones that depend most on insect-pollination (Chagnon et al. 1989) and produce the largest strawberries (Free 1993), such that only one flower per plant was analysed.

To avoid mixing up the flower treatments (oilseed rape or control), we used one of the treatments in each pot. To keep track of the position in the foraging bout and the number of hand-pollination treatments each flower within a pot had been through, we colour marked hand-pollinated flowers with one piece of drinking straw (one separate colour for each position in the foraging bout) per performed foraging bout. The first strawberry flowers opened and were hand-pollinated on the 1st May and we continued hand-pollination until the 21st May. A few days there was a lack of flowers, such that we could not accomplish any entire foraging bouts with six visits. When fewer than six flowers were available or when a few flowers remained to be pollinated after all complete simulated foraging bouts had been accomplished, we made shorter simulated foraging bouts.

Measures of strawberry pollination

Ripe strawberries were harvested continuously and were brought to the lab immediately for assessment. Strawberries were considered ripe when they were completely red and could easily be detached from the pedicle. In the lab we took the following measurements on each strawberry: weight, number of malformations (groups of undeveloped achenes), the number of developed and undeveloped achenes and ripening time (number of days from the flower opened until the strawberry was harvested). We continued harvesting until the 19th June, when only two poorly pollinated primary strawberries (and several from later flowers) remained, one control strawberry and one oilseed rape treated. These two strawberries were excluded when we tested weight, malformation and ripening time, but not the number of pollinated achenes, which were easily identified.

STATISTICAL ANALYSES

We used R.3.3.2 for windows and analysed data with general or generalized linear models from the R package 'stats' (R Core Team 2016). We extracted model estimates that we used to plot the results and to calculate effect sizes using package 'effects' (Fox 2003).

Effects of oilseed rape pollen and strawberry pollen

In separate analyses we tested how number of developed achenes, weight, number of malformations and ripening time were affected by an interaction between treatment (i.e. whether the simulated foraging bout was an oilseed rape or a control bout) and the flower's position in the simulated foraging bout (1-6). Because the total number of achenes sets the upper limit for the number of developed achenes and for strawberry weight, we aimed to use this as an offset variable in these two models. However, on four poorly pollinated strawberries (one oilseed rape and three control, all from the first position in the foraging bout) we could not count the total number of achenes, because the undeveloped achenes were embedded in surrounding strawberry tissue. Since accounting for the number of achenes resulted in the exclusion of these strawberries we ran the models both with and without the offset variable. For weight and ripening time we assumed Gaussian distribution of the residuals and for developed achenes and malformations we used quasi-Poisson to handle overdispersion. We removed non-significant ($P \geq$ 0.05) terms hierarchically to establish the best model explaining each of our response variables.

Effects of developed achenes on weight, malformations and ripening time

To verify our assumption that well-pollinated strawberries are larger (Abbott & Webb 1970; Nitsch 1950), have fewer malformations and ripen faster than poorly pollinated ones, we performed three additional tests using number of developed achenes as predictor variable and

weight, malformations and ripening time, respectively, as response variable. For weight and ripening time we assumed Gaussian distribution of residuals and for malformations we used quasi-Poisson to handle overdispersion.

RESULTS

Effects of oilseed rape pollen

Strawberries had on average (mean ± sd) 138.5 ± 73.5 developed achenes, weighed 13.3 ± 4.2 grams, had 7.0 ± 4.0 malformations and ripened in 32.4 ± 2.3 days. Oilseed rape pollen had no effect on any of these variables, neither via the interaction with the flower's position in the foraging bout (*developed achenes*: $T_{64} = 0.92$; $P = 0.36$/ (with offset) $T_{60} = 1.57$; $P = 0.12$, *weight*: $T_{62} = 1.07$; $P = 0.29$/ (with offset) $T_{60} = -0.31$; $P = 0.76$, *malformations*: $T_{62} = 0.17$; $P = 0.86$, *ripening time*: $T_{62} = -0.95$; $P = 0.35$, Fig. 1), nor as an independently acting factor (*developed achenes*: $T_{65} = 0.69$; $P = 0.49$/ (with offset) $T_{61} = 0.40$; $P = 0.69$; *weight*: : $T_{63} = 1.25$; $P = 0.22$/ (with offset) $T_{61} = -1.23$; $P = 0.22$, *malformations*: : $T_{63} = 0.09$; $P = 0.93$, *ripening time*: $T_{63} = 0.20$; $P = 0.84$, Fig. 1).

Effects of strawberry pollen

When excluding the offset variable, thereby including strawberries with unknown number of achenes, the number of previously visited flowers in the simulated foraging bout (1-6) had a positive effect the number of developed achenes (12% increase per visited flower; $T_{66} = 3.19$; $P = 0.002$, Fig. 1) and strawberry weight (0.87 grams increase per visited flower; $T_{64} = 3.04$; $P = 0.003$, Fig. 1), but this effect became marginally non-significant (*developed achenes*: $T_{62} = 1.95$; $P = 0.056$) or disappeared (*weight*: $T_{62} = -1.24$; $P = 0.22$) when using the number of achenes as offset variable. Increasing number of previously visited flowers in the simulated foraging bout had a negative impact on ripening time (0.44 days shorter per visited flower; $T_{64} = -2.81$; $P = 0.007$, Fig. 1), but no effect on the number of malformations ($T_{64} = -0.84$; $P = 0.41$, Fig. 1).

Effects of developed achenes on other measurements

The number of developed achenes was positively related to strawberry weight (4.2 grams increase per 100 achenes; $T_{64} = 7.92$; $P < 0.001$, Fig. 2), unrelated to the number of malformations ($T_{64} = -0.30$; $P = 0.77$, Fig. 2) and negatively related to ripening time (-1.9 days per 100 achenes; $T_{64} = -5.72$; $P < 0.001$, Fig. 2).

DISCUSSION

Oilseed rape pollen had no effect on strawberry pollination, weight, malformations or ripening time. However, the number of developed achenes increased (significantly or marginally non-significantly so, depending on analysis) with the number of previously visited flowers in the simulated foraging bout, possibly because of higher conspecific pollen loads (Quesada et al. 2001) or higher pollen diversity (Paschke et al. 2002) on the bee body. Increasing number of developed achenes in turn, had a positive influence on strawberry size and reduced the

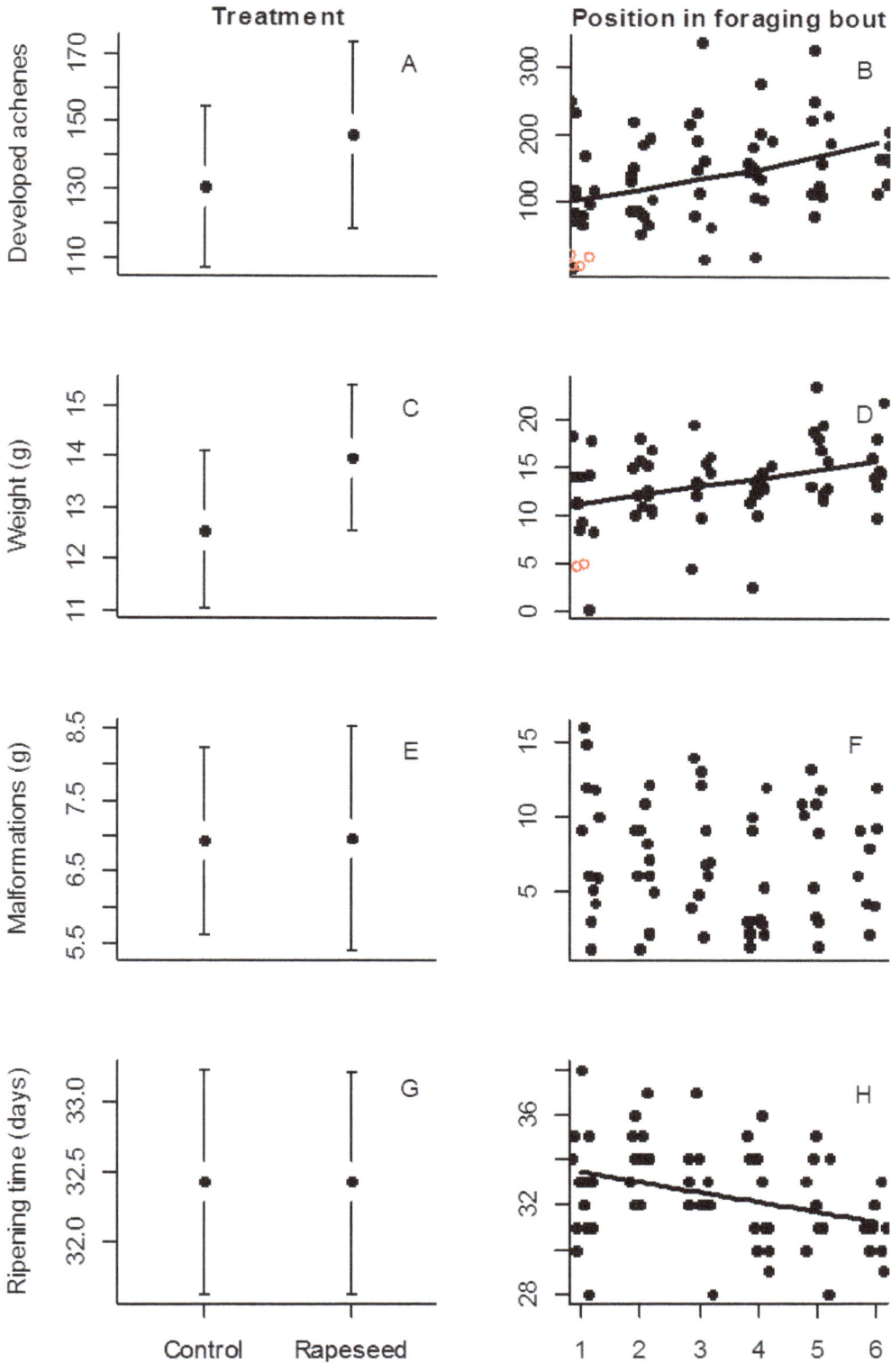

FIGURE I. The panel shows (A,B) the number of developed achenes, (C,D) strawberry weight, (E,F) number of malformations and (G,H) ripening time in relation to (A,C,E,G) oilseed rape treatment and (B,D,F,H) the flower's position in the simulated foraging bout. In the left column raw mean values for the two treatments are shown with error bars showing 95% confidence intervals. In the right column individual strawberries are shown in black and trend lines are added for significant effects. When accounting for the total number of achenes, the flower's position in the foraging bout had a marginally non-significant effect on number of developed achenes ($P = 0.056$) and no effect on weight ($P = 0.22$). Strawberries that lacked data on the number of achenes, and consequentially were excluded from the latter analyses, are shown as red circles.

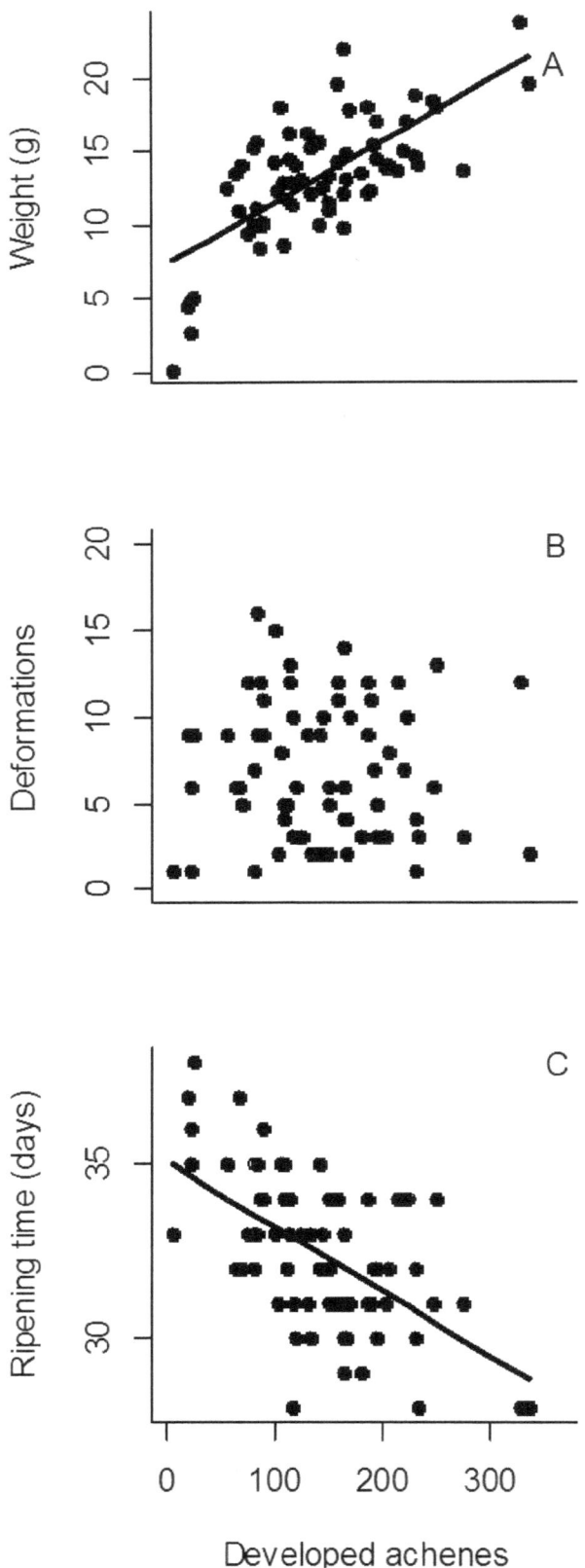

FIGURE 2. (A) Strawberry weight, (B) number of malformations and (C) ripening time in relation to the number of developed achenes. Dots show data on individual strawberries and trend lines are shown if significant.

ripening time, but had no impact on that the number of malformations under these experimental condition.

We found no evidence for oilseed rape pollen to have any negative impact on strawberry pollination, but this does not exclude that oilseed rape increases pollen limitation in strawberry fields by other means. Pollen limitation could occur as a result of competition for pollinators (cf. Holzschuh et al. 2011). We are not aware of any studies showing that competition for pollinators between co-flowering crops reduces yields, but intraspecific competition over pollinators can affect flower-visitation at both field (among coffee plants; Veddeler et al. 2006) and landscape scales (among sunflower fields within 2 km; Riedinger et al. 2014), suggesting that it also affects co-flowering crops, which attract the same pollinators. Pollen limitation could also be the result of pollen loss, i.e. the deposition of pollen on the stigma of a competing species (Morales & Traveset 2008). The effect of losing pollen in this way seems to be of higher importance for pollination than the negative consequences from receiving foreign pollen (Campbell & Motten 1985; Feinsinger & Tiebout 1991). Yet pollen loss is a far less studied phenomenon (Morales & Traveset 2008) and we are not aware of any studies assessing how it affects co-flowering crops.

From a farmer's perspective it may be important to differentiate between pollen limitation caused by lack of pollinators and loss of pollen, because the outcome of potential actions that the farmers commit likely depends on what mechanism is responsible for the effect. Whereas competitive pollinator dilution could be avoided by boosting pollinator densities, pollen loss may instead be aggravated if the same actions are taken and may instead require separation between the two crops. Honeybees can be used to increase pollinator densities, but since they commonly forage more than a kilometer away from the nest (Steffan-Dewenter & Kuhn 2003), spatial separation of the strawberry and oilseed rape fields would be difficult if using honeybees as pollinators. Solitary bees may provide a solution to this problem, because they have shorter foraging ranges than honeybees (Beekman & Ratnieks 2000; Gathmann & Tscharntke 2002), and their nests could possibly be placed to maximize the distance to the oilseed rape (e.g. in the middle of a strawberry field or on a side if the strawberry field where no oilseed rape is grown). Farmers could also grow later flowering strawberry varieties, which flower after the oilseed rape. This would allow the usage of honeybees without risking increased pollen loss and may be particularly beneficial for strawberry yields in landscapes with high proportion oilseed rape where farmers can take advantage of increased bumblebee densities (e.g. Riedinger et al. 2014).

In agreement with previous studies on cloudberry (Kortesharju 1993) and raspberry (Jennings & Topham 1971), we found that well-pollinated strawberries ripened significantly faster than poorly pollinated strawberries. This finding is interesting and should be further investigated, because shorter ripening time means that the critical time when strawberries are exposed to pests and diseases, and the need for water and suitable weather conditions, is minimized.

There are potential shortcomings of this study. First, effects of pollination treatment on the number of pollinated achenes and strawberry weight were only significant when including all strawberries, i.e. without accounting for the total number of achenes. While the estimated effect of pollination treatment on developed achenes remained near-significant ($P = 0.056$) when accounting for the number of achenes and thereby also incidentally excluding some of the less pollinated strawberries, the effect on weight completely disappeared. Increasing amount and diversity of strawberry pollen is likely to have a positive influence on strawberry weight, but if this is not the case, then the lack of effect from oilseed rape pollen should also be carefully interpreted. Second, the fact that no strawberries in our study completely lacked malformations suggest that pollination was generally poor, e.g. because of the low number of "bee visits" or because hand-pollination with dead bees that do not collect pollen actively and do not move around in the same way as living bees do not adequately mimic pollination by live bees. Both these differences could be relevant for the interpretation of the results, e.g. if pollination saturates when the number of pollinator visits increases. Thus, additional studies using live bees in realistic settings are needed. Third, previous studies have shown that insect-pollination reduces the number of malformed strawberries (Free 1993). However, in contrast to our other pollination proxies, the number of malformations in this study was unrelated to the number of developed achenes. A possible reason is that the number of malformations is an unprecise estimate of pollination success; in this study strawberries were generally small and poorly pollinated, and some badly pollinated strawberries had one or a few large malformations instead of many smaller.

Conclusion

Although strawberries are partly self-pollinated (Free 1993), the present and previous studies show that pollinators can increase the value of strawberry yields by several means. Not only does increased pollination result in larger strawberries and higher fruit set (Free 1993), but also in longer shelf-life (Klatt et al. 2014), a more desired fruit shape (Free 1993; Klatt et al. 2014) and, as shown in this paper, shorter ripening time. The present study suggests that foreign pollen from oilseed rape is unlikely to interact with these outcomes, but some caution is needed because increased number of previously visited strawberry flowers also had no significant impact on weight and developed achenes when accounting for the total number of achenes. Other potential effects of oilseed rape on strawberry pollination (e.g. pollen loss and competition for pollinators) still remain to be explored.

ACKNOWLEDGEMENTS

We acknowledge M. Saez for help in the greenhouse, O. Olsson for providing us with a script to plot the model estimates, T. Jack for checking the language, and the Royal Physiographic Society for financing the study. HS was financed by the Swedish Research Council FORMAS.

AUTHOR CONTRIBUTIONS

LH and BKK formulated the idea, IG and LH developed the methods with input from HGS, IG and LH collected the data, IG and LH analysed the data, LH and IG wrote the manuscript with contribution from BKK and HGS. LH and IG contributed equally to this work.

REFERENCES

Abbott AJ & Webb RA (1970). Achene spacing of strawberries as an aid to calculating potential yield. Nature 225:663-664. doi: 10.1038/225663b0

Aizen MA, Garibaldi LA, Cunningham SA, Klein AM (2009). How much does agriculture depend on pollinators? Lessons from long-term trends in crop production. Annals of Botany 103:1579-1588. doi: 10.1093/aob/mcp076

Albano S, Salvado E, Duarte S, Mexia A, Borges PAV (2009). Pollination effectiveness of different strawberry floral visitors in Ribatejo, Portugal: selection of potential pollinators. Part 2. Advances in Horticultural Science 23:246-253. doi: 10.1400/121241

Beekman M, Ratnieks FLW (2000). Long-range foraging by the honey-bee, Apis mellifera L. Functional Ecology 14:490-496. doi: 10.2307/2656543

Blaauw BR, Isaacs R (2014). Flower plantings increase wild bee abundance and the pollination services provided to a pollination-dependent crop. Journal of Applied Ecology 51:890-898. doi: 10.1111/1365-2664.12257

Bommarco R, Kleijn D, Potts SG (2013). Ecological intensification: harnessing ecosystem services for food security. Trends in Ecology & Evolution 28:230-238. doi: 10.1016/j.tree.2012.10.012

Brown B, Mitchell R (2001). Competition for pollination: effects of pollen of an invasive plant on seed set of a native congener. Oecologia 129:43-49. doi: 10.1007/s004420100700

Campbell DR, Motten AF (1985). The Mechanism of competition for pollination between two forest herbs. Ecology 66:554-563. doi: 10.2307/1940404

Carvalheiro LG, Seymour CL, Nicolson SW, Veldtman R (2012). Creating patches of native flowers facilitates crop pollination in large agricultural fields: mango as a case study. Journal of Applied Ecology 49:1373-1383. doi: 10.1111/j.1365-2664.2012.02217.x

Carvalheiro LG et al. (2011). Natural and within-farmland biodiversity enhances crop productivity. Ecology Letters 14:251-259. doi: 10.1111/j.1461-0248.2010.01579.x

Chagnon M, Gingras J, De Oliveira D (1989). Effect of honey bee (Hymenoptera: Apidae) visits on the pollination rate of strawberries. Journal of Economic Entomology 82:1350-1353. doi: 10.1093/jee/82.5.1350

Chagnon M, Gingras J, de Oliveira D (1993). Complementary aspects of strawberry pollination by honey and indigenous bees (Hymenoptera). Ecology and Behaviour 86:416-420. doi: 10.1093/jee/86.2.416

Cussans J, Goulson D, Sanderson R, Goffe L, Darvill B, Osborne JL (2010). Two bee-pollinated plant species show higher seed production when grown in gardens compared to arable farmland. PLoS ONE 5:e11753. doi: 10.1371/journal.pone.0011753

Eilers EJ, Kremen C, Smith Greenleaf S, Garber AK, Klein AM (2011). Contribution of pollinator-mediated crops to nutrients in the human food supply. PLoS ONE 6:e21363. doi: 10.1371/journal.pone.0021363

FAOSTAT (2017). Food and agriculture organisation of the United Nations - Statistics Division. [online] URL: http://faostat3.fao.org/download/Q/QV/E. (accessed February 2017).

Feinsinger P, Tiebout HM (1991). Competition among plants sharing hummingbird pollinators: Laboratory experiments on a mechanism. Ecology 72:1946-1952. doi: 10.2307/1941549

Fox J (2003). Effect displays in R for generalised linear models. Journal of Statistical Software 8:1-27. doi: 10.18637/jss.v008.i15

Free JB (1993). Insect pollination in crops. Academic Press Inc. Ltd, London.

Garibaldi LA et al. (2013). Wild pollinators enhance fruit set of crops regardless of honey bee abundance. Science 339:1608-1611. doi: 10.1126/science.1230200

Gathmann A, Tscharntke T (2002). Foraging ranges of solitary bees. Journal of Animal Ecology 71:757-764. doi: 10.1046/j.1365-2656.2002.00641.x

Ghazoul J (2006). Floral diversity and the facilitation of pollination. Journal of Ecology 94:295-304. doi: 10.2307/3599633

Holzschuh A, Dormann CF, Tscharntke T, Steffan-Dewenter I (2011). Expansion of mass-flowering crops leads to transient pollinator dilution and reduced wild plant pollination. Proceedings of the Royal Society B 278:3444–3451. doi: 10.1098/rspb.2011.0268

Jennings DL, Topham PB (1971). Some consequences of rapsberry pollen dilution for its germination and for fruit development. New Phytologist 70:371-380. doi: 10.1111/j.1469-8137.1971.tb02535.x

Kanchan S, Chandra J (1980). Pollen allelopathy - a new phenomenon. New Phytologist 84:739-746. doi: 10.1111/j.1469-8137.1980.tb04786.x

Klatt BK (2013). Bee pollination of strawberries on different spatial scales – from crop varieties and fields to landscapes. Fakultät für Agrarwissenschaften. Thesis. Göttingen: Georg-August-Universität Göttingen.

Klatt BK, Holzschuh A, Westphal C, Clough Y, Smit I, Pawelzik E, Tscharntke T (2014). Bee pollination improves crop quality, shelf life and commercial value. Proceedings of the Royal Society B 281:20132440:. doi: 10.1098/rspb.2013.2440

Klein AM, Vaissière BE, Cane JH, Steffan-Dewenter I, Cunningham SA, Kremen C, Tscharntke T (2007). Importance of pollinators in changing landscapes for world crops. Proceedings of the Royal Society B 274:303-313. doi: 10.1098/rspb.2006.3721

Kortesharju J (1993). Ecological factors affecting the ripening time of cloudberry (Rubus chamaemorus) fruit under cultivation conditions. Annales Botanici Fennici, 30:263-274. doi: http://www.jstor.org/stable/23726462

Kovács-Hostyánszki A, Haenke S, Batáry P, Jauker B, Báldi A, Tscharntke T, Holzschuh A (2013). Contrasting effects of mass-flowering crops on bee pollination of hedge plants at different spatial and temporal scales. Ecological Applications 23:1938-1946. doi: 10.1890/12-2012.1

Morales CL, Traveset A (2008). Interspecific pollen transfer: magnitude, prevalence and consequences for plant fitness. Critical Reviews in Plant Sciences 27:221-238. doi: 10.1080/07352680802205631

Nitsch JP (1950). Growth and morphogenesis of the strawberry as related to auxin. American Journal of Botany 37:211-215. doi: 10.2307/2437903

Paschke M, Abs C, Schmid B (2002). Effects of population size and pollen diversity on reproductive success and offspring size in the narrow endemic Cochlearia bavarica (Brassicaceae). American Journal of Botany 89:1250-1259. doi: 10.3732/ajb.89.8.1250

Quesada M, Fuchs EJ, Lobo JA (2001). Pollen load size, reproductive success, and progeny kinship of naturally pollinated flowers of the tropical dry forest tree Pachira quinata (Bombacaceae). American Journal of Botany 88:2113-2118

R Core Team. (2016). R: a language and environment for statistical computing. R Foundation for Statistical Computing. Vienna, Austria. https://www.R-project.org/

Ricketts TH, Regetz J, Steffan-Dewenter I, Cunningham SA, Kremen C, Bogdanski A, Gemmill-Herren B, Greenleaf SS, Klein AM, Mayfield MM, Morandin LA, Ochieng' A, Viana BF (2008). Landscape effects on crop pollination services: are there general patterns? Ecology letters 11:499-515. doi: 10.1111/j.1461-0248.2008.01157.x

Riedinger V, Renner M, Rundlöf M, Steffan-Dewenter I, Holzschuh A (2014). Early mass-flowering crops mitigate pollinator dilution in late-flowering crops. Landscape Ecol:425-435. doi: 10.1007/s10980-013-9973-ySmith MR, Singh GM, Mozaffarian D, Myers SS (2015). Effects of decreases of animal pollinators on human nutrition and global health: a modelling analysis. The Lancet 386:1964-1972. doi: 10.1016/S0140-6736(15)61085-6

Stanley D, Stout J (2014). Pollinator sharing between mass-flowering oilseed rape and co-flowering wild plants: implications for wild plant pollination. Plant Ecology 215:315-325. doi: 10.1007/s11258-014-0301-7

Steffan-Dewenter I, Kuhn A (2003). Honeybee foraging in differentially structured landscapes. Proceedings of the Royal Society B 270:569-575. doi: 10.1098/rspb.2002.2292

Tuck SL, Winqvist C, Mota F, Ahnström J, Turnbull LA, Bengtsson J (2014). Land-use intensity and the effects of organic farming on biodiversity: a hierarchical meta-analysis. Journal of Applied Ecology 51:746-755. doi: 10.1111/1365-2664.12219

Veddeler D, Klein A-M, Tscharntke T (2006). Contrasting responses of bee communities to coffee flowering at different spatial scales. Oikos 112:594-601. doi: 10.1111/j.0030-1299.2006.14111.x

20

ATTRACTIVENESS OF THE DARK CENTRAL FLORET IN WILD CARROTS: DO UMBEL SIZE AND HEIGHT MATTER?

Victor H. Gonzalez[1]*, Peter Cruz[2], Nadiyah Folks[3], Sarah Anderson[4], Dillon Travis[5], John M. Hranitz[6], & John F. Barthell[7]

[1]*Undergraduate Biology Program and Department of Ecology and Evolutionary Biology, Haworth Hall, 1200 Sunnyside Ave., University of Kansas, Lawrence, Kansas, 66045, USA*
[2]*Montclair State University, Montclair, New Jersey, 07043, USA*
[3]*University of Texas at El Paso, El Paso, Texas, 79902*
[4]*University of Kansas, Lawrence, Kansas, 66045, USA*
[5]*Boston University, Boston, Massachusetts, 02215, USA*
[6]*Biological and Allied Health Sciences, Bloomsburg University, Bloomsburg, PA, 17815, USA*
[7]*Department of Biology and Office of Provost & Vice President for Academic Affairs, University of Central Oklahoma, Edmond, Oklahoma, 73034, USA*

Abstract—The function of the dark central floret (DCF) in the wild carrot, *Daucus carota* L. (Apiaceae), is uncertain. It has been suggested that it is a vestigial structure without a function, that it serves as a long or short distance signal to attract pollinators, or that it might function as a defense mechanism against herbivores. We experimentally assessed the role of the umbel size and height in the attractiveness of the DCF to insects in a coastal population of *D. carota* in western Turkey. We did not find differences in the number of insect visits between umbels with a DCF and umbels in which the DCF was removed when they were of average diameter (10 cm) and were placed either at the average inflorescence height (120 cm) or at 147 cm above ground. Similarly, we did not find differences in the number of insect visits before and after the removal of the DCF from an umbel or between umbels of small (5–7 cm) and large (11–13 cm) diameters. However, umbels of average diameter with DCF received more insect visits than those without it when we placed them at 81 cm above ground. These results suggest that umbel height, not diameter *per se*, influences the attractiveness of the DCF in the studied population. Thus, our study supports the hypothesis that DCF function depends on ecological context, reliant on both the visitor community and the predominant flower phenotype.

Keywords: Bees; Coleoptera; Diptera; Pollination; Turkey

INTRODUCTION

The wild carrot, *Daucus carota* L. (Apiaceae), is native to Europe, Asia and North Africa, and it has been introduced to North America, Australia, New Zealand, and South Africa (Lamborn & Ollerton 2000). As in other umbellifers, the flowers of this species are in a compound inflorescence (umbel) at the end of a stem. Each umbel consists of many umbellets, which are composed of several individual florets. In *D. carota*, one or several of the florets in the central umbellet are pink or dark purple and thus stand out among the white florets of the remaining umbellets. The function of the dark central florets (DCF), which not only vary in numbers but also in their presence within a population, has been a matter of debate for many years. Some authors have suggested that the

DCF is a vestigial structure without a function (Darwin 1888), that it serves as a long or short distance signal to attract or deter flower visitors, or that it might function as a defense mechanism against herbivores, such as the gall midge *Kiefferia pericarpiicola* (Bremi) (Diptera: Cecidomyiidae) (e.g., Eisikowitch 1980; Lamborn & Ollerton 2000; Goulson et al. 2009; Polte & Reinhold 2013).

Such a diversity of explanations attributed to the DCF appears to be a reflection of the generalized pollination system of *D. carota*, varying in function depending on the local pollinator availability and composition (Ollerton et al. 2007; Goulson et al. 2009; Polte & Reinhold 2013). For example, the number of floral visitors recorded for *D. carota* ranges from 20 species in Europe to more than 300 species in North America. In addition, different taxa of floral visitors respond differently to the removal of the DCF among locations and even between years in the same population (e.g., Bohart & Nye 1960; Westmoreland & Muntan 1996; Lamborn & Ollerton 2000; Goulson et al. 2009).

*Corresponding author: victorgonzab@gmail.com

The adaptive value of the DCF might also depend on other factors, such as the height and size of the umbel. In general, flowers or inflorescences of taller plants naturally attract more bees and other pollinators than those of short plants (e.g., Gumbert & Kunze 1999; Lortie & Aarssen 1999), and large umbels tend to receive higher visitation rates than small umbels (Thomson 1988). In fact, Goulson et al. (2009) noted that adults of the dermestid beetle *Anthrenus verbasci* (Linnaeus) are more abundant in larger umbels of *D. carota*, as well as in umbels with greater numbers of dark florets.

Herein, we sought to answer the following questions: Does the DCF have a role in the attraction of floral visitors in a coastal population of *D. carota* in western Turkey? Do umbel size and height have a role in the attractiveness of the DCF to visitors?

MATERIALS AND METHODS

From June 21 to 27, 2016, we conducted observations on a coastal population of *D. carota* (40°04'30.67" N, 26°21'36.41" E, 11 m a.s.l.) located next to the Dardanos Dormitory at Çanakkale Onsekiz Mart University (ÇOMÜ) in the Republic of Turkey. Umbels ranged from 6.0 to 16.5 cm ($\bar{x} = 10.2 \pm 2.10$, $N = 106$) in diameter and from 64.8 to 211.2 cm ($\bar{x} = 126.6 \pm 29.7$, $N = 106$) in height, the latter measured from the ground to the base of the umbel. Umbel diameter was positively associated with height (Spearman's correlation, $r_s = 0.283$, $P = 0.003$). To explore the role of umbel size and height in the attractiveness of the DCF, we assessed insect visitation in the following four experiments (Table 1), each of which controlled for one or both independent variables:

Experiment 1. We compared umbels of average diameter (10 cm) placed at the average population's height (120 cm).

Experiment 2. We compared umbels of average diameter (10 cm) placed at 81 and 147 cm above the ground, representing heights below and above the average population's height.

Experiment 3. We compared small (5–7 cm) versus large (11–13 cm) umbels placed at the average population's height.

For experiments 1–3, we set up one or two parallel transects 0.4–0.5 m apart. Along the transects, we established treatments using water picks (plastic tubes filled with water and covered with a rubber cap that has a hole in the center through which the flower stem is inserted). We tied pairs of water picks onto stems of plants, each 4–5 m apart along the length of the transect. The number of transects and pairs of water picks varied among experiments (Table 1). In Experiment 1, we set up a single transect consisting of 10 pairs of water picks. In Experiment 2, we set up two transects, each consisting of five pairs of water picks, placed at alternated heights (five pairs at 81 cm, five pairs at 147 cm above the ground). In Experiment 3, we set up two transects, each consisting of six pairs of water picks.

About 30–40 minutes before we started the experiments, we placed umbels in the water picks and then randomly removed, with forceps, the DCF from one of the umbels (treatment) of each pair. The umbel with the intact DCF served as a control. In Experiment 3, we set up two additional types of pairs (Table 1), one in which we left the DCF intact in both umbels (small and large) and one in which we removed the DCF from both of them, thus serving as positive and negative controls, respectively. Along a transect, we repeated each type of pair three times.

We selected undamaged umbels without galls. If spiders or beetles were present on the umbels, we removed them with forceps or a brush before we placed them in the water picks. We conducted Experiments 1–3 for two days each and used a new umbel each time. Two observers recorded hourly insect visitation to umbel pairs, for 2.5 min at 10:00 through 14:00 hours, when activity was highest and before umbels began to wilt. To avoid disturbing insect visitors, we conducted observations about 1 m away from the umbels. We considered a visit only to be when insects landed directly on the umbels.

Table 1. Summary of experiments conducted on a coastal population of *Daucus carota* L. in western Turkey. We completed Experiments 1–3 on two days, each day using a new umbel, and Experiment 4 on the same day and same umbel, after the removal of the dark central floret. — = not applicable, as the same umbel was compared before and after the removal of the DCF.

Experiment	Description	Function	#Transects	# Pairs
1	Average size umbels (10 cm) at average population height (120 cm)	Baseline comparisons	1	10
2	Average size umbels below (81 cm) and above (147 cm) average population height (120 cm)	Testing effect of height	2	5 (below) 5 (above)
3	Umbel of small (5–7 cm) and large (11–13 cm) diameters with (W) and without (Wo) the dark central floret	Testing effect of umbel diameter	2	3 (small/W vs. large/W) 3 (small/W vs. large/Wo) 3 (small/Wo vs. large/W) 3 (small/Wo vs. large/Wo)
4	Average size umbels at average population height; umbels not removed from plants ($N = 41$)	Control for possible handling effects	1	—

If an insect moved from one to another umbel and returned to the original umbel during the observation period, we recorded it as two separate visits.

Experiment 4. To control for possible handling effects of the umbels, we selected 41 umbels (one umbel per plant) of average diameter and height and compared insect visitation before and after the removal of the DCF. We chose umbels the day before the experiment and marked the base of the plant with a piece of non-adhesive plastic ribbon. We began observations at 10:35 by recording the number of insect visits during one minute per umbel. Immediately after this observation period, we removed the DCF and brushed off arthropods, such as spiders, ants, and hemipterans, from the umbel. Then, about 80 min later, we again recorded the number of insect visits for another minute starting with the first umbel.

For Experiments 1–3, we used generalized linear models with Poisson distribution to examine the effect of the presence of the DCF and umbel height and diameter on the number of insect visits per umbel. For Experiment 4, we used a Sign test to compare the number of visits before and after removal of the DCF on the same umbel. We considered a P-value of $\leq$ 0.05 to be statistically significant. Finally, to estimate the prevalence of umbels with DCF in the studied population, we randomly selected and then examined 230 umbels.

RESULTS AND DISCUSSION

At our study site, beetles (e.g., Dermestidae), flies (e.g., Syrphidae), small bees (e.g., *Hylaeus* sp., *Andrena* sp., *Lasioglossum* spp.) and wasps (e.g., Ichneumonidae) visited umbels of *D. carota*. During the observation period (2.5 min), the number of visits per observation ranged from 0 to 19 ($\bar{x}$

$= 1.76 \pm 2.65$, $N = 264$). The percentage of umbels with a DCF in the studied population was 76.1% ($N = 230$) and we only observed three umbels with galls in 124 umbels examined (2.8%).

In Experiment 1, the total number of insect visits between umbels with (control) and without DCF (treatment) were similar, Wald $\chi^2 = 0.346$, df $= 1$, $P = 0.557$ (Fig. 1A). In Experiment 2, the total number of insect visits differed between heights above and below the population's average height (Wald $\chi^2 = 12.269$, df $= 1$, $P = 0.000$), and between the control and treatment (Wald $\chi^2 = 4.670$, df $= 1$, $P = 0.031$); however, the interaction between height and presence of DCF was not significant (Wald $\chi^2 = 1.312$, df $= 1$, $P = 0.252$). We recorded a higher number of visits in umbels with the DCF and in umbels placed at 81 cm above ground, below population's average (Fig. 1B). In Experiment 3, the total number of insect visits was similar between umbels of small and large diameters (Wald $\chi^2 = 0.623$, df $= 1$, $P = 0.430$) and between the control and treatment (Wald $\chi^2 = 2.502$, df $= 1$, $P = 0.114$), but the interaction between umbel size and presence of DCF was significant (Wald $\chi^2 = 4.291$, df $= 1$, $P = 0.038$). In Experiment 4, the number of insect visits before and after the DCF was removed from an umbel was similar (Sign test, Z $= -1.278$, $P = 0.201$, $N = 41$).

Our results suggest context-dependent effects of the DCF, as umbel height influenced the attractiveness of the DCF in the studied population of *D. carota*. At least one study, using pan-traps, demonstrated that height, even as small as 70 cm above ground, may play a significant role in the kinds of bees collected, especially in their average body size (Gonzalez et al. 2016). In natural systems, distinctions have even been made between large- and small-bodied bees and the plant

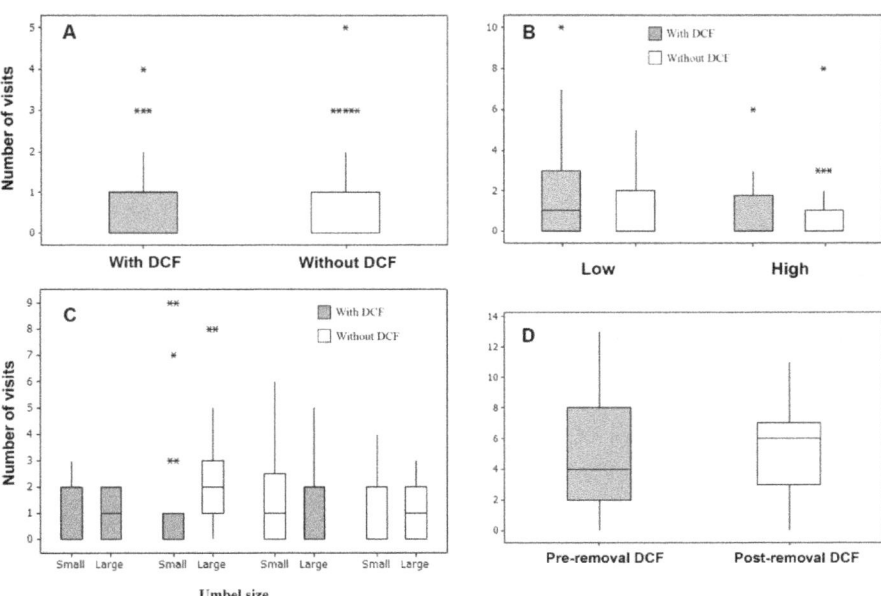

FIGURE 1.   Number of insect visits recorded from umbels of *Daucus carota* L with and without the dark central floret (DCF) in Experiments 1–4 (A–D, respectively). A = comparison between pairs of umbels of average population's diameter (10 cm) placed at average population height (120 cm). B = comparisons among pairs of umbels of average diameter placed below (81 cm) and above (147 cm) population's average height. C = comparisons among pairs of umbels of small (5–7 cm) and large (11–13 cm) diameters placed at average population's height. D = comparison between umbels of average diameter and height before and after the removal of DCF. Boxplots display median, quartiles, and extreme values, the latter indicated by an asterisk.

communities they forage among (Frankie et al. 1983; Barthell et al. in preparation). These patterns may relate not only to the species-specific characteristics of the bees but the energetics of size in the foraging dynamics of bee species within plant communities (Schaefer et al. 1979).

If bees and other pollinators tend to fly in the horizontal stratum and they are naturally attracted to flowers or inflorescences of taller plants (Gumbert & Kunze 1999; Lortie and Aarssen 1999; Cane et al. 2000), then short umbels likely experience fewer insect visits than tall umbels. Thus, the presence of the DCF in short umbels might increase their attractiveness to potential visitors in this community. We did not assess if the presence of the DCF varies with plant height in our study population. However, anecdotal observations in one coastal population of *D. carota* at Kalloni Bay, Island of Lesvos, Greece, revealed that the DCF was frequently absent from short umbels (< 100 cm above ground), on the same plants having umbels of average or above average height with the DCF. The co-occurrence of umbels with and without the DCF on the same plant suggests that environmental factors (e.g., light intensity, arid conditions, arthropod visitation) and plant developmental factors affect the expression of flower traits (e.g., Gonzáles et al. 2016).

Finally, we cannot refute the hypothesis that the DCF may mediate visitation by certain taxa, as we did not address this aspect in our study. Our conflicting results for umbels at different heights support the hypothesis that DCF function depends on ecological context, reliant on both the visitor community and the predominant flower phenotype (Goulson et al. 2009; Polte & Reinhold 2013).

ACKNOWLEDGEMENTS

We are indebted to Jeff Ollerton, Jane Stout, and an anonymous reviewer for comments and suggestions that improved this manuscript. Our colleagues Özge Can Niyaz and Cüneyt Aki from Çanakkale Onsekiz Mart University for their unconditional help and support during our field studies in Turkey. This work was supported by the National Science Foundation's REU program (DBI 1560389).

REFERENCES

Bohart GE, Nye WP (1960) Insect Pollinators of Carrots in Utah. Utah Agricultural Experimental Station Bulletin 419:1–16.

Cane JH, Minckley RL, Kervin LJ (2000) Sampling bees (Hymenoptera: Apiformes) for pollinator community studies: pitfalls of pan-trapping. Journal of the Kansas Entomological Society 73: 225–231.

Darwin C (1888) The different forms of flowers on plants of the same species, 3rd edn. John Murray, London, UK.

Eisikowitch D (1980) The role of dark flowers in the pollination of certain umbelliferae. Journal of Natural History 14:737–742.

Frankie GW, Haber WA, Opler PA, Bawa KS (1983) Characteristics and organization of the large bee pollination system in the Costa Rican dry forest. In: Jones CE, Little RJ (eds) Handbook of Experimental Pollination Biology. Van Nostrand-Reinhold Company Inc., New York, pp 411–447.

Gonzáles WL, Suarez LH, Gianoli E (2016) Genetic variation in the reduction of attractive floral traits of an annual tarweed in response to drought and apical damage. Journal of Plant Ecology 9:629–635.

Gonzalez VH, Park KE, Çakmak I, Hranitz JM, Barthell JF (2016) Pan traps and bee size in unmanaged urban habitats. Journal of Hymenoptera Research 51:241–247.

Goulson D, McGuire K, Munro EE, Adamson S, Colliar L, Park KJ, Tinsley MC, Gilburn AS (2009) Functional significance of the dark central floret of *Daucus carota* (Apiaceae) L.; is it an insect mimic? Plant Species Biology 24:77–82.

Gumbert A, Kunze J (1999) Inflorescence height affects visitation behavior of bees—A case study of an aquatic plant community in Bolivia. Biotropica 31:466–477.

Lamborn E, Ollerton J (2000) Experimental assessment of the functional morphology of inflorescences of *Daucus carota* (Apiaceae): testing the 'fly catcher effect'. Functional Ecology 14:445–454.

Lortie CJ, Aarssen LW (1999) The advantage of being tall: higher flowers receive more pollen in *Verbascum thapsus* L. (Scrophulariaceae). Ecoscience 6:68–71.

Ollerton J, Killick A, Lamborn E, Watts S, Whiston M (2007) Multiple meanings and modes: on the many ways to be a generalist flower. Taxon 56:717–728.

Polte S, Reinhold K (2013) The function of the wild carrot's dark central floret: attract, guide or deter? Plant Species Biology 28:81–86.

Schaffer WM, Jensen DB, Hobbs DE, Gurevitch J, Todd JR, Schaffer VM (1979) Competition, foraging energetics, and the cost of sociality in three species of bees. Ecology 60:976–987.

Thomson JD (1988) Effects of variation in inflorescence size and floral rewards on the visitation rates of traplining pollinators of *Aralia hispida*. Evolutionary Ecology 2:65–76.

Westmoreland D, Muntan C (1996) The influence of dark central florets on insect attraction and fruit production in Queen Anne's Lace (*Daucus carota* L.). American Midland Naturalist 135:122–129.

21

INSECT POLLINATION IMPROVES YIELD OF SHEA (*VITELLARIA PARADOXA SUBSP. PARADOXA*) IN THE AGROFORESTRY PARKLANDS OF WEST AFRICA

Jane C. Stout[1]*, Issa Nombre[2], Bernd de Bruijn[3], Aoife Delaney[1], Dzigbodi Adzo Doke[4], Thomas Gyimah[5], Francois Kamano[6], Ruth Kelly[1], Peter Lovett[7], Elaine Marshall[6], Adama Nana[8,] Latif Iddrisu Nasare[4], Japheth Roberts[5], Prudence Tankoano[8], Cath Tayleur[6,9], David Thomas[6], Juliet Vickery[9,10], Peter Kwapong[11]

[1]*School of Natural Sciences, Trinity College Dublin, Dublin, Republic of Ireland*
[2]*Laboratoire de Biologie et Ecologie Végétales, Université Ouaga I Pr Joseph KI-ZERBO, Institut des sciences, 01 BP 1757 Ouagadougou 01, Burkina Faso*
[3]*Vogelbescherming Nederland - BirdLife in The Netherlands, P.O. Box 925, 3700 AX Zeist, The Netherlands*
[4]*Faculty of Natural Resources and Environment, University for Development Studies, Tamale, Ghana*
[5]*Ghana Wildlife Society, P.O. Box 13252, Accra, Ghana*
[6]*BirdLife International, David Attenborough Building, Pembroke Street, Cambridge, CB2 3QZ, UK*
[7]*Form International, Bevrijdingsweg 3, 8051 EN Hattem, The Netherlands*
[8]*Naturama, 01 B.P. 6133, 01, Ouagadougou, Burkina Faso*
[9]*RSPB Centre for Conservation Science, The Royal Society for the Protection of Birds, The Lodge, Sandy, Bedfordshire SG19 2DL, UK*
[10]*University of Cambridge, David Attenborough Building, Pembroke Street, Cambridge, CB2 3QZ, UK*
[11]*Department of Conservation Biology and Entomology, University of Cape Coast and International Stingless Bee Centre, Cape Coast, Ghana*

Abstract—Pollinator decline, driven primarily by habitat degradation, has the potential to reduce the quantity and quality of pollinator-dependent crops produced across the world. *Vitellaria paradoxa*, a socio-economically important tree which grows across the sub-Saharan drylands of Africa, produces seeds from which shea butter is extracted. However, the habitats in which this tree grows are threatened with degradation, potentially impacting its ability to attract sufficient pollinators and to produce seeds. The flowers of *V. paradoxa* are insect-pollinated, and we investigated flower visitors in six sites in southern Burkina Faso and northern Ghana and tested whether plants were capable of fruit set in the absence of pollinators. We found that the majority of flower visitors (88%) were bees, most frequently small social stingless bees (*Hypotrigona gribodoi*), but native honey bees (*Apis mellifera adansonii*) were also common visitors to flowers early in the morning. The number of fruit produced per inflorescence was significantly lower when insects were excluded during flowering by bagging, but any fruits and seeds that were produced in bagged treatments were of similar weight to un-bagged ones. We conclude that conservation of habitat to protect social bees is important to maintain pollination services to *V. paradoxa* and other fruit-bearing trees and cultivated crops on which local livelihoods depend.

Keywords: Bees, Tree pollination, Fruit set, Livelihoods, Stingless bees, Tropical crops

INTRODUCTION

Pollinator decline, driven by agricultural intensification causing the fragmentation, degradation and loss of habitat, as well as climate change, parasites/disease and other factors (Goulson et al. 2015), can reduce pollination success and thus yield in many crop and wild plant species (Klein et al. 2007; Ollerton et al. 2011). Both the quantity and quality of yield can be affected via a reduction in the number and/or weight of fruits/seeds produced and the nutritional or commercial value of fruits, nuts and oils (Bommarco et al. 2012; Brittain

et al. 2014). Conversely, an increased number and/or diversity of pollinators can improve yields (Garibaldi et al. 2016). Many studies investigating the role of pollinators in crop production have focussed on herbaceous temperate food crops (but see Klein et al. 2003; Macias-Macias et al. 2009, Kudom and Kwapong 2010 etc.), and the role of pollinators in the production of tropical tree crops has been comparatively understudied (Kwapong et al. 2014, but see Carvalheiro et al. 2010; Freitas et al. 2014), particularly in Africa (Rodger et al. 2004).

In the sub-Saharan drylands of Africa, some trees are deliberately maintained in a landscape which is also used for the cultivation of crops and/or animals in agroforestry parkland systems (Boffa 1999). These systems have been maintained by the practice of shifting cultivation, where

*Corresponding author: stoutj@tcd.ie

cropping and grazing have alternated with fallow periods, for thousands of years (Gallagher et al. 2016). However, the semi-arid sub-Saharan drylands are a rapidly degrading habitat, under pressure from agricultural intensification, fuelwood demand and climate change (Bodart et al. 2013; Kandji et al. 2006). Reduced habitat diversity has been linked with insect pollinator decline worldwide (Kennedy et al. 2013), and degraded agroforestry parklands, which contain fewer species of tree and other vegetation due to reduced periods of fallow, may not be able to maintain the pollinator richness required for pollination of insect-dependent species (Tornyie and Kwapong 2015). Since 94% of tropical plant species are animal pollinated (Ollerton et al. 2011), this has implications for the reproduction of the majority of both crop and wild species in these habitats. Quantification and documentation of pollination deficits has been recognised as a priority issue for these areas (Gemmill-Herren et al. 2014). Although several native tree species in these agroforestry parklands produce edible or medicinal fruits, including *Parkia biglobosa* (African Locust bean), *Adansonia digitata* (Baobab) and *Tamarindus indica* (Tamarind), one of the most financially important to local communities is *Vitellaria paradoxa* (Shea/Karité).

Vitellaria paradoxa (Sapotaceae) grows in 21 sub-Saharan countries, with *V. paradoxa paradoxa* in West Africa and *V. paradoxa nilotica* in East Africa (Naughton et al. 2015), and "shea butter", extracted from the seeds of both subspecies, is the primary edible oil for 80 million people, and is growing in economic importance as a major export product, worth an estimated US$120 million annually (Naughton et al. 2015). In addition, the wood from this tree has a range of local uses (fuel as firewood or charcoal, building poles, making local utensils etc.), and the fruit pulp provides food for local communities during the "hungry season". The vast majority of shea butter production, from collection of the fruits to production of the oil, is carried out by women, and local trade in shea butter provides income to support education and diet in 18.4 million families (Pouliot 2012; Schreckenberg et al. 2006). *Vitellaria paradoxa* trees are usually not planted or sown, but naturally regenerate and, once established, certain saplings are selected and protected from damage by agricultural practices because of the value of the shea fruit. However, permanent cultivation and/or grazing, with reduced or non-existent fallow periods, prevents naturally regenerating shea seedlings attaining a size at which farmers will select for protection and recruitment into the parkland populations. In addition, a lack of tree planting, increased intensification and mechanisation of cropping, uncontrolled tree felling for fuel, and increased urbanisation greatly reduce habitat diversity and contribute to degradation of shea parklands (Boffa 2015; Elias 2013; Lovett & Haq 2000).

Vitellaria paradoxa flowers are hermaphrodite, predominantly outcrossing, and insect pollination has been noted to result in modest increases in fruit set (Klein et al. 2007; Okullo et al. 2003). Recent studies in a village in southern Burkina Faso have suggested that *Apis mellifera* (honey bees) are the primary pollinating species, and showed increases in pollination success when *A. mellifera* hives were nearby (Lassen et al. 2016). At least four taxa of smaller stingless and solitary bees are thought to compensate for *A.*

mellifera in their absence (Lassen et al. 2016). However, there is a higher diversity of potential pollinator species in these habitats, including other bees, Diptera, Lepidoptera and Coleoptera, as well as several species of bird, and it is not clear which taxa visit shea flowers in addition to *A. mellifera* across the region. Furthermore, given geographic variation in genetic structure, microclimatic conditions, agricultural intensification and yields among sites (Boffa 2015; Gaisberger et al. 2017; Lovett & Haq 2000; Naughton et al. 2015), and the lack of applied pollination studies in this part of the world (Rodger et al. 2004), further study is required to confirm the findings of Lassen et al. (2016). In addition, it is not clear whether pollination is limiting fruit/seed set and weight, i.e. whether increased pollination could result in improved yields, particularly given the differences in yields recorded in different land-uses (Lamien et al. 2004). Pollinator limitation is common in many plant species (Burd 1994), caused by either insufficient pollinator visitation resulting in suboptimal pollen export and import, or caused by inappropriate pollen deposition (self or heterospecific).

To address these knowledge gaps, the current study tested the following hypotheses:

1. Flowers are predominantly visited by a small proportion of the available pollinating fauna, principally by honey bees and other wild social and solitary bees.

2. Shea yields (in terms of number of fruit set per inflorescence, fruit weight and seed weight) are

    i. lower when pollinators are excluded from flowers compared with open pollinated flowers i.e. flowers are dependent on animal-pollination, and

    ii. higher with pollen supplementation by hand compared with open pollinated flowers i.e. flowers are pollen limited.

MATERIALS AND METHODS

Study species

The flowers of *V. paradoxa* are produced during the dry season (December to April, depending on geographic location) in dense inflorescences at the end of usually leafless branches (Fig. 1). Inflorescences bear variable numbers of flowers (mean 31.9 ± 22.4 SD per inflorescence, $N = 330$ from 44 trees, authors' personal observations). Flowers are actinomorphic, approximately 15mm in diameter, have 8-10 creamy-white petals, and are protogynous, with the style (occasionally two styles) and fertile stigmas protruding from the buds before petals open (Hall et al. 1996) (Fig. 1). Nectar is produced at the base of the flower and is protected by petaloid staminodes which open early in the morning to allow access to flower visitors (Lassen 2016). Low volumes of nectar are produced, and small nectar standing crops have been recorded (mean 0.25 µl ± 0.58 SD per flower, $N = 20$ from 4 trees, authors' personal observations). After flowering, the ovary develops into a fruit containing one (occasionally two) seeds. Each inflorescence typically produces a small number of fruits (typically 2-3, rarely > 10, personal observations).

FIGURE 1:   *V. paradoxa* inflorescence, open male-stage flowers (with dehiscing anthers), and female-stage buds (with protruding stigmas, marked with the dashed circle).

Study sites

Six sites in northern Ghana (Kanfaiyili, Damongo, Zini 1 and Zini 2) and southern Burkina Faso (Torem 1 and Torem 2) were selected opportunistically in consultation with local communities (Fig. 2, Tab. 1). In each site, 8-10 mature, flowering *V. paradoxa* subsp. *paradoxa* trees, 10-50 m apart, were selected within an area of < 0.5 km² as focal trees.

Insect surveys

In all sites, insects visiting *V. paradoxa* flowers were captured via hand-netting (Fig. 3a). Ten minutes was spent at each tree and all individuals seen to visit flowers were captured using long-handled nets. Netting was conducted early (06:00-07:30 hrs GMT), during the middle (11:30-12:30) and late (16:30-18:00) in the day in order to maximise the chances of capturing all insect species visiting flowers. Netting was conducted on eight separate days in each of the Zini sites, on two days in each of Kanfaiyili and Damongo, and Torem 1, and on four days in Torem 2.

In four sites (Kanfaiyili, Damongo, Torem 1 and Torem 2), insects were also sampled using pan traps to survey the flower visitor fauna present in the sites (Westphal et al. 2008). Each trap consisted of three 1.2 m plastic pipes (60 mm diameter) driven 200mm into the soil, 1m apart from each other in a triangular pattern (Fig. 3b). Small plastic cups (35 mm deep, 70 mm diameter), painted with fluorescent white, yellow or blue paint (which have previously been shown to attract a range of insect taxa), were set into the top of each pipe and half-filled with water and a drop of detergent to break the surface tension (Droege et al. 2010). Traps were left open for 24 hours, and then the contents of each cup were strained and insects were stored in 70% alcohol until they could be identified. This was conducted four times in Kanfaiyili and Damongo and five times in each of the Torem sites.

In two sites (Zini 1 and Zini 2), direct observations of flower visitation were conducted to quantify visitation rates.

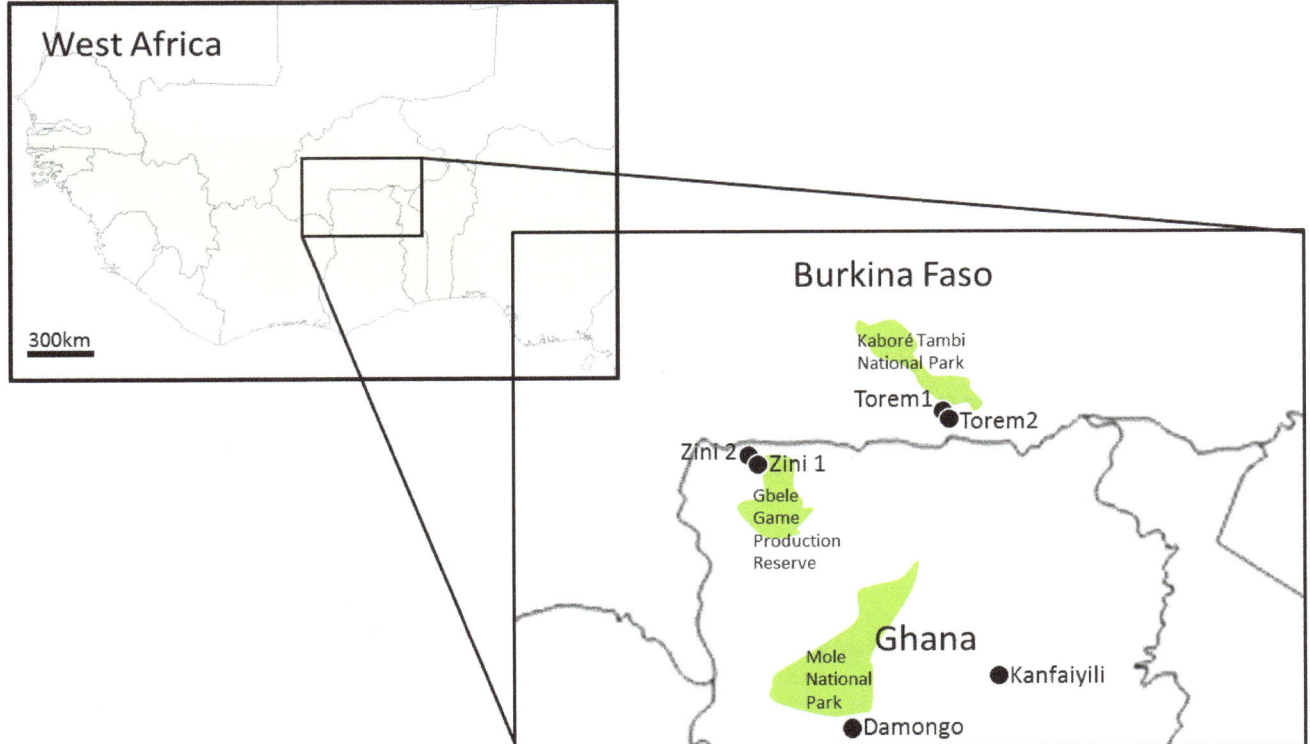

FIGURE 2:   The *Vitellaria paradoxa* (shea) zone in West Africa (Naughton et al. 2015) (main map,) and sampling sites (inset). Protected areas are shaded on inset map. For details see Tab. 1.

TABLE I: Study sites and data collected at each site.

Country	Site Name	Location (Grid Reference)	Site Description	Insect surveys			Pollination experiments		
				Pan trapping	Hand netting	Insect visitation rates	open	bagged	Supplemental hand pollination
Ghana	Kanfaiyili, Tamale	9.50688, -0.90753	Agricultural land 13km from centre of Tamale, little natural vegetation, 73km from Mole National Park	√	√		√	√	√
	Damongo	9.09127, -1.82695	Former fallow, beside school on edge of town, some natural vegetation, 8.6km from Mole National Park	√	√		√	√	√
	Zini 1	10.833333, -2.382556	Agricultural land located approx. 2 km away from dwellings, with some natural vegetation in the adjacent fallow land, 50km from Gbele Game Production Reserve		√	√	√	√	
	Zini 2	10.870583, -2.413556	Agricultural land located approx. 1 km away from dwellings with little natural vegetation, 55km from Gbele Game Production Reserve		√	√	√	√	
Burkina Faso	Torem 1	11.2170731, -1.1897331	Agricultural land within savannah landscape, 4.5km from Kaboré Tambi National Park, and 2km to Pô, chief town of Nahouri Province.	√	√		√	√	√
	Torem 2	11.206331, -1.1892506	As above, 5.5km from Kaboré Tambi National Park	√	√		√	√	√

Five inflorescences were observed over 10 minutes on each tree. Observations were conducted early (06:00-07:30), during the middle (11:30-12:30) and late (16:30-18:00) in the day in order to determine when visitors were most active. Observations were made on seven separate days in each site at approximately weekly intervals during February and March.

All insects were identified to order, except bees, which were identified to species.

Pollination treatments

On each tree, three pollination treatments were applied to entire inflorescences ($N = 3$ per treatment per tree) which were marked with coloured tape. These treatments were:

1. Open pollination ("open") – no manipulation of flowers

2. Pollinator exclusion ("bagged") – inflorescences were bagged whilst still in bud using bridal veil material (mesh

FIGURE 3:   Insect trapping methods a) hand netting using long-handled nets to sample insects directly from flowers, b) pan trapping using plastic cups painted with fluorescent yellow, white and blue, half-filled with water.

size ~1 mm) – any buds with already protruding styles were removed before bagging and bags were removed when flowering was completed.

3. Supplemental hand pollination ("hand pollination") – pollen was applied to protruding stigmas directly from the anthers of flowers from different trees.

Treatments were applied during January/February 2016 when trees started to flower and fruit formation was monitored until maturity in June 2016. Due to logistical constraints, hand pollination was only performed on each inflorescence once during flowering, and any untreated flowers were removed from inflorescences. Fruits from all treatments were counted in the middle of May 2016 before fruit dehiscence, immediately weighed, and then pulp was removed and seeds counted and weighed.

Data analysis

Differences in 'fruit number' (number of fruit per inflorescence), 'fruit weight' (mean fruit weight per inflorescence) and 'seed weight' (mean seed weight per inflorescence) between "bagged", "open" and "hand pollinated" treatments were assessed using Generalized Linear Mixed Models (GLMMs). Thus, we compared both the quantity (fruit number) and quality of fruit (fruit and seed weight) in the presence and absence of pollinators ('open' vs. 'bagged' treatments). The comparison between 'open' and 'hand pollinated' treatments was used to assess the role of pollen limitation in determining fruit quantity and quality in the field, with differences between open and hand pollinated treatments representing the shortfall between current productivity and the potential maximum productivity with no pollen limitation. In some cases it was not possible to weigh fruits or seeds as fruit had already fallen from the trees prior to the return site visits, hence for fruit number $N = 464$, whilst for fruit weight $N = 137$ and for seed weight $N = 134$. For each response variable ('fruit number', 'fruit weight' and 'seed weight') a separate GLMM was constructed with

treatment as a predictor variable (fixed factor) and 'Tree ID' nested within 'Site' as a nested random factor to account for correlation between measures both within trees and sites. GLMMs were initially fitted with a Gaussian response distribution and residuals were assessed for normality using a Shapiro-Wilk test. In all three cases, model residuals were not normally distributed, therefore models were recalculated using alternative response distributions. 'Fruit number' models were first refitted using a Poisson distribution with a log link function appropriate for count data, but model residuals showed over-dispersion and therefore the model was refitted using a negative binomial model with a log link function. 'Fruit weight' and 'seed weight' models were refitted using a 'gamma' distribution with a log link function suitable for non-negative continuous data. GLMMs were fitted using the package 'glmmADMB' in R 3.3.3 (Fournier et al. 2012; R Core Team 2017). Where GLMMs indicated significant differences between treatments, post-host analysis of the pairwise differences between treatment combinations (i.e. open vs. bagged, open vs. hand pollinated and bagged vs. hand pollinated) were calculated using the Tukey Honest Significant Difference test in the package 'multcomp' (Hothorn et al. 2008).

RESULTS

Insect surveys

Over a total of 26 ten-minute hand-netting periods, a total of 280 insects were captured whilst foraging on shea flowers (Tab. 2); 247 (88.2%) individuals were bees, from six species (*Apis mellifera adansonii*, *Ceratina moerenhouti*, *Compsomelissa nigrinervis*, *Hypotrigona gribodoi*, *Meliponula ferruginea* and *Meliponula beccari*). *Hypotrigona gribodoi* were most frequently captured (182 individuals), followed by *Apis mellifera adansonii* (48 individuals). The Zini sites had the greatest diversity of bees (five species), whilst no bees were captured on flowers at Torem I.

TABLE 2: Insects captured at each site using Pan trapping and Hand netting (K = Kanfaiyili, D = Damongo, T1 = Torem 1, T2 = Torem 2, Z = Zini – both sites combined).

Site	Pan trapping				Hand netting				
	K	D	T1	T2	K	D	T1	T2	Z
Number of trapping sessions	4	4	5	5	2	2	2	4	16
Bees									
Apis mellifera adansonii	0	1	0	0	11	6	0	1	26
Hypotrigona gribodoi	2	7	8	3	6	2	0	1	173
Meliponula ferruginea	0	0	0	0	2	2	0	0	2
Meliponula beccari	0	0	0	0	0	0	0	1	4
Compsomelissa	0	0	1	3	0	0	0	0	2
Xylocopa olivacea	1	0	0	0	0	0	0	0	0
Amegilla calens	2	2	8	4	0	0	0	0	0
Lassioglossum duponti	0	4	1	1	0	0	0	0	0
Lipotriches natalensis	0	0	3	3	0	0	0	0	0
Pseudoanthidium truncatum	0	0	0	1	0	0	0	0	0
Ceratina moerenhouti	0	0	0	0	0	0	0	1	0
Bees total	9	22	27	16	19	10	3	4	207
Wasps	5	13	7	5	0	1	1	3	3
Flies	24	65	26	8	1	0	0	6	0
Beetles	4	12	10	10	0	2	1	0	0
Ants	0	5	2	1	0	0	0	0	4
Bugs	0	8	1	3	2	0	0	1	0
others	11	15	14	18	1	1	4	2	0
All insects total	53	140	87	61	23	14	9	16	214
Insects per trapping session	13	35	17	12	11.5	7	4.5	4	13.4
number of bee spp	3	4	5	6	3	3	0	4	5
% bees	17	16	31	26	83	71	33	25	97

TABLE 3: Shea flower visitation in the two sites in Zini by *Apis mellifera adansonii* and other bees early (06.00-07.30), during the middle (11.30-12.30) and late (16.30-18.00) in the day.

		Total number of bees observed		Average number of visits per inflorescence per hour
		Apis mellifera	Other bees	
Zini1				
	Early	79	223	0.014
	Middle	0	37	0.0017
	Late	8	122	0.0060
Zini2				
	Early	56	151	0.0096
	Middle	0	21	0.00097
	Late	5	82	0.0041

A total of 341 insects were captured in pan traps over the four sites (Tab. 2). Seventy-four (21.7%) individuals were bees, from eight species (*Amegilla calens, Apis mellifera adansonii, Compsomelissa nigrinervis, Hypotrigona gribodoi, Lassioglossum duponti, Lipotriches natalensis, Pseudoanthidium truncatum* and *Xylocopa olivacea*). Other insects were only identified to Order, including Hymenoptera (wasps and ants), Diptera (flies), Coleoptera (beetles), Hemiptera (bugs) and other unidentified specimens (Tab. 2). The highest abundance of insects was captured at Damongo, but Torem 2 had the greatest species richness of bees (six species), whilst Kanfaiyili had only three bee species.

TABLE 4: Fruit set (number of fruits per inflorescence), fruit weight (mean per inflorescence) and seed weight (mean per inflorescence) following pollination treatments ("Bagged" = pollinator exclusion; "Open" = no manipulation; "Hand pollinated" = supplemental hand pollination; N = number of inflorescences, fruits and seeds respectively).

Site	Treatment	Fruit number			Fruit weight (g)			Seed weight (g)		
		N	Mean	Range	N	Mean	Range	N	Mean	Range
Damongo	Bagged	30	0.27	0 - 2	1	9.96	9.96 - 9.96	1	0.32	0.32 - 0.32
	Open	29	1.21	0 - 6	6	21.31	10.66 - 28.24	6	7.74	3.87 - 9.86
	Hand pollinated	30	1.40	0 - 6	9	16.63	8.17 - 25.64	9	6.56	3.38 - 10.37
Kanfiayili	Bagged	30	0.93	0 - 4	8	11.27	3.93 - 17.32	8	5.09	1.04 - 8.43
	Open	30	3.17	0 - 16	9	10.47	5.24 - 16.22	9	5.32	2.72 - 8.16
	Hand pollinated	28	2.25	0 - 12	10	9.37	3.37 - 23.98	10	5.63	2.31 - 16.90
Torem 1	Bagged	30	0.50	0 - 3	5	23.2	14.75 - 36.82	5	7.56	5.77 - 8.98
	Open	30	1.37	0 - 11	3	20.32	16.40 - 22.47	3	8.2	7.83 - 8.68
	Hand pollinated	30	2.53	0 - 10	8	20.85	14.54 - 40.71	8	7.82	4.25 - 14.60
Torem 2	Bagged	30	2.20	0 - 6	3	13.41	10.96 - 16.78	3	6.55	6.31 - 6.93
	Open	30	2.47	0 - 6	12	18.02	11.04 - 34.46	12	7.36	4.56 - 11.35
	Hand pollinated	29	3.72	0 - 14	14	19.24	9.29 - 27.19	14	7.13	4.07 - 10.55
Zini1	Bagged	27	0.48	0 - 3	9	16.58	0.70 - 31.75	6	8.13	5.90 - 9.70
	Open	27	2.00	0 - 7	20	25.56	16.10 - 40.00	20	8.15	2.80 - 11.60
Zini2	Bagged	27	0.19	0 - 1	5	21.5	14.30 - 27.20	5	5.98	2.30 - 8.10
	Open	27	0.85	0 - 3	15	24.24	13.90 - 38.00	15	8.42	2.30 - 14.10

Observations of visitation rates at the two Zini sites confirmed that bees were the most frequent visitors to flowers. During a total of 71.5 hours of observations, 784 flower visitors were recorded: 148 *A. mellifera* and 636 other bees (Tab. 3). *A. mellifera* were mainly active early in the day (06:00-07:30), were never seen on flowers in the middle of the day, and rarely seen later in the day. Other bees were also most active early in the day, less active in the middle of the day, but were also reasonably active later in the day. Total visitation rates (bees per inflorescence per hour) were more than twice as frequent in the morning compared to the afternoon (0.059 visitors per hour in the morning vs. 0.025 visitors per hour in the afternoon).

Pollination treatments

On average, across all sites, open pollinated inflorescences produced 1.86 ± 0.18 (mean ± s.e.) fruits per inflorescence, whilst pollinator exclusion (bagged inflorescences) produced 0.78 ± 0.09 fruits. Fruit set was marginally higher with supplemental hand pollination, with 2.47 ± 0.26 fruits per inflorescence (Tab. 4).

Fruit number was significantly higher in 'open' and 'hand pollinated' treatments than in 'bagged' treatments (β = 0.936 ± 0.129, P < 0.001; and β = 1.107 ± 0.139, P < 0.001 respectively). There was no significant difference in fruit number between 'open' and 'hand pollinated' treatments (β = -0.135 ± 0.176, P = 0.717) (Fig. 4a). Fruit weight and seed weight did not differ significantly between treatments (χ = 3.864, P = 0.145 and χ = 1.144, P = 0.564) (Fig. 4b & 4c). Some variation was observed among sites, with bagged flowers at Torem I and open flowers at Kanfiayili producing relatively more fruit than the same treatments in other sites (Appendix I).

DISCUSSION

Visitors to *V. paradoxa* flowers during our study were primarily bees, despite the presence of other flower-visiting insects in the shea parklands during the flowering season. Unlike Lassen et al. (2016), we did not observe sunbirds visiting shea flowers, nor the flowers of the hemi-parasitic plants that grow on shea trees (Zwarts 2015), during observation periods. It is possible that the phenology, nectar chemistry or structure of the flowers place constraints on which insects can forage on *V. paradoxa* flowers. Those insects active later in the day, or with a proboscis which is too short to access the nectaries, may be prevented from utilising shea as a forage resource, and thus not be frequent visitors or effective pollinators (Nienhuis & Stout 2009; Stang et al. 2006). In addition, the nectar chemistry, and secondary compounds present in the nectar (Meda et al. 2005), may also affect which species visit flowers (Adler 2000; Tiedeken et al. 2016).

Honey bees (*Apis mellifera adansonii*) and six species of stingless bee were recorded as the most frequent visitors, and like Lassen et al. (2016), we found that honey bees preferred to visit early in the morning, whilst temperatures were lower. Very little nectar was found in flowers, even though we sampled shortly after dawn. Later in the day, the very low nectar volumes available in flowers, as well as higher temperatures, might explain the absence of honey bees. However, without further understanding of patterns of nectar secretion, visual and olfactory floral traits, or quality of nectar

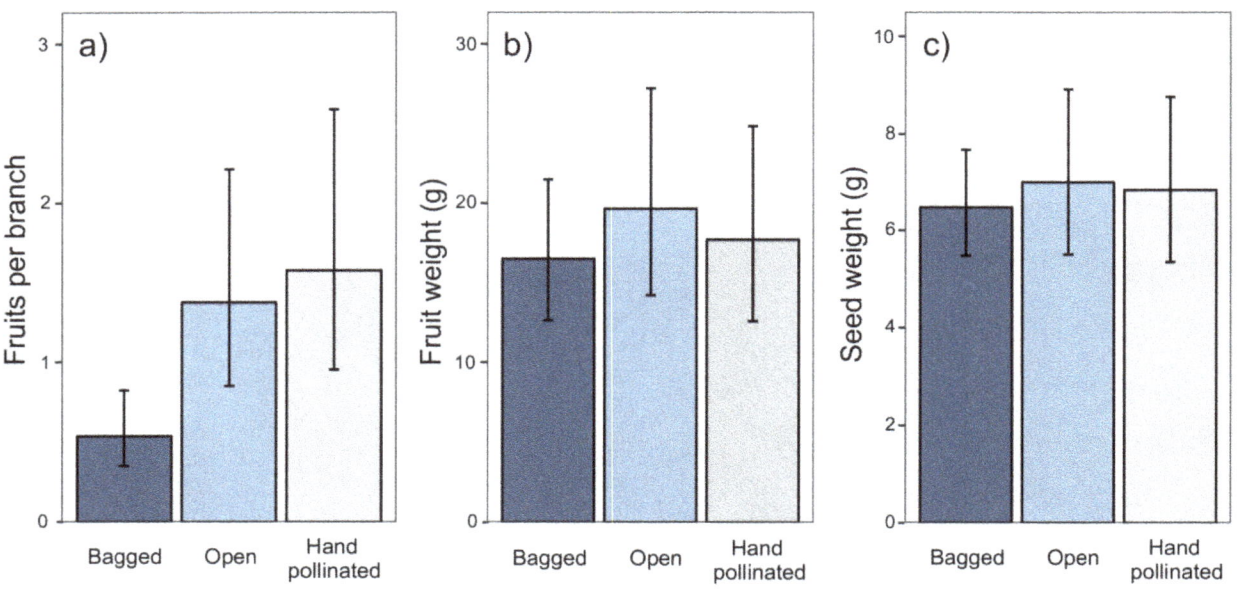

FIGURE 4:   Fitted mean ± 95% of confidence intervals based on GLMM outputs for a) fruit set, b) fruit weight and c) seed weight from bagged, open and hand-pollinated shea flowers.

and pollen rewards, in both shea and other co-flowering species, it is not possible to speculate on what is driving visitor behaviour. We did not assess the behaviour or relative pollination efficiency of the different bee visitors in this study. However, some assessment of the ability of different species to pick-up and deposit pollen appropriately (spatially and temporally), in addition to the frequency of their visits, would be important to determine which pollinators are the most effective (Kasina et al. 2013). It is possible that the stingless *Hypotrigona gribodoi* bees, whilst more frequent visitors than honey bees, are less efficient as pollinators due to their smaller body size. Furthermore, observations of behaviour of different species would held to determine how pollination service varies spatially within trees and within populations (Kwapong et al. 2014).

We confirmed that *V. paradoxa* flowers require insect visits for fruit production (Lassen et al. 2016; Yidana 2004). Fruit set per inflorescence more than doubled in the open compared with bagged treatments. Fruit can fall to the ground or be taken by birds/bats and so it is possible that our fruit counts for this treatment are under-representative of actual fruit set. Some fruit were produced from the bagged treatments, particularly in Torem 1, and we assume that this was due to within-bag pollen movement facilitated by wind or animals, pollen entering through the mesh of the bags, stigmas extruding outside of bags, or due to incomplete protogyny (Silva & Goring 2001). Flowers do appear to be strongly protogynous though, with the stigma becoming receptive before anthers dehisce, making within-flower self-fertilization unlikely. However, on a single inflorescence there can be many closely-packed flowers at different stages of maturity (Fig. 1). Thus within-inflorescence selfing can occur. Indeed, within a single tree, insect pollination can facilitate geitonogamy (within-plant selfing) – this might be particularly common on

trees with an abundance of flowers, or on trees isolated from neighbours.

Although hand pollination was associated with higher mean fruit set in the Torem sites and at Damongo, the same was not true at Kanfaiyili where open flowers produced more fruit per inflorescence than at the other sites, and overall there was no statistical difference in fruit set between and hand and open pollinated inflorescences. This could be an artefact of the methods used: hand pollination was performed on each tree only once during flowering and we removed non-treated flowers, thus potentially reducing the total number of flowers on hand-pollinated inflorescences. It may have been better to mark treated flowers and repeat treatments throughout the flowering season so that all flowers on an inflorescence were treated with outcross pollen. Also it is possible that we treated non-receptive stigmas, as 27.5% of hand-pollinated inflorescences produced zero fruit (data not shown). Alternatively, we may have just failed to detect pollen limitation because hand pollinations were not done at the scale of the whole tree (Wesselingh 2007). Furthermore, fruit set may be limited by other biotic and abiotic resources, which vary greatly from site to site, including soil, nutrients, water availability, pesticide use, and pest and pollinator density. Given the variation between sites and the methodological limitations, and potential other constraints on fruit production, it would be worth repeating these tests for pollen limitation.

Since we found no evidence for reduced fruit or seed weight as a result of pollinator exclusion, we could also be tempted to conclude that whilst pollinator visitation influences yield quantity, it has no impact on quality. However, we did not perform germination tests to determine biological "quality", nor did we examine the oil content of the seeds to determine economic quality according to treatment. These are questions that require further investigation.

The lack of bees captured on flowers at Torem I may simply have been due to low sampling effort (only two netting sessions were conducted). Bees were captured at this site in pan traps, and given fruits were produced from the open pollination treatments at this site, we can assume that flowers were visited at some point. More comprehensive surveying of visitors, including potential nocturnal pollinators, would be beneficial, although logistically difficult.

All six sites used in this study were within actively cultivated systems, except the one on the edge of the small town of Damongo, which was an old fallow - the current shea tree distribution at this site shows that these were once-farmed parklands. This site was also within 10km of the largest protected area in Ghana (Mole National Park), and in an area which has experienced a decline in farming due to recent economic focus on timber extraction. These factors may explain the elevated number of insects captured in pan-traps at this site. However, we did not see an increase in the number of flower visitors, or in fruit set, at Damongo. Although previous studies have shown that proximity to protected areas can increase pollination services to tropical trees (e.g. Freitas et al. 2014), our study design did not allow us to test this here. While three sites (Damongo, and the two Zini sites) were located relatively near to protected areas, and these areas may provide resources for bees, small stingless bees are unlikely to travel this far to forage (Araújo et al. 2004). Honey bees can travel over large distances to forage, and so the spatial pattern of resources at a landscape scale may influence visitation and pollination of shea. The proximity to urban centres, habitat structure and land- use (including fallow periods, ploughing methods, tree density, pesticide use etc.) could all influence bee abundance and deserve further consideration. Furthermore, how the pollinators of *V. paradoxa* respond to other flowering plants in the parklands, and what limits their populations (natural enemies and response to environmental fluctuations, as well as anthropogenic activity) should all be addressed in order to improve management recommendations for maximising pollination services in the shea parklands.

In conclusion, we have demonstrated that bees are very important for the pollination and fruit set of *V. paradoxa* across several sites in the agroforestry parklands of West Africa. Continued clearing of natural habitat and less regeneration via fallows could potentially damage bee populations (Tornyie & Kwapong 2015). Beekeeping, with both honey bees and stingless bees, as well as enhancing plant diversity in the parkland as a food source for bees, could therefore present a win-win opportunity for local communities, enhancing both pollination services to *V. paradoxa* and other useful fruit-bearing trees and other crops (Kudom & Kwapong 2010; Kasina et al. 2013; Kiatoko et al. 2014), and providing honey to supplement household incomes. Furthermore, increased semi-natural habitat may have wider biodiversity benefits, for example in supporting nesting and foraging sites for resident and migrant birds (Zwarts 2015). However, more research is needed to understand the pollination ecology of these sub-Saharan ecosystems, including both cultivated and wild plant species, as well as community-level interactions (Rodger et al. 2004; Gemmill-Herren et al. 2014).

ACKNOWLEDGEMENTS

This work was funded by BirdLife International and Vogelbescherming Nederland (BirdLife in The Netherlands) and the RSPB, and was part of a project investigating socio-economic and ornithological characteristics of the Shea parkland. We are grateful to Abukari Andani, James Braimah and Gnané Frank and their families for access to sites and for their assistance with field work at Kanfiayilli, Damongo and Torem respectively.

APPENDICES

Additional supporting information may be found in the online version of this article:
APPENDIX I.   Fruit set per treatment per site

REFERENCES

Adler LS (2000) The ecological significance of toxic nectar. Oikos 91:409-420.

Araújo ED, Costa M, Chaud-Netto J, Fowler HG (2004) Body size and flight distance in stingless bees (Hymenoptera: Meliponini): inference of flight range and possible ecological implications. Brazilian Journal of Biology 64:563-568.

Bodart C, Brink A, Donnay F, Lupi A, Mayaux P, Achard F (2013) Continental estimates of forest cover and forest cover changes in the dry ecosystems of Africa between 1990 and 2000. Journal of Biogeography 40:1036-1047.

Boffa J-M (1999) Agroforestry parklands in sub-Saharan Africa. Food and Agriculture Organization of the United Nations, Rome.

Boffa J-M (2015) Opportunities and challenges in the improvement of the shea (*Vitellaria paradoxa*) resource and its management. World Agroforestry Centre Occasional paper 24, Nairobi.

Bommarco R, Marini L, Vaissière B (2012) Insect pollination enhances seed yield, quality, and market value in oilseed rape. Oecologia 169:1025-1032.

Brittain C, Kremen C, Garber A, Klein A-M (2014) Pollination and Plant Resources Change the Nutritional Quality of Almonds for Human Health. PLoS ONE 9:e90082. doi: 10.1371/journal.pone.0090082.

Burd M (1994) Bateman principle and plant reproduction - the role of pollen limitation in fruit and seed set. Botanical Review 60:83-139.

Carvalheiro LG, Seymour CL, Veldtman R, Nicolson SW (2010) Pollination services decline with distance from natural habitat even in biodiversity-rich areas. Journal of Applied Ecology 47:810-820.

Droege SAM, Tepedino VJ, Lebuhn G, Link W, Minckley RL, Chen Q, Conrad C (2010) Spatial patterns of bee captures in North American bowl trapping surveys. Insect Conservation and Diversity 3:15-23.

Elias M (2013) Influence of agroforestry practices on the structure and spatiality of shea trees (*Vitellaria paradoxa* C.F. Gaertn.) in central-west Burkina Faso. Agroforestry Systems 87:203-216. doi: 10.1007/s10457-012-9536-2.

Fournier D, Skaug H, Ancheta J, Ianelli J, Magnusson A, Maunder M, Nielsen A, Sibert J (2012) AD Model Builder: using automatic differentiation for statistical inference of highly parameterized complex nonlinear models. Optimization Methods and Software 27:233-249.

Freitas BM, Pacheco Filho AJS, Andrade PB, Lemos CQ, Rocha EEM, Pereira NO, Bezerra ADM, Nogueira DS, Alencar RL, Rocha RF, Mendonça KS (2014) Forest remnants enhance wild pollinator visits to cashew flowers and mitigate pollination deficit in NE Brazil. Journal of Pollination Ecology 12(4): 22-30.

Gaisberger H, Kindt R, Loo J, Schmidt M, Bognounou F, Da SS, Diallo OB, Ganaba SS, Gnoumou A, Lompo D, Lykke AM, Mbayngone E, Nacoulma BMI, Ouedraogo M, Ouédraogo O, Parkouda C, Porembski S, Savadogo P, Thiombiano A, Zerbo G, Vinceti B (2017) Spatially explicit multi-threat assessment of food tree species in Burkina Faso: A fine-scale approach. PLOS ONE 12:e0184457. doi: 10.1371/journal.pone.0184457.

Gallagher DE, Dueppen SA, Walsh R (2016) The archaeology of shea butter (*Vitellaria paradoxa*) in Burkina Faso, West Africa. Journal of Ethnobiology 36:150-171.

Garibaldi LA, Vaissière BE, Gemmill-Herren B, Hipólito J, Freitas BM, Ngo HT, Azzu N, Sáez A, Åström J, An J, Blochtein B, Buchori D, García FJC, Oliveira da Silva F, Devkota K, Ribeiro MdF, Freitas L, Gaglianone MC, Goss M, Irshad M, Kasina M, Filho AJSP, Kiill LHP, Kwapong P, Parra GN, Pires C, Pires V, Rawal RS, Rizali A, Saraiva AM, Veldtman R, Viana BF, Witter S, Zhang H (2016) Mutually beneficial pollinator diversity and crop yield outcomes in small and large farms. Science 351:388-391.

Gemmill-Herren B, Kwapong PK, Aidoo K, Martins D, Kinuthia W, Gikungu M, Eardley CD (2014) Priorities for research and development in the management of pollination services for agricultural development in Africa. Journal of Pollination Ecology 12:40-51..

Goulson D, Nicholls E, Botías C, Rotheray EL (2015) Bee declines driven by combined stress from parasites, pesticides, and lack of flowers. Science 347:1255957.

Hall JB, Aebischer DP, Tomlinson HF, Osei-Amaning E, Hindle JR (1996) *Vitellaria paradoxa*: a monograph School of Agricultural and Forest Sciences publication, no. 8., University of Wales, Bangor, UK.

Hothorn T, Bretz F, Westfall P (2008) Simultaneous Inference in General Parametric Models. Biometrical Journal 50:346-363.

Kandji S, Verchot L, Mackensen J (2006) Climate change and variability in the Sahel region: impacts and adaptation strategies in the agricultural sector. World Agroforestry Centre (ICRAF), Nairobi (Kenya)

Kasina M, Kraemer M, Nderitu J, Martius C, Wittmann D (2013) The importance of African honey bees (*Apis mellifera* L.) as pollinators of high value crops in Kenya: A case of Butternut Squash (*Cucurbita Moschata* Duchesne ex Poir.) pollination. East African Journal of Agriculture and Forestry 79:143-149.

Kennedy CM et al. (2013) A global quantitative synthesis of local and landscape effects on wild bee pollinators in agroecosystems. Ecology Letters 16:584-599.

Kiatoko N, Raina SK, Muli E, Mueke J (2014) Enhancement of fruit quality in *Capsicum annum* through pollination by *Hypotrigona gribodoi* in Kakamega, Western Kenya. Entomological Science 17:106-110.

Klein A-M, Steffan-Dewenter I, Tscharntke T (2003) Fruit set of highland coffee increases with the diversity of pollinating bees. Proceedings of the Royal Society B: Biological Sciences 270:955-961.

Klein AM, Vaissière BE, Cane JH, Steffan-Dewenter I, Cunningham SA, Kremen C, Tscharntke T (2007) Importance of pollinators in changing landscapes for world crops. Proceedings of the Royal Society Series B - Biological Sciences 274:303-313.

Kudom AA, Kwapong PK (2010) Floral visitors of *Ananas comosus* in Ghana: A preliminary assessment. Journal of Pollination Ecology 2:27-32.

Kwapong PK, Frimpong-Anin K, Ahedor B (2014) Pollination and yield dynamics of cocoa tree. Research and Reviews in Biosciences 8:337-342.

Lamien N, Ouédraogo SJ, Boukary DO, Guinko S (2004) Productivité fruitière du karité (*Vitellaria paradoxa* Gaertn. C. F., Sapotaceae) dans les parcs agroforestiers traditionnels au Burkina Faso. Fruits 59:1-7.

Lassen KM (2016) Pollination strategies to increase productivity of the African fruit trees *Vitellaria paradoxa* subsp. *paradoxa* and *Parkia biglobosa*. PhD, University of Copenhagen.

Lassen KM, Nielsen LR, Lompo D, Dupont YL, Kjær ED (2016) Honey bees are essential for pollination of *Vitellaria paradoxa* subsp. *paradoxa* (Sapotaceae) in Burkina Faso. Agroforestry Systems:1-12.

Lovett PN, Haq N (2000) Evidence for anthropic selection of the Sheanut tree (*Vitellaria paradoxa*). Agroforestry Systems 48:273-288.

Macias-Macias O, Chuc J, Ancona-Xiu P, Cauich O, Quezada-Euán JJG (2009) Contribution of native bees and Africanized honey bees (Hymenoptera:Apoidea) to Solanaceae crop pollination in tropical México. Journal of Applied Entomology 133:456-465.

Meda A, Lamien CE, Romito M, Millogo J, Nacoulma OG (2005) Determination of the total phenolic, flavonoid and proline contents in Burkina Fasan honey, as well as their radical scavenging activity. Food Chemistry 91:571-577.

Naughton C, Lovett P, Mihelcic J (2015) Land suitability modeling of shea (*Vitellaria paradoxa*) distribution across sub-Saharan Africa. Applied Geography 58:217-227.

Nienhuis CM, Stout JC (2009) Effectiveness of native bumblebees as pollinators of the alien invasive plant *Impatiens glandulifera* (Balsiminaceae) in Ireland. Journal of Pollination Ecology 1:1-11.

Okullo JB, Hall JB, Masters E (2003) Reproductive biology and breeding systems of *Vitellaria paradoxa*. In: Teklehaimanot Z (ed) Improved Management of Agroforestry Parkland Systems in Sub-Saharan Africa. EU/INCO Project Contract IC18-CT98-0261, Final Report, University of Wales Bangor, UK, pp 66-84.

Ollerton J, Winfree R, Tarrant S (2011) How many flowering plants are pollinated by animals? Oikos 120:321-326.

Pouliot M (2012) Contribution of "Women's Gold" to West African Livelihoods: The Case of Shea (*Vitellaria paradoxa*) in Burkina Faso. Economic Botany 66:237-248.

R Core Team (2017) R: A language and environment for statistical computing. R Foundation for Statistical Computing, Vienna, Austria.

Rodger JG, Balkwill K, Gemmill B (2004) African pollination studies: where are the gaps? International Journal of Tropical Insect Science 24:5-28.

Schreckenberg K, Awono A, Degrande A, Mbosso C, Ndoye O, Tchoundjeu Z (2006) Domesticating indigenous fruit trees as a contribution to poverty reduction. 16:35-51.

Silva NF, Goring DR (2001) Mechanisms of self-incompatibility in flowering plants. Cell. Mol. Life Sci. 58:1988-2007.

Stang M, Klinkhamer PGL, van der Meijden E (2006) Size constraints and flower abundance determine the number of interactions in a plant-flower visitor web. Oikos 112:111-121.

Tiedeken EJ et al. (2016) Nectar chemistry modulates the impact of an invasive plant on native pollinators. Functional Ecology 30:885-893.

Tornyie F, Kwapong PK (2015) Nesting ecology of stingless bees and potential threats to their survival within selected landscapes in the northern Volta region of Ghana. African Journal of Ecology 53:398-405.

Wesselingh RA (2007) Pollen limitation meets resource allocation: towards a comprehensive methodology. New Phytologist 174 26-37.

Westphal C et al. (2008) Measuring bee diversity in different European habitats and biogeographical regions. Ecological Monographs 78:653-671.

Yidana JA (2004) Progress in developing technologies to domesticate the cultivation of shea tree (*Vitellaria paradoxa* L.) in Ghana. Agricultural and Food Science Journal of Ghana 3:249-267.

Zwarts L (2015) Tree preference of insectivorous birds in the *Vitellaria* zone, West Africa. A&W-report 2152, Feanwâlden, The Netherlands

22

Estimating pollinator performance of visitors to the self-incompatible crop-plant *Brassica rapa* by single visit deposition and pollen germination: a comparison of methods

R Patchett[1], G Ballantyne[2] & PG Willmer[1]*

[1]*Sir Harold Mitchell Building, School of Biology, University of St Andrews, St Andrews KY16 9TH*
[2]*School of Applied Sciences, Edinburgh Napier University, Edinburgh, EH11 4BN*

Abstract—Estimating the pollen-deposition effectiveness of flower visitors is fundamental to understanding their performance as pollinators. While estimates of visitation rates, pollen loads, and single visit deposition (SVD) are all useful proxies for performance, and so help to reveal the relative effectiveness of different visitors, none take into account the breeding system of the plants, or the quality of pollen deposited. Here we compare the performance of visitors to the self-incompatible plant *Brassica rapa* (turnip) using SVD and pollen germination. We also report the first use of the staining of *Brassica rapa* stigma papilla cells (known to reveal a specific reaction to self-pollen) to compare self-pollen deposition between insect visitors. We found that most of the pollen grains deposited by insect visitors (and therefore counted by SVD methods) were non-germinating self-pollen. A smaller proportion of grains were outcrossed and so germinated. There was also a significant positive relationship between environmental conditions (wind speed) and pollen deposition, but not pollen germination.

Both methods identified *Bombus* spp. as the best-performing visitors on turnip flowers, followed by *Eristalis* spp., whereas performance estimates for *Episyrphus balteatus* and 'other hoverflies' were no higher than controls for both methods. This study provides further insight into the methodology for estimating pollinator performance, especially in plants when only cross-pollen can germinate.

Keywords: Pollination, Pollinator effectiveness, Pollen deposition, Pollen germination, Brassicaceae

Introduction

Understanding plant-pollinator interactions is vital, as pollinators play a key role in ecosystem services that maintain biodiversity. Thirty-five per cent of global food production relies on insect pollinators (Klein et al. 2007), and approximately 87% of flowering plant species globally are entomophilous (Ollerton et al. 2011). With many insect pollinator populations in decline (e.g. Potts et al. 2010) and a heavy dependency on a small number of pollinator species for crop pollination (Kleijn et al. 2015), a deeper quantitative insight into pollinator performance on crops is essential. Surveys of visitation patterns and rates can produce valuable large datasets quickly, but lack information on visit quality. More time-consuming surveys of single visit pollen deposition (SVD; *sensu* Ne'eman et al. 2010) provide a measure of visit quality and can distinguish conspecific deposition, but lack information on the viability of the pollen deposited (Ballantyne et al. 2015; Ballantyne et al. 2017). Here we compare SVD measures with counts of the number of pollen grains germinating after a single visit to a self-incompatible species, providing a direct measure of cross-pollen deposition. Outcrossed pollen is essential for fertilisation in self-incompatible species, and self-pollen deposition can be deleterious via stigma-clogging (Shore &

Barrett 1983; Galen et al. 1989; Gross 2005).

Counting the number of germinated pollen grains after a single visit is the ideal method for estimating pollinator performance in self-incompatible plants, but is only suitable for species where the reproductive systems are well-known. Wist & Davis (2013) aimed to compare SVD with pollen germination in the apparently self-incompatible *Echinacea angustifolia* (Asteraceae), but found it to be self-compatible, confounding pollen germination results. There are thus no studies to date that focus on a fully self-incompatible plant species.

Turnip, *Brassica rapa*, was chosen as an important crop species that is also fully self-incompatible through the sporophytic self-incompatibility (SSI) mechanism (Hiscock & McInnis 2003). Callose plugs are deposited by pollen tubes as they grow, and stigmatic papillae produce callose in response to the presence of self-pollen (Currier 1957). Callose can be stained with aniline blue and viewed with fluorescence microscopy (Kearns & Inouye 1993; Ästergaard et al. 2002). This allows pollen tubes of successfully germinated pollen grains, and also papilla cells that have reacted to self-pollen, to be identified and quantified. The average number of successfully germinated pollen grains after single visits from insect groups is then a measure of pollinator performance. For the first time ever, the papilla cell response can then be used to show the presence of self-pollen, so that self-pollen deposition can be compared between visitors.

*Corresponding author: pgw@st-andrews.ac.uk

This study asks the question: how do pollinator performance estimates compare between SVD and pollen germination methods? A positive relationship between SVD and pollen germination will indicate that SVD estimates are unlikely to be confounded by self-pollen deposition.

MATERIALS AND METHODS

Plants, visitors and study site

A row of *Brassica rapa* ssp. *rapa* (Brassicaceae) was grown from seed at Earlshall Castle garden, Fife (56 22.8' N, 2 52.1' W). Thirty-three plants were individually labelled for identification. The field site had a variety of flowering plant species in its vicinity, but the plot was maintained so that only *Brassica rapa* flowered within the experimental row. Data were collected through the main flowering period (28th August to 18th September 2015) between 10:00 and 15:00 on each suitable day, whenever dry and calm weather conditions permitted. Temperature, humidity and wind speed were recorded at 30-minute intervals throughout the sampling period.

Flower visitors were identified by photography or catch and release methods (Appendix I). *Bombus* were identified to species (although *B. terrestris* and *B. lucorum* are difficult to distinguish in the field (Falk 2015), so we hereafter refer to *B. terrestris/lucorum*); but later grouped by genus for analysis. For hoverflies, *Eristalis* were identified to genus, and *Episyrphus balteatus* to species. All others were small syrphids and grouped as 'other hoverflies'. Visits from *Apis mellifera*, a *Lasioglossum* species and a *Sphecodes* species were also recorded, but were too infrequent to include in the analysis.

Measuring pollinator performance

Inflorescences were covered with mesh bags 24 hours prior to sampling, and a single petal was removed to identify flowers that were open prior to bagging. Flowers that opened and dehisced whilst in the bag were then used for SVD or pollen germination sampling. No data were collected from undehisced flowers or flowers greater than one day old based on preliminary tests that confirmed cross-pollen grains were best able to germinate on flowers that had dehisced and were <I day old.

For SVD measurements, individual open and virgin flowers were observed until their first insect visit. The time, visitor identity, visit duration and foraging behaviour (feeding on pollen or nectar) were recorded for each visit. After the visitor had left, the flower was carefully dissected using fine forceps, and the stigma's receptive surfaces were then dabbed onto a cube of fuchsin gel on a microscope slide, which was then melted under a coverslip. Stigmas were checked using a hand lens to ensure all pollen had been removed. Pollen grains were identified under a light microscope as either conspecific or heterospecific (Fig.I). This process was repeated ($N = 39$) on unvisited flowers to control for pollen deposition due to wind or to handling.

For pollen germination, the SVD method was followed until a visit was completed, but the stigma was not disturbed. Instead, the anthers were removed to prevent any additional

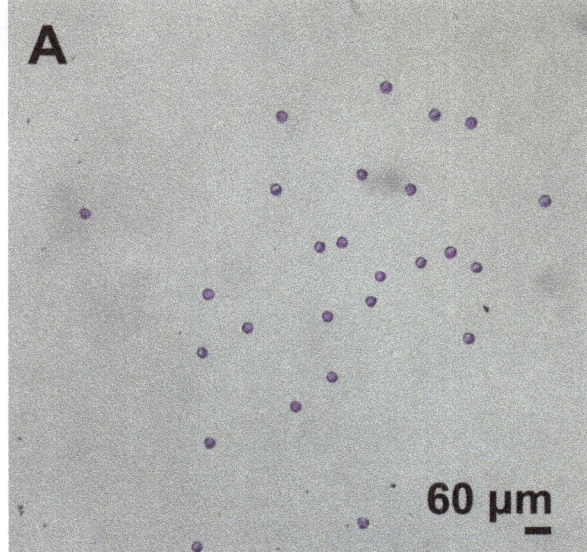

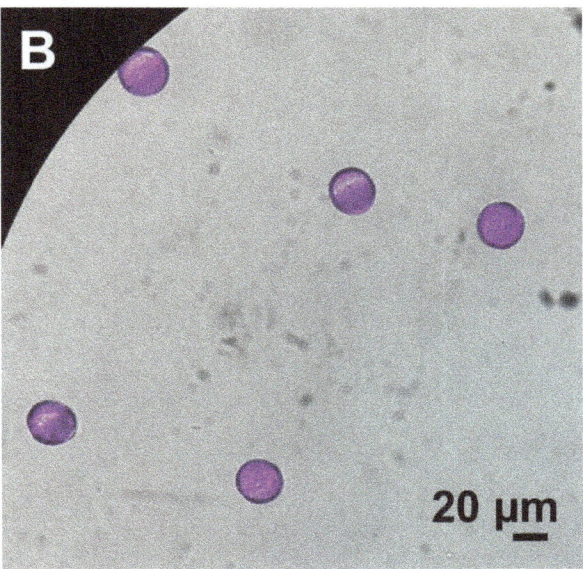

FIGURE I. *Brassica rapa* pollen grains in fuchsin gel viewed with light microscopy at (A) 100 × magnification and at (B) 400 × magnification.

pollen deposition and the inflorescences were re-bagged for 24 hours. Then bags were removed and the flowers dissected to access the pistils (stigma, style and ovaries), which were placed into Eppendorf tubes containing 1.0 ml of FAA fixative (I part formalin: I part acetic acid: 18 parts 50% ethanol [Kearns & Inouye, 1993]). After 18 hours the pistils were transferred to new tubes containing 1.0 ml of 70% ethanol, until they were counted. For microscopy, they were removed from the ethanol and softened in 1.0 ml of I M NaOH for 6 hours at room temperature, then washed with distilled water, placed on a microscope slide with a drop of decolourised aniline blue (DAB) (0.0005% w/v) and squashed under a coverslip. The DAB stock solution mixed 0.01% w/v aniline blue (Sigma-Aldrich) with 0.I M K_2HPO_4 buffer; this buffer is used to decolourise aniline blue, or it may interfere with the fluorescence (Kearns & Inouye 1993).

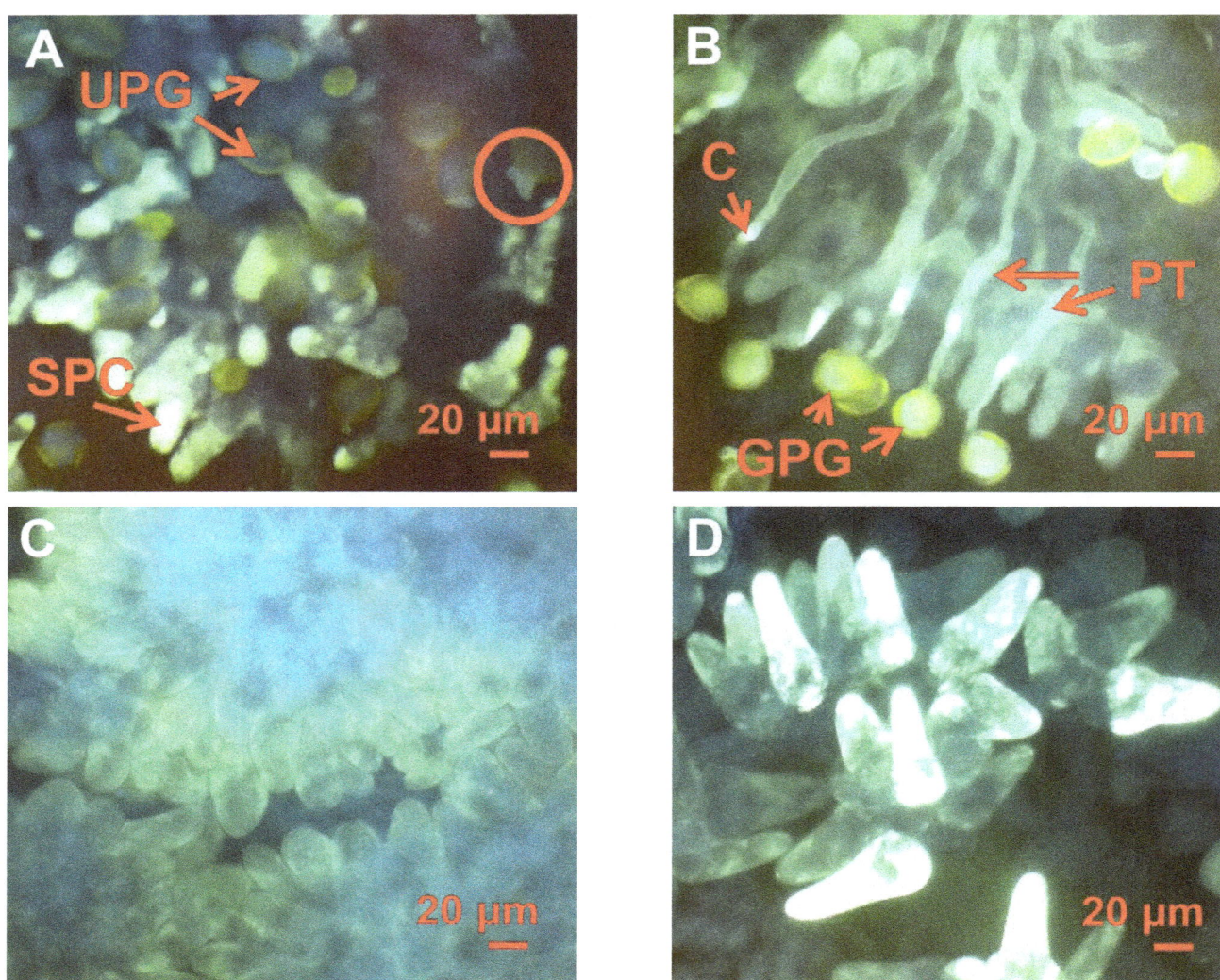

FIGURE 2.   Fluorescence microscopy images of *Brassica rapa* stigmas (375 × magnification). (A) Ungerminated pollen grains (UPG), stained papilla cells (SPC) and a germinated pollen grain with a pollen tube that has not penetrated the stigma (circled in red). (B) Germinated pollen grains (GPG) with pollen tubes (PT) that have penetrated the stigma, and a callose deposit within a pollen tube (C). (C) Papilla cells on a virgin stigma. (D) Stained papilla cells after contact with self-pollen.

We used fluorescence microscopy (Leitz Ortholux II with a mercury vapour UV light source, and Leitz filter cube A; 375 × magnification with a water-immersion objective lens) to count germinated pollen grains (those producing pollen tubes that penetrate the stigma), non-germinated pollen grains, and stained papilla cells (see Fig. 2). Some pollen grains germinated but produced pollen tubes that failed to penetrate the stigma (Fig. 2A). These were easily identified: the tubes were very short and narrow compared to those from successfully germinated grains, and did not contain the callose deposits that were observed within the penetrating pollen tubes. Pollen grains that produced these failed tubes are typical of self-pollen deposition (Sulaman et al. 1997) and were not counted as having successfully germinated. This process was repeated ($N = 27$) with unvisited flowers (controls) as before, and with manual self- and cross-pollination tests to confirm the self-incompatibility of the plants.

Statistical analysis

Generalised linear models (GLMs) and mixed models (GLMMs) were used to analyse the data, with R version 3.2.1 (R Core Team 2015). The Lme4 package (Bates et al. 2015) was used for GLMMs. Random effects (intercept) considered in models were: *Plant* to account for between plant differences (e.g. position within the row, and possible biological differences between plants) and because each plant was used more than once for data collection; *Individual,* because in some cases the same insect would visit two or three flowers that were being watched for data collection so that more than one data point was collected from the same visitor; and *Date* was included to account for between-day variations in unmeasured abiotic conditions, which could affect pollen germination and tube growth. Fixed effects considered in models were: *visitor, visit length, time of day,* and *wind speed.* The interactions '*visitor*visit length*' and '*visitor*wind*' were not tested in GLMM's (i.e. alongside random effects) due to insufficient data. Instead, they were tested in GLMs and found to be non-significant. The full models vary between analyses depending on what specification allowed for a robust model to be constructed. Stepwise model simplification was carried out by removing non-significant terms and comparing models using Akaike's information criterion (AIC) to produce minimum adequate models (Crawley 2007) (see Appendix II for details of full and minimum models).

The model validation procedure in Thomas et al. (2013) was followed, checking for over-dispersion and for patterns in the deviance residuals. In GLMs the dispersion parameter (theta) was calculated by dividing the residual deviance by the residual degrees of freedom. A theta value between 0.75 and 1.5 is deemed acceptable (Zuur et al. 2009; Thomas et al. 2013). Theta in GLMMs was calculated using the 'blmeco' package (Korner-Nievergelt et al. 2015). Negative binomial and Poisson distributions were used in the models (see Appendix II for detail on specific models). *Temperature* and *humidity* co-varied with *time of day* (Pearson correlation coefficients: 0.54, $P < 0.001$, and -0.45, $P < 0.001$ respectively) so both were omitted from the analyses.

Model statistics and sample sizes are reported in Tab. 1, rather than in-text. Standard errors are back-transformed from GLMs, and are therefore asymmetrical.

RESULTS

Single visit pollen deposition

A total of 101 SVD data points were collected from insect visits, plus 39 control data points. *Episyrphus* accounted for 31%, of observations, *Eristalis* 28%, *Bombus* 23% and 'other hoverflies' 18%. More than 99.9% of the pollen counted was conspecific. *Bombus* and *Eristalis* deposited significantly more pollen grains than found on control stigmas, whereas single visits from *Episyrphus* and 'other hoverflies' deposited no more pollen than on control stigmas (Fig. 3A; Tab. 1). Wind speed had a marginally significant positive effect on pollen deposition (Tab. 1).

Pollen germination

Manual cross- and self-pollination tests confirmed that the *Brassica rapa* plants were sporophytically self-incompatible; Fig. 4A shows that cross-pollen germinated and penetrated the stigma, whilst self-pollen did not. Self-pollination correlated with staining of papilla cells whereas cross-pollination did not (Fig. 4B). It is possible that some papilla cells did stain under the cross-pollination treatment, and that detection of these could have been masked by the staining of germinated pollen grains and pollen tubes. The control stigmas in Fig. 5 show a slightly higher number of stained papilla cells than those from cross-pollination in Fig. 4A, which is to be expected since some self-pollen is likely to end up on the stigmas of unvisited flowers.

A total of 71 pollen germination samples were collected from single visits (plus 27 control stigmas). Episyrphus accounted for 32%, of observations, Eristalis 28%, Bombus 24% and 'other hoverflies' 15% (i.e. giving similar proportions to the SVD study). Only one germinated pollen grain was found, on one control stigma; this is much lower

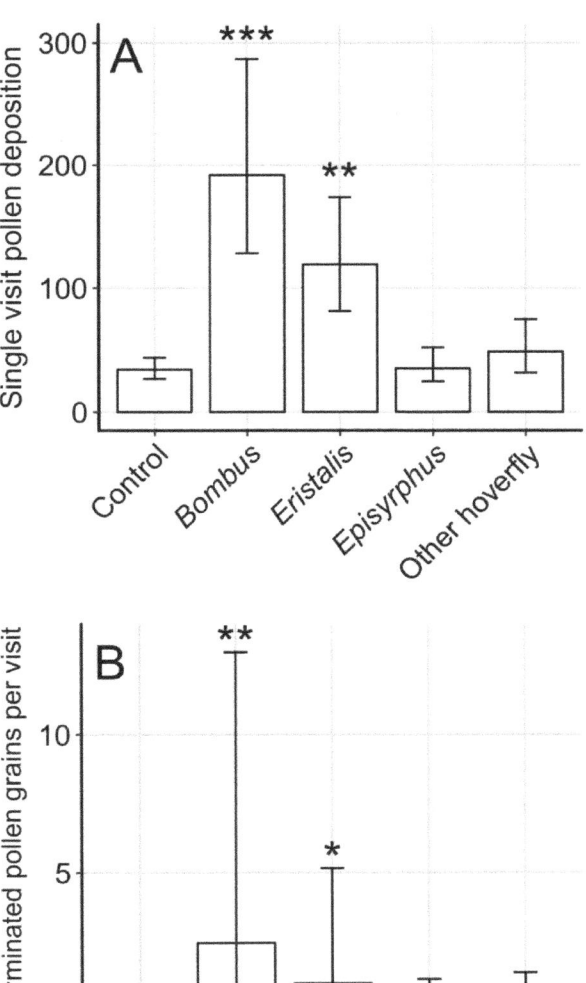

FIGURE 3. Pollinator performance of insect visitors on *Brassica rapa* (turnip) showing that *Bombus* and *Eristalis* were the most effective individual pollinators estimated by (A) single visit pollen deposition (SVD) and (B) pollen germination. Asterisks show significance in comparison to the controls: '***' = $P < 0.001$; '**' = $P < 0.01$; '*' = $P < 0.05$. Estimates (mean ± 1 SE) are produced from final model (see Tab. 1 for statistics and sample sizes).

than for SVD methods. This difference is most likely because ungerminated pollen is washed off stigmas during preparation for fluorescence microscopy. Only germinated pollen remains attached to the stigma. The numbers of germinated pollen grains detected after single visits from Bombus and Eristalis were also low, but significantly greater than on control stigmas, whereas the numbers found after single visits from *Episyrphus* and 'other hoverflies' were no different from controls (Fig. 3B; Tab. 1). Single visits from *Bombus*, *Eristalis* and *Episyrphus* led to significantly greater numbers of stained papilla cells than found on control stigmas (Fig. 5; Tab. 1).

TABLE I.   Pollinator performance final model results for SVD and pollen germination methods.

	Sample size	Estimate	SE	Z-value	P value
Field data					
SVD					
Control (Intercept)	39	3.533	0.246	14.361	<0.001 ***
Bombus	23	1.724	0.403	4.282	<0.001 ***
Episyrphus	31	0.047	0.373	0.125	0.9
Eristalis	28	1.249	0.381	3.275	<0.01 **
Other hoverflies	19	0.351	0.43	0.815	0.415
Wind (kph) (mean centred)		0.154	0.073	2.107	0.035 *
Pollen germination					
Control (Intercept)	27	-3.606	1.641	-2.198	0.028 *
Bombus	17	4.495	1.672	2.688	<0.01 **
Episyrphus	23	1.983	1.727	1.148	0.251
Eristalis	20	3.574	1.672	2.138	0.033 *
Other hoverflies	11	2.119	1.777	1.193	0.233
Papilla cells					
Control (Intercept)	27	3.226	0.228	14.181	<0.001 ***
Bombus	17	1.49	0.317	4.707	<0.001 ***
Episyrphus	23	0.721	0.285	2.531	0.014 *
Eristalis	20	1.32	0.314	4.193	<0.01 **
Other hoverflies	11	0.655	0.347	1.887	0.059
Manual Pollination					
Germinated pollen					
Cross-pollination (Intercept)	7	5.348	0.44	12.16	<0.001 ***
Self-pollination	15	-6.195	0.676	-9.17	<0.001 ***
Papilla cells					
Cross-pollination (Intercept)	7	1.925	0.378	5.092	<0.001 ***
Self-pollination	15	3.358	0.452	7.427	<0.001 ***

DISCUSSION

Both the SVD and pollen germination methods identified *Bombus* as the most effective pollinators of *Brassica rapa*, followed by *Eristalis*. This is not surprising, as Rader et al. (2009) found similar patterns for *Brassica rapa*, and Ali et al. (2011) reported that *Apis* (there were no *Bombus* in their geographical area) were more effective than *Eristalis*, which were more effective than *Episyrphus*. Jauker et al. (2012) found that hoverflies were individually less effective pollinators of *Brassica rapa* than bees, and in particular that hoverflies were poor at delivering cross-pollen, which agrees with the pollen germination data in this study.

The self-pollen deposition in self-incompatible plants is an obvious concern in using SVD to assess pollinator performance, and analysing pollen germination helps to address this issue. Our study does not explicitly separate self-pollen and un-germinated cross-pollen; however it is likely that much of the pollen recorded on *Brassica rapa* by SVD is self-pollen, given the match of papilla cells staining in response to self-pollen deposition and the number of stained papilla cells found after single visits (e.g. single visits from *Bombus* had the greatest pollen deposition, but also the greatest number of stained papilla cells). But it is also likely that some of the pollen counted with SVD was low quality cross-pollen, since this included all deposition by insect visitors, by wind, and by handling, whereas counting germinated pollen on the stigma of self-incompatible plants included only viable cross-pollen. Here, the number of pollen grains that germinated was two orders of magnitude lower than the total number of grains recorded by SVD, highlighting the importance of including pollen quality as a component of PE.

SVD is a reasonable proxy for estimating performance in our study, even though it includes high numbers of pollen grains that do not germinate. However, SVD is unlikely be a good proxy in all situations, since the effect of pollen quality is likely to vary depending on the breeding system of the plant species, plant and flower density, visitor species and abiotic conditions. Heat stress reduces pollen viability for example, so the proportion of pollen grains germinating can vary with temperature (Orueta 2002; Galen & Stanton 2003; Cross et al. 2003).

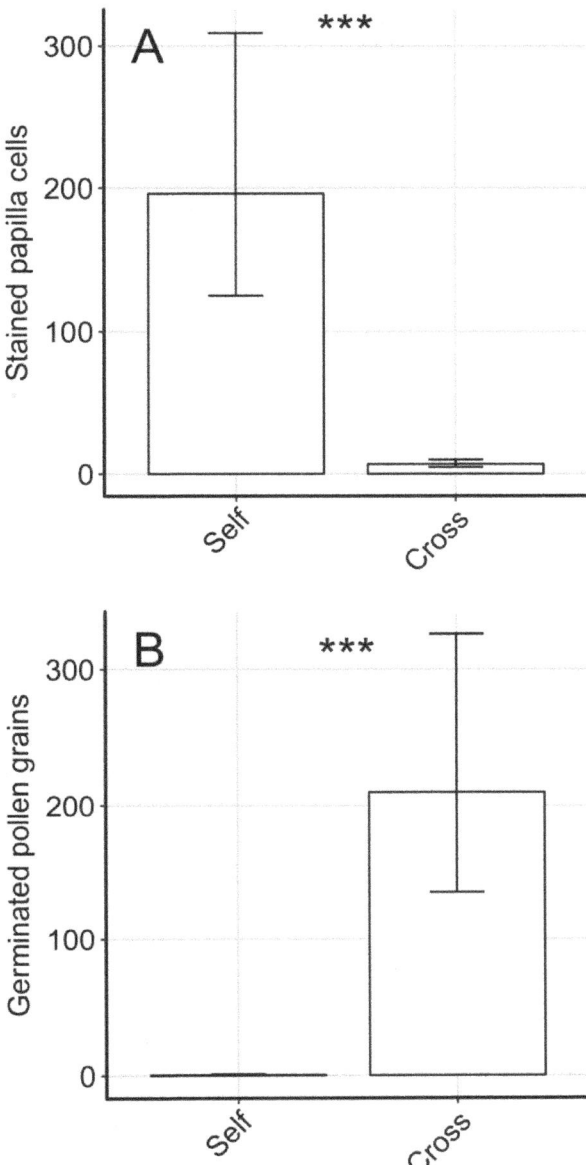

FIGURE 4. Results of manual self- and cross-pollination tests showing that the *Brassica rapa* plants were self-incompatible, and showing that papilla cells of the stigma stain in response to self-pollen, but not cross-pollen. (A) Pollen germination response to self- and cross-pollen. Germinated pollen grains are those that have produced pollen tubes that penetrate the stigma (i.e. 'GPG' in Fig. 2B). (B) Papilla cell response to cross- and self-pollen. An example of the stained papilla cells is shown in Fig. 2D. Asterisks show significance in difference between self- and crosses: '***' = $P < 0.001$. Estimates (mean $\pm$ 1 SE) are produced from final model (see Tab. 1 for statistics and sample sizes).

Frier et al. (2016) argued that using SVD to estimate PE on *Lonicera caerulea* (Caprifoliaceae) could incorrectly imply that some visitors were ineffective pollinators, since high variance in intra-floral self-pollen deposition would limit detection of pollen deposited by some visitor groups. The precise origin of the self-pollen in our study is unknown, and could either be intra- or inter-floral. It is likely that an insect visit leads to deposition from both sources (i.e. geitonogamous pollination from inter-floral but within-plant

movements by insects, and intra-floral pollen transfer from anthers to stigma during a visit). However both sources of self-pollen are important in terms of sexual interference (Barrett 2002), so it may be unnecessary to know the source of self-pollen in the context of pollinator effectiveness, but simply to know whether it will germinate. In addition, the amount of high quality pollen required for seed production varies greatly between plant species (Cruden 2000), and ideally this should be considered when assessing pollinator performance.

SVD can quantify heterospecific pollen deposition, whilst adding on pollen germination analysis does not, since the preparative stages can remove some un-germinated conspecific and heterospecific grains. Heterospecific pollen deposition can have the same stigma-clogging effect in any plant as self-pollen has in self-incompatible plants (Traveset & Richardson 2006; Brown et al. 2013) and is therefore informative; for example it is often used in assessing the effect of invasive plants on native plant communities (Larson et al. 2006; Bartomeus et al. 2008). The low levels of heterospecific pollen deposition reported here have been observed elsewhere (e.g. Moragues & Traveset 2005; Bartomeus et al. 2008; Willmer et al. 2017), although levels can be rather variable (e.g. Montgomery & Rathcke 2012; Fang & Huang 2013). Choosing between simple SVD and the addition of pollen germination analysis therefore depends on the question being asked; knowing the proportion of heterospecific pollen grains deposited may be more useful than the number germinated in some circumstances.

Pollen deposition due to wind, as seen here (Tab. 1), is not unexpected, but it is notable that other PE studies have not detected it. The effect of wind is likely to vary between plant species; for example, it might have a greater effect on self-pollen deposition in plant species with small pollen grains, or species that have anthers close to the stigma. The control stigmas in SVD studies take into account pollen deposition due to the wind and other factors, setting the baseline for comparisons after insect visits; control stigmas must therefore be taken regularly enough to account for the variation in wind speed throughout the data collection period so that the differences in pollen deposition between visitor species (or group) are not confounded by changing wind.

It is important not to over-generalise from this study, which concerns a single crop-plant species at one site, with data gathered during periods of peak visitation and good weather. Ideally it should be repeated on more SSI species to see if the results are typical, i.e. to determine if SVD and pollen germination regularly identify the same order of PE between visitors. It should also be carried out on plants with other self-incompatibility mechanisms, although this will be more difficult to achieve because both self- and cross-pollen germinate on the stigma in species with gametophytic (Newbiggin et al. 1993), cryptic (e.g. Jones 1994) or late-acting self-incompatibility (Gibbs 2014). Using the papilla cell callose response as a proxy for self-pollen deposition is restricted to species where the self-incompatibility mechanism acts at a papillate stigma, and even then callose deposition is not guaranteed and must be tested. There are relatively few studies that have observed the papilla cell

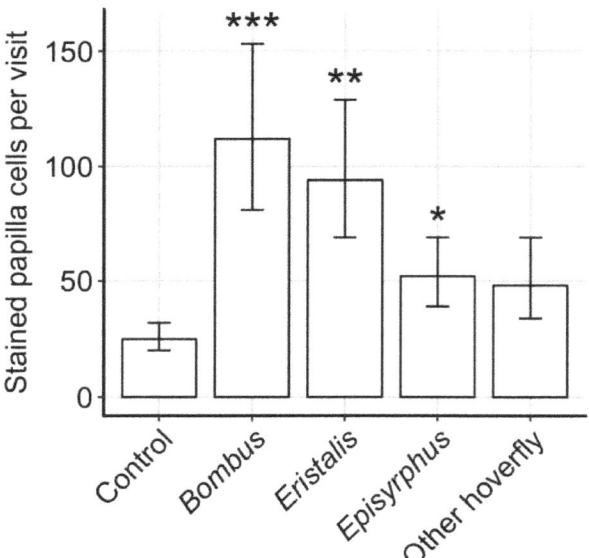

FIGURE 5. Results of papilla cell response to single insect visits, showing that insect visitors transfer self-pollen onto the stigma during a visit. Asterisks show significance in comparison to the control: '***' = $P < 0.001$; '**' = $P < 0.01$; '*' = $P < 0.05$. Estimates (mean ± 1 SE) are produced from final model (see Tab. 1 for statistics and sample sizes).

response outside of the Brassicaceae; (see Friedman & Barrett (2008) for Asteraceae; Pontieri & Sage (1999) for Saururaceae; and Sedgley (1979) for Lauraceae). Nevertheless, using the papilla cell response in a suitable study system may prove valuable in researching the ecology of self-pollen deposition.

A number of variables that were not measured in this study could affect pollen deposition and pollen germination rates between flowers. The number of inflorescences on a plant, the number of open flowers in an inflorescence, variation in nectar production per flower and resultant flower density are all likely to affect the amount of self-pollen moved between individual plants, and between flowers of the same plant, and there is also likely to be within-plant variation in pollen germination rates. In addition, we did not record visitation rates in this study, but it is a key component in other measures of pollinator performance (e.g. in calculating Pollinator Importance in Ballantyne et al. (2015)). However, the main focus of this study, the methods comparison, is unlikely to be biased by these factors.

Future research should compare SVD and pollen germination methods across self-incompatibility mechanisms and should ideally include comparisons with a direct measure of fitness, such as fruit or seed set. Using the pollen germination method at the community-level is unlikely to be achievable, although it is an excellent technique for single species studies. A novel approach may be required to include pollen quality in community-level pollinator performance studies. SVD has already proven to be achievable at the community level, and this study helps to validate it as a suitable method for assessing pollinator performance.

ACKNOWLEDGEMENTS

Advice about fluorescence microscopy and callose staining was gratefully received from Dr Peter Gibbs.

APPENDICES

Additional supporting information may be found in the online version of this article:
    APPENDIX I.   Insect visitor sample sizes and taxonomic groupings.
    APPENDIX II.   Model selection analyses.

REFERENCES

Ästergaard L, Petersen M, Mattsson O, Mundy J (2002) An *Arabidopsis* callose synthase. Plant Molecular Biology 49:559–566.

Ballantyne G, Baldock KCR, Willmer PG (2015) Constructing more informative plant-pollinator networks: visitation and pollen deposition networks in a heathland plant community. Proceedings of the Royal Society B 282:e20151130.

Ballantyne G, Baldock KCR, Rendell L, Willmer PG (2017) Pollinator importance networks and the crucial value of bees in a highly speciose plant community. Scientific Reports 7:8389.

Barrett, SCH (2002) Sexual interference of the floral kind. Heredity 88:154-159.

Bartomeus I, Bosch J, Vilà M (2008) High invasive pollen transfer, yet low deposition on native stigmas in a *Carpobrotus*-invaded community. Annals of Botany 102:417–424.

Bates D, Maechler M, Bolker B, Walker S (2015) Fitting linear mixed-effects models using lme4. Journal of Statistical Software 67:1–48.

Brown BJ, Mitchell RJ, Graham SA (2003) Competition for pollination between an invasive species (purple loosestrife) and a native congener. Ecology 80:2328–2336.

Crawley, M (2007) The R book. John Wiley and Sons Ltd, Chichester.

Cross RH, McKay SB, McHughen G, Bonham-Smith PC (2003) Heat-stress effects on reproduction and seed set in *Linum usitatissimum* L. (flax). Plant, Cell and Environment 26:1013–1020.

Cruden RW (2000) Pollen grains: why so many? Plant Systematics and Evolution 222: 143-165.

Currier H (1957) Callose substance in plant cells. American Journal of Botany 44:478–488.

Falk S (2015) Field guide to the bees of Great Britain and Ireland. Bloomsbury, London.

Fang Q, Huang SQ (2013) A directed network analysis of heterospecific pollen transfer in a biodiverse community. Ecology 94:1176–1185.

Friedman J, Barrett SCH (2008) High outcrossing in the annual colonizing species *Ambrosia artemisiifolia* (Asteraceae). Annals of Botany 101:1303–1309.

Frier SD, Somers CM, Sheffield CS (2016) Comparing the performance of native and managed pollinators of Haskap (*Lonicera caerulea*: Caprifoliaceae), an emerging fruit crop. Agriculture, Ecosystems and Environment 219:42–48.

Galen C, Gregory T, Galloway LF (1989) Costs of self-pollination in a self-incompatible plant, *Polemonium viscosum*. American Journal of Botany 76:1675–1680.

Galen C, Stanton ML (2003) Sunny-side up: flower heliotropism as a source of parental environmental effects on pollen quality and performance in the snow buttercup, *Ranunculus adoneus*. American Journal of Botany 90:724–729.

Gibbs PE (2014) Late-acting self-incompatibility – the pariah breeding system in flowering plants. New Phytologist 203:717–734.

Gross C (2005) Pollination efficiency and pollinator effectiveness. In: Dafni A, Kevan PG, Husband BC (eds.). Practical pollination biology. Enviroquest Ltd, Cambridge, Ontario, pp 83-146.

Hiscock SJ, McInnis SM (2003) Pollen recognition and rejection during the sporophytic self-incompatibility response: *Brassica* and beyond. Trends in Plant Science 8:606–613.

Jones KN (1994) Nonrandom mating in *Clarkia gracilis* (Onagraceae): a case of cryptic self-incompatibility. American Journal of Botany 81:195–198.

Kearns C, Inouye D (1993) Techniques for pollen biologists. University of Colorado Press, Niwot, Colorado, USA.

Kleijn D, Winfree R, Bartomeus I, Carvalheiro L, Henry M, Isaacs R, Klein A-M, Kremen C, M'Gonigle LK, Rader R, Ricketts TH, Williams NM, Adamson NL, Ascher JS, Báldi A, Batáry P, Benjamin F, Biesmeijer JC, Blitzer EJ, Bommarco R, Brand MR, Bretagnolle V, Button L, Cariveau DP, Chifflet R, Colville JF, Danforth BN, Elle E, Garratt MPD, Herzog F, Holzschuh A, Howlett BG, Jauker F, Jha S, Knop E, Krewenka KM, Le Féon V, Mandelik Y, May E, Park MG, Pisanty G, Reemer M, Riedinger V, Rollin O, Rundlöf M, Sardiñas HS, Scheper J, Sciligo AR, Smith HG, Steffan-Dewenter I, Thorp R, Tscharntke T, Verhulst J, Viana BF, Vaissière BE, Veldtman R, Westphal C, Potts SG (2015) Delivery of crop pollination services is an insufficient argument for wild pollinator conservation. Nature Communications 6:e7414.

Klein AM, Vaissière BE, Cane JH, Steffan-Dewenter I, Cunningham SA, Kremen C, Tscharntke T (2007) Importance of pollinators in changing landscapes for world crops. Proceedings of the Royal Society B 274:303–313.

Korner-Nievergelt F, Roth T, Felten S, Guelat J, Almasi B, Korner-Nievergelt P (2015) Bayesian data analysis in ecology: using linear models with R. Elsevier, London, UK.

Larson DL, Royer RA, Royer MR (2006) Insect visitation and pollen deposition in an invaded prairie plant community. Biological Conservation 130:148–159.

Montgomery BR, Rathcke BJ (2012) Effects of floral restrictiveness and stigma size on heterospecific pollen receipt in a prairie community. Oecologia 168:449–458.

Moragues E, Traveset A (2005) Effect of *Carpobrotus* spp. on the pollination success of native plant species of the Balearic Islands. Biological Conservation 122:611–619.

Ne'eman G, Jürgens A, Newstrom-Lloyd L, Potts SG, Dafni A (2010) A framework for comparing pollinator performance: effectiveness and efficiency. Biological Reviews 85:435-451.

Newbigin E, Anderson M, Clarke A (1993) Gametophytic self-incompatibility systems. The Plant Cell. 5:1315–1324.

Ollerton J, Winfree R, Tarrant S (2011) How many flowering plants are pollinated by animals? Oikos 120:321–326.

Orueta D (2002) Thermal relationships between *Calendula arvensis* inflorescences and *Usia aurata* bombyliid flies. Ecology 83:3073–3085.

Pontieri V, Sage, TL (1999) Evidence for stigmatic self-incompatibility, pollination induced ovule enlargement and transmitting tissue exudates in the paleoherb, *Saururus cernuus* L. (Saururaceae). Annals of Botany 84:507–519.

Potts SG, Biesmeijer JC, Kremen C, Neumann P, Schweiger O, Kunin WE (2010) Global pollinator declines: trends, impacts and drivers. Trends in Ecology and Evolution 25:345–353.

Sedgley M (1979) Structural changes in the pollinated and unpollinated avocado stigma and style. Journal of Cell Science 38:49–60.

Shore JS, Barrett SCH (1983) The effect of pollination intensity and incompatible pollen on seed set in *Turnera ulmifolia* (Turneraceae). Canadian Journal of Botany 62:298–1303.

Sulaman W, Arnoldo MA, Yu K, Tulsieram L, Rothstein SJ, Goring DR (1997) Loss of callose in the stigma papillae does not affect the *Brassica* self-incompatibility phenotype. Planta 203:327–331.

R Core Team (2015) R: a language and environment for statistical computing. R Foundation for Statistical Computing, Vienna, Austria.

Thomas R, Vaughan I, Lello J (2013) Data analysis with R statistical software. Eco-explore, Cardiff, UK.

Traveset A, Richardson DM (2006) Biological invasions as disruptors of plant reproductive mutualisms. Trends in Ecology and Evolution 21:208–216.

Willmer PG, Cunnold H, Ballantyne G (2017) Insights from measuring pollen deposition – quantifying the pre-eminence of bees as flower visitors and effective pollinators. Arthropod-Plant Interactions 11:411-425.

Wist TJ, Davis AR (2013) Evaluation of inflorescence visitors as pollinators of *Echinacea angustifolia* (Asteraceae): comparison of techniques. Journal of Economic Entomology 106:2055–2071.

Zuur A, Ieno E, Walker N, Saveliev A, Smith G (2009) Mixed effects models and extensions in ecology. Springer, New York.

23

LOW OVERNIGHT TEMPERATURES ASSOCIATED WITH A DELAY IN 'HASS' AVOCADO (*PERSEA AMERICANA*) FEMALE FLOWER OPENING, LEADING TO NOCTURNAL FLOWERING

David Pattemore[1,2]*, Max N. Buxton[1], Brian T. Cutting[1], Heather McBrydie[1], Mark Goodwin[1], Arnon Dag[3]

[1]*The New Zealand Institute for Plant & Food Research Limited, Ruakura Research Centre, Hamilton 3210, New Zealand*
[2]*School of Biological Sciences, University of Auckland, Auckland, New Zealand*
[3]*Gilat Research Center, Agricultural Research Organization, 85280, Israel*

Abstract—Avocado (*Persea americana*) has synchronously protogynous flowers: flowers open first in female phase before closing and opening the next day in male phase. Cultivars are grouped based on whether the flowers typically first open in female phase in the morning (type A), or in the afternoon (type B). However, it is known that environmental factors can alter the timing of flower opening, with cold temperatures being shown to affect the timing of flowering. The aim of this study was to investigate how low spring temperatures in New Zealand affect the flowering cycle of commercial avocado cultivars, focusing primarily on the receptive female phase of 'Hass', a type A cultivar. Time-lapse photography was used to assess flower opening times of 'Hass' over three years. Decreasing minimum overnight temperatures were associated with a delay in the timing of 'Hass' female flower phases and resulted in nocturnal flowering of both male and female phase flowers. We recorded insects visiting female flowers at night, and some nocturnal flower visitors collected were carrying avocado pollen. Our study suggests that nocturnal pollination needs to be considered for avocados grown in temperate regions. Furthermore, as the timing of the female phase of 'Hass' varied significantly with overnight temperature, the activity patterns of potential pollinators need to be considered to ensure adequate pollinator activity across the range of times in which 'Hass' flowers are receptive.

Keywords: Pollination, pollinators, floral biology, phenology, honey bees, flies

INTRODUCTION

Avocado (*Persea americana*) is an evergreen subtropical fruit tree native to Central America and Mexico and is now grown commercially in tropical and temperate regions globally (Knight Jr 2002). Although self-compatible, avocado flowers show several adaptations for maximising cross-pollination. Flowers are protogynous, opening first as functionally female (pistillate) before closing and opening on the following day as functionally male (staminate), thus preventing self-pollination within a flower (Ish-Am & Eisikowitch, 1993; Evans et al. 2010). Opportunity for geitonogamy is further reduced through synchronous flower opening within a tree and within cultivars, with cultivars grouped in two broad types based on the timing of flower phases. Type 'A' cultivars typically open as female in the morning, then close before opening as male in the afternoon of the following day (e.g. 'Hass'), while type 'B' cultivars open first as female in the afternoon and then close before opening as male in the morning of the following day (e.g. 'Bacon', 'Zutano', and 'Fuerte') (Alcaraz et al. 2011). This dichogamy promotes cross-pollination between the two different cultivar 'types', with one releasing pollen at the same time as the other cultivar is functionally female. While self-pollination can produce fruit, there is evidence of preferential retention of cross-pollinated fruits leading to a high proportion of these fruits at harvest (Gazit & Gafni 1986; Goldring et al. 1987; Degani et al. 1997). Therefore, cultivar types are usually inter-planted to maximise opportunities for cross-pollination in commercial orchards.

Flowering in avocado is known to be sensitive to environmental conditions (Davenport, 1986), with low temperatures being shown to delay and/or lengthen typical flowering cycles (Ish-Am & Eisikowitch, 1991), omit the female stage of flowering (Ish-Am & Eisikowitch, 1991), and/or reduce yield (Sedgley 1977; Sedgley & Grant 1983). Ish-Am & Eisikowitch (1991) reported that this delay in flowering did occasionally lead to flowers remaining open overnight, but only for type B flowers and type A staminate flowers. In regions where temperatures are cooler, a clear understanding of how temperature can influence the timing of the receptive female phase is important for determining which pollinators may be of particular importance and for developing management strategies to improve pollination.

With relatively sticky pollen (Gazit 1976), avocado flowers require an insect vector for pollination, rather than being able to rely on wind pollination. In most regions, including New Zealand, commercial honey bee (*Apis mellifera*) hives are introduced into orchards to provide this service (Ish-Am & Eisikowitch, 1993; Avocado Industry Council, 2006). However, honey bees are not the only insects

*Corresponding author: David.Pattemore@plantandfood.co.nz

to visit avocado flowers (Read et al. 2017), and some species, such as bumble bees (*Bombus* spp.), can achieve greater levels of cross-pollination than honey bees for avocado (Ish-Am et al. 1998). As different insect species vary in their daily activity patterns, complementarity in these activity patterns may play an important role in optimising fruit set, especially if the timing of flowering also varies during the day (Read et al. 2017).

Avocado growers in New Zealand experience climatic conditions that are considered marginal for avocado production, with relatively cold and variable spring weather (Appendix 1). The implications of this for the timing of avocado flowering and the probability of fruit set are poorly understood. The aim of this study was to assess correlations between overnight temperatures and the timing of the 'Hass' avocado flower phases in New Zealand, with a focus on the receptive female phase. This study also aimed to record which nocturnal insects visited the flowers and to determine if these visitors carried avocado pollen, in order to assess the potential for a new paradigm of nocturnal pollination of avocado.

MATERIALS AND METHODS

Time-lapse images of flowering

Data were collected between late October to early December over three years (2011, 2012 and 2013) from three avocado orchards in the Bay of Plenty and Waikato regions of New Zealand: Katikati in 2011 and 2013 (37°30'50"S 175°58'30"E), Aongatete in 2012 (37°36'32"S 175°55'14"E) and Maungatautari in 2011 (37°58'21"S 175°34'40"E). Nikon D5100 Digital SLR cameras were used to take time-lapse images of 'Hass' avocado flowers (every 5 minutes in 2011, and every 10 minutes in 2012 & 2013), for multiple consecutive days as they cycled through their flowering sequence. Cameras were powered by paired 36-amp hour, 12-volt hour deep cycle marine batteries, with voltage transformed to 220V AC and distributed via 30 m power cables. Batteries were changed every 2-3 days. This power setup was required to keep the cameras operating long enough to capture full sequences of flowering, but then also limited the range and number of trees that could be photographed at any given time. Due to electrical faults and weather conditions the system shut down on occasion before the regular battery replacement and camera check was performed, leading to lost opportunities to capture full flowering cycles.

All cameras were set up to take images of clusters of flowers within an inflorescence over several days and were moved to new inflorescences when the majority of flowers in view had finished. Inflorescences were selected based on a visual assessment of the number of flower buds that were close to opening in the female phase. Only complete sequences of flower opening and closing in either male or female phase were used for analyses.

In 2011, four cameras were used to collect time-lapse images of flowers between 31 October and 2 December. 35 inflorescences on 14 'Hass' trees were filmed, with between 1 and 14 flowers filmed per inflorescence. In 2012, three cameras were used between 24 September and 21 November.

26 inflorescences on 5 'Hass' trees were filmed, with between 1 and 16 flowers filmed per inflorescence. In both 2011 and 2012, only inflorescences on the northern aspect of focal trees were used to keep aspect constant. Although multiple trees were filmed in these years, the logistics of obtaining the images resulted in very low replication per tree in the final dataset. As all 'Hass' trees are clones of single original tree, variation between trees is less important that variation between sites and years. So, the effect of tree identity was not analysed further in this study. Temperature was recorded with HortPlus temperature and humidity loggers placed within a Henshall radiation screen (Henshall 1989) and positioned near the centre of each study block.

In 2013, three cameras were used to collect time-lapse images of flowers between 7 October and 15 November, with cameras placed on the northern, western and south-eastern sides of the tree to assess whether flower opening differed between sides of a tree. A single 'Hass' tree was photographed in 2013, with a total of 16 inflorescences filmed, with between 2 and 17 flowers filmed per inflorescence. Temperature measurements were made at the four quadrants around the tree (north, south, east and west) using thermistors (3 * 50mm tubes) mounted in Henshall screens. All temperature sensors were scanned at 60-second intervals and the average was recorded each hour using a Campbell CR10 data logger.

Images were reviewed and the flower stage was recorded for each flower in view that changed state during the entire time-lapse sequence. Flower stage was assigned based on a predetermined phenology scale of five stages (Fig. 1). We report here on the timing of Stage 1, when flowers first showed signs of opening, which was hypothesised to be most likely correlated to overnight temperatures; Stage 2, when the stigma was first accessible; and Stage 5, when the stigma was no longer accessible on the closing female flower.

All occasions when a flower was recorded reaching Stage 1, Stage 2, or Stage 5 were identified. For the first image of a flower at Stage 1, Stage 2 or Stage 5, the date and time (in minutes since that day's sunrise) were recorded and matched with the minimum temperature recorded the previous night by the relevant sensor (e.g. the southern sensor was used for the south-eastern flowers). All occasions of a flower remaining open overnight in either male or female phase were recorded.

We used binomial generalised linear mixed models in R version 3.2.4 (package lme4), to test if there was a significant association between overnight minimum temperatures and whether the female flowers remained open overnight or not. The response variable was a binary variable indicating whether the flower was open overnight or not, and the main fixed effect variable was the minimum overnight temperature. We ran two models, one with all data from the three years and including year as a random effect, and the second as a generalised linear model just on the data from 2013 with the side of the tree also included as a fixed effect. P-values were generated using the Satterthwaite method of denominator synthesis, implemented within the package lmerTest (this method results in non-integer degrees of freedom). The significance of the 'overnight temperature' term was determined with a likelihood ratio test between models with and without this term. Due to small sample sizes (18 individual flowers in

FIGURE I. Avocado female flower stages. A = Stage 0, unopened flower bud; B = Stage I, female flowers show first opening movement; C = Stage 2, stigma first accessible on flower; D = Stage 3, female flower fully open; E = Stage 4, female flower showing first closing movements; F = Stage 5, stigma no longer accessible on closing female flower

2011, 29 in 2012, and 75 in 2013), GLMMs run on the dataset with the timing of each stage as response variables did not converge. So, for these data we ran an analysis of variance on the datasets for each of Stages I, 2 and 5, with year as a blocking factor, using the command 'aov' in R version 3.2.4 with alpha = 0.05. For each year we then separately assessed the significance of Pearson product-moment correlation coefficients between the previous night's minimum temperature and the mean time (minutes since sunrise) to reach each of Stage I, Stage 2 and Stage 5, using the command 'cor.test' in R. We then followed this same process for each of three sides of the tree in 2013. To reduce the probability of type I errors due to the multiple comparisons, we used a Bonferroni-corrected alpha of 0.006.

Nocturnal flower visitors

Nocturnally active insects were collected from two 'Hass' avocado orchards in the Bay of Plenty Region, New Zealand. Collections were made on 10 November 2012 at the Maungatautari site, 29 and 30 October 2013 at the Katikati site, and 6 and 8 November 2014 at a second Katikati site (37°30'41.06"S 175°57'14.44"E) on evenings when both female- and male-phase 'Hass' flowers remained open after dark and insects were active. Both orchards also had type 'B' pollenisers planted at a ratio of about 1:8 (polleniser: 'Hass'). Sampling took place between approximately 2100 h and 0000 h.

When possible, insects were caught directly into plastic containers (~50 mL volume) from avocado flowers which they were visiting. Some samples were captured using nets, including insects that were active in proximity to, but not interacting directly with, avocado flowers, as the sampling method with head lamps frequently disrupted insect flower visitors causing them to fly off. Sample containers were immediately placed onto ice to slow and euthanize insects quickly, minimizing opportunity for attached avocado pollen to become dislodged or consumed. Samples were then stored at -20°C until processing, where each sampled insect was inspected under a stereomicroscope and adhered avocado pollen was identified morphologically and counted. Insect specimens were subsequently preserved and identified to the lowest taxonomic level possible.

RESULTS

In total, the timing of the flowering cycles of 122 female-phase and 165 male-phase flowers were documented between 2011 – 2013. Overnight minimum temperature significantly predicted whether female 'Hass' flowers stayed open overnight the following night or not, both when considering the entire dataset ($P = 0.00027$, Table I) or when modelling the data in 2013 including the sides of the tree ($P = 0.00013$, Table 2), although the side of the tree was not a significant factor (Table 2).

TABLE 1.   Output from fitting a binomial Generalised Linear Mixed Model to data from 2011-2013 relating whether female 'Hass' flowers remained open overnight to the overnight minimum temperature from the previous night (fixed effect), taking into account variation by year (random effect).

Variable	Estimate	Std. Error	Z value	P value	ΔAIC	X^2 value	$P\text{-}value$
(intercept)	1.83497	0.99982	1.835	0.06646			
Overnight temperature	-0.30948	0.09119	-3.394	0.00069	11.21	13.248	0.00027

TABLE 2.   Output from fitting a binomial Generalised Linear Model to data from 2013 relating whether a female 'Hass' flower remained open overnight to the overnight minimum temperature from the previous night and the side of the tree.

Variable	Estimate	Std. Error	Z value	P value	ΔAIC	X^2 value	$P\text{-}value$
(intercept)	2.99829	1.09005	2.751	0.005949			
Overnight temperature	-0.39092	0.11569	-3.379	0.000727	12.65	14.651	0.00013
SE side of tree	0.02378	0.70802	0.034	0.973205			
W side of tree	0.42807	0.66913	0.640	0.522341			

Overnight minimum temperatures were significantly correlated with the timing of Stage 1 ($F = 66.45$, $P < 0.0001$), Stage 2 ($F = 64.41$, $P < 0.0001$) and Stage 5 ($F = 27.51$, $P < 0.0001$), and Year had a significant effect for Stage 1 ($F = 3.65$, $P = 0.029$) and Stage 2 ($F = 3.49$, $P = 0.034$), but not Stage 5 ($F = 1.48$, $P = 0.23$).

The timing of Stages 1 (Fig. 2A), 2 (Fig. 2C) and 5 (Fig. 2E) were correlated with overnight minimum temperature in both 2011 ($P < 0.0001$, $P = 0.00036$ and $P < 0.0001$, respectively) and 2013 (all $P < 0.0001$). In 2013 this correlation was apparent for both Stage 1 and Stage 2 on the northern ($P = 0.0045$ and $P = 0.001$ respectively) and western ($P = 0.0001$ and $P < 0.0001$) sides of the tree (Figs. 2B & 2D).

Using the Bonferroni-corrected alpha of 0.006, no significant correlation between temperature and flower stage timing was detected in 2012 for Stage 1 ($P = 0.024$, Fig. 2A), Stage 2 ($P = 0.044$, Fig. 2C), or Stage 5 ($P = 0.9277$, Fig. 2E), or in 2013 on the south-eastern side of the tree for Stage 1 and 2 ($P = 0.0065$ and $P = 0.1032$ respectively, Figs. 2B & 2D), or Stage 5 in 2013 for any of the sides, north, west or south-east ($P = 0.0162$, $P = 0.0444$ and $P = 0.2113$ respectively, Fig. 2F). These results show that the timing of Stage 1 and Stage 2 was delayed by approximately 30 minutes for every 1°C degree drop in previous overnight minimum temperature, with the latest mean Stage 2 recorded 9 – 10 h after sunrise following nights with minimum temperatures below 8°C (Fig. 2A – D).

Declining overnight temperature had a stronger effect on the timing of Stage 5 when the flowers were no longer able to be pollinated, with an approximate 70-minute delay for every 1°C drop in overnight minimum temperature (Fig. 2E), although the data showed strong bimodality, with a cluster of data points in all years showing Stage 5 occurring 23 – 26 h after sunrise (i.e. the following morning). These data points represent female 'Hass' flowers closing the following morning after a nocturnal flowering event.

The proportion of female phase flowers remaining open overnight increased with decreasing overnight minimum temperatures (Fig. 3). When minimum overnight temperatures dropped to 4-6°C, c. 50% of the flowers remained open overnight; when minimum overnight temperatures were 13-15°C, only c. 10% of the flowers remained open overnight (Fig. 3). In contrast, over 80% of male phase flowers remained open overnight when overnight temperatures ranged between 7°C and 15°C, and only dropped to 60% when overnight temperatures were 4-6°C (Fig. 4).

The nocturnal female flowering indicated in the time-lapse data was confirmed through night-time visits to these orchards following low-temperature events. Furthermore, it was observed through both orchard visits and through the time-lapse images that male phase 'Hass' avocado flowers remained open overnight on almost all nights studied, including during the warmest conditions recorded in this study.

A total of 161 individual arthropods were caught from avocado flowers at night, representing eight different orders: Araneae, Blattodea, Coleoptera, Diptera, Heteroptera, Hymenoptera, Lepidoptera and Neuroptera (Table 3, Appendix 2), with the largest diversity of flower visitors in Coleoptera, Diptera and Lepidoptera (Fig. 5A). Coleoptera, Diptera and Neuroptera individuals carried more pollen grains on average and at greater frequencies than individuals from other orders (Table 3, Figs. 5B & 5C), while despite having the highest species richness the average number of pollen grains carried by Lepidoptera was low. Of the Diptera, Anisopodids (Wood gnats) and Tipulids (Crane flies) had multiple individuals carrying relatively high numbers of pollen grains (Table 3), as did the Noctuid moth *Rhapsa scotosialis*, the Coleopteran *Costelytra zealandica* (Grass Grub beetle), and the Neuropteran *Micromus tasmaniae* (Tasmanian lacewing; Table 3). One resting *Apis mellifera* (Hymenoptera) individual was also collected and carried 120 pollen grains but was not included in the tables or analyses because it was not active when caught.

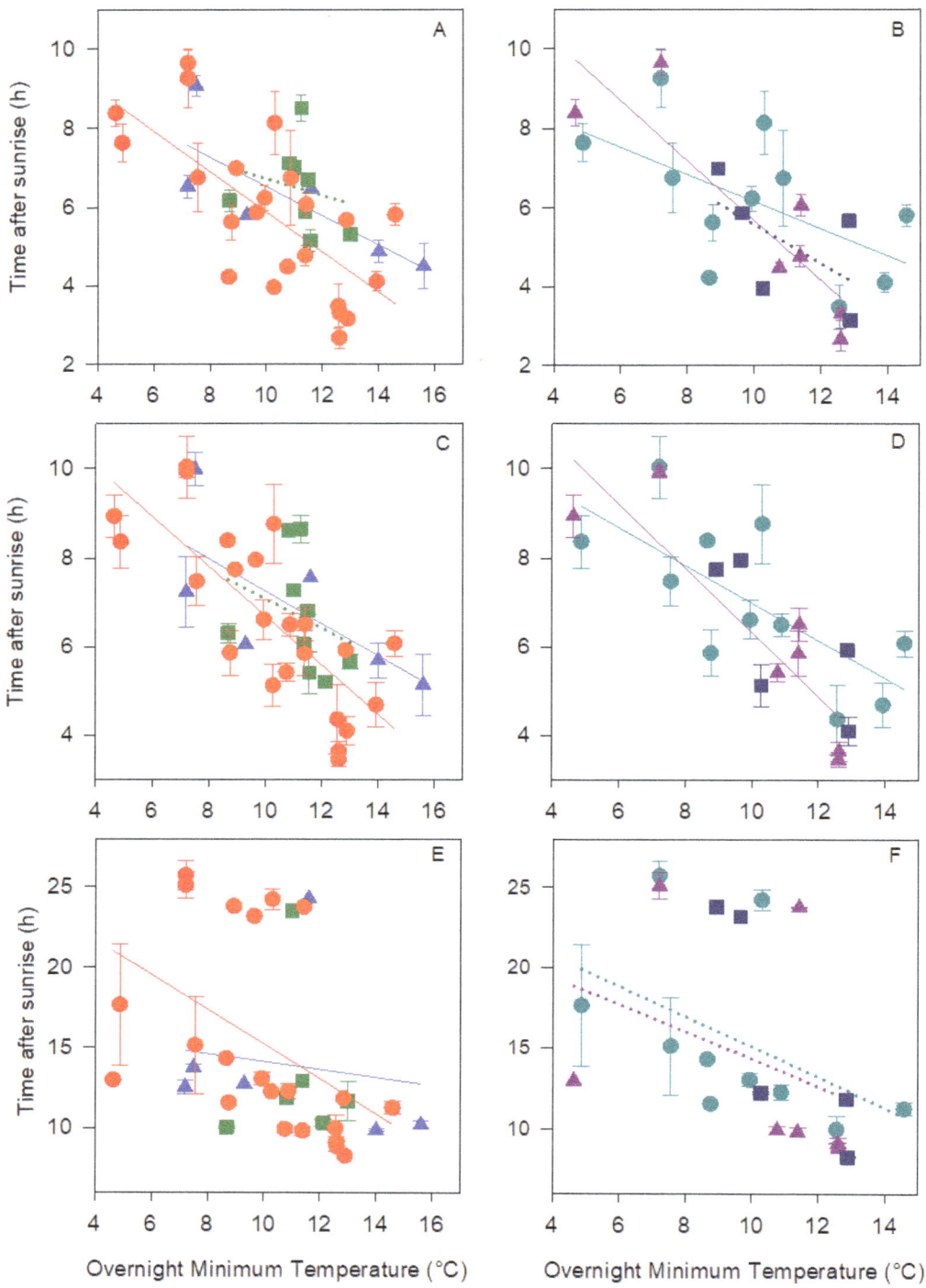

FIGURE 2: The relationship between overnight minimum temperatures and the time to avocado flowering Stages 1 (A), 2 (C) and 5 (E) for 2011, 2012 and 2013, and the time to Stages 1 (B), 2 (D) and 5 (F) in regards to the location of the flowers on the tree (2013 only). Datapoints show times averaged by day of observations (mean +/- SEM). Sunset occurred between 13-16 hours after sunrise during this study. The cluster of points between 23-26 hours in Fig. 2E & 2F indicate flowers closing at or after dawn the following day. Significant correlations (P < 0.006) are indicated by solid lines; dotted lines indicate correlations that had P-values between 0.05 – 0.006.

DISCUSSION

This study demonstrates that low overnight minimum temperatures are associated with the delay in the female flower phase in 'Hass' avocados such that the flowers remain open overnight. Furthermore, a greater proportion of flowers remained open as functionally female overnight as the prior nights' minimum temperatures decreased. Over the temperature ranges recorded in this study, male phase 'Hass' flowers remained open almost every night, similar to findings in previous studies (Ish-Am & Eisikowitch 1991), which resulted in our study in significant nocturnal male-female phase overlap within the single cultivar following cold nights.

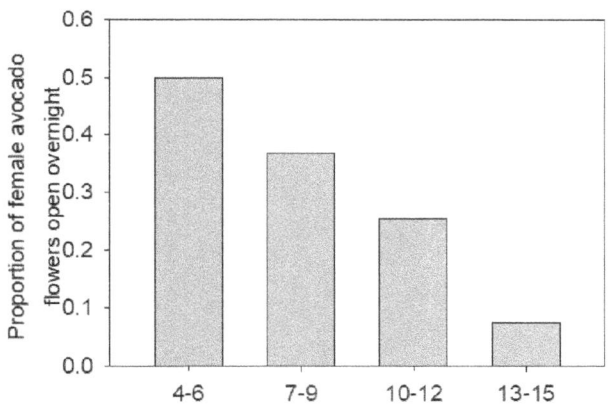

FIGURE 3: The relationship between overnight minimum temperature ranges and the proportion of female 'Hass' avocado flowers that remained open overnight. Temperatures are grouped into classes, each encompassing a range of 3°C.

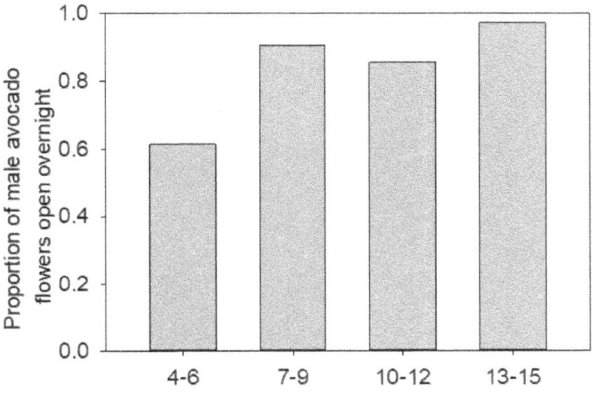

FIGURE 4: The relationship between overnight minimum temperature ranges and the proportion of male 'Hass' avocado flowers that remained open overnight. Temperatures are grouped into classes, each encompassing a range of 3°C.

To obtain sufficient replication of sequences of the full cycle of flowering required a camera system powered by a central battery bank, which limited how far apart cameras could be placed. Our approach in 2011 of shifting the cameras and battery bank frequently resulted in just 18 full sequences from 11 trees. In subsequent years we focused more on obtained full sequences of flowering. While this prevented us from assessing the influence of tree identity in this study, tree identity is less important in commercial horticulture with crops like 'Hass' avocado where every tree is a clone of the original parent 'Hass' tree. As the influence of the location of the orchard and the year are likely to have a much stronger influence on the pattern of flowering than tree identity, further work is needed to validate the relationship we have documented here in other regions of the world where 'Hass' are grown.

This nocturnal flowering suggests nocturnal pollination could be important in temperate regions with strong climatic fluctuations during the flowering period. The question remains whether these nocturnal female phase flowers are capable of being pollinated, fertilised and producing viable seed or fruit. Of particular importance is whether pollen deposited on stigmas at night are able to germinate and fertilise the ovules. Pollen growth and ovule fertilisation in many food crops appears to be sensitive to temperature stress, and these processes are often considered to be the most vulnerable in plant stress tolerance (Zinn et al. 2010). Temperature extremes have been shown to alter typical flower and fruit development in avocado (Gazit 1976; Sedgley & Annells 1981; Whiley & Winston 1987; Loupassaki et al. 1997; Alcaraz et al. 2011). While low temperatures have been shown to negatively affect pollen germination and pollen tube growth rates in a number of crops (e.g. Hedhly et al. 2005; Pham et al. 2015; Milatović et al. 2016), the results for avocado have not been as clear with some studies showing a negative effect in particular laboratory conditions (Loupassaki et al. 1997), while others report a difference in the effect of temperature on pollen tube growth between cultivars but no strong positive or negative trend across the temperature ranges tested (Alcaraz et al. 2011).

However, while cold overnight temperature may trigger a process that leads to nocturnal flowering, the temperatures may actually be significantly warmer the following night when the female flowers remain open. In this study, the weather conditions on the nights insects were collected from female flowers at night were mild (~10°C -15°C minimum), rather than cold (<10°C). Further work is required to understand the variation in the activity patterns of these nocturnal flower visitors with temperature. Likewise, more work is required to understand how pollen germination and pollen tube growth are affected by nocturnal conditions and whether that might inhibit fertilisation. While pollen tube growth can be slowed by low temperatures initially, warmer temperatures in the following day may lead to a resumption in pollen tube growth.

Pollination of flowers can take place only when the female flowers are receptive and there is pollen available for transfer. When female-phase 'Hass' flowers start to open in the late afternoon and are thus only receptive during evening, overnight and early the next morning, this will reduce or eliminate pollination by honey bees (*Apis mellifera*), whose activity peaks in the middle of the day or early afternoon (Corbet et al. 1993). It is therefore important that future research should determine whether these late-opening flowers can set fruit successfully in the absence of honey bee visitation, as growers primarily rely on honey bees to manage pollination. It will also be important to understand what proportion of the flowering season is dominated by these colder temperatures in each growing region, and what potential pollinators may be visiting these flowers. If fruit set can occur at these times and the plants encounter these conditions on a significant proportion of the days during flowering, it will be important to ensure that there is adequate pollinator activity across the range of times that these 'Hass' flowers are receptive, especially in the late afternoon, dusk, overnight and dawn.

Eight different invertebrate orders were captured from avocado flowers at night. Coleoptera, Diptera and Lepidoptera were the most frequently caught floral visitors, but it was coleopteran, dipteran and neuropteran individuals

TABLE 3: Arthropod visitors to avocado flowers carrying at least one avocado pollen grain, the total number caught ('Caught'), the number of individuals that were carrying avocado pollen ('Carrying'), and the mean number of avocado pollen grains carried with standard errors when the number carrying pollen is greater than 1 ('mean ± SEM'). Only species found to carry pollen are represented in this table; a full list of floral visitors is available in Appendix 2.

Order	Family	Species	Caught	Carrying	Mean ± SEM
Araneae	Unidentified	sp.	I	I	I
Coleoptera	Scarabaeidae	*Costelytra zealandica*	14	10	6.5 ± 2.53
	Cerambycidae	*Oemona hirta*	I	I	5
	Chrysomelidae	*Diachus auratus*	I	I	30
		Eucolaspis sp.	I	I	17
Diptera	Anisopodidae	*Sylvicola* sp.	8	6	30.4 ± 2.53
	Bibionidae	*Dilophus nigrostigma*	I	I	26
		sp.	I	I	171
	Chironomidae	*Chironomus* sp.	2	2	3.5 ± 0.5
	Drosophilidae	*Drosophila* sp.	I	I	13
	Mycetophilidae	*Anomalomyia* sp.	I	I	3
		Macrocera scoparia	I	I	6
	Tipulidae	sp.	13	8	19.5 ± 6.56
Heteroptera	Miridae	*Diomocorus* sp.	15	3	0.4 ± 0.24
Hymenoptera	Ichneumonidae	*Netelea ephippiata*	I	I	4
Lepidoptera	Geometridae	*Cleora scriptaria*	I	I	11
		Declana floccosa	2	2	1.5 ± 1.5
		Pseudocoremia suavis	8	2	4 ± 3.86
	Noctuidae	*Rhapsa scotosialis*	10	5	11.4 ± 7.89
	Tineidae	*Opogona omoscopa*	3	I	0.3 + 0.00
	Unidentified	sp. A	2	I	0.5 ± 0.5
Neuroptera	Hemerobiidae	*Micromus tasmaniae*	12	5	7.1 ± 5.11

that carried the greatest number of pollen grains on average. This is an important distinction to make, as not all floral visitors behave as pollinators: visitation does not necessarily infer pollination (King et al. 2013; Popic et al. 2013; although see Vázquez et al. 2005). Species such as *Costelytra zealandica* (Coleoptera), *Micromus tasmaniae* (Neuroptera), along with Tipulidae and *Sylvicola* species (Diptera) may be especially important, as these were both frequently caught and often carried a high number of pollen grains. Compared with diurnal pollination, nocturnal pollination is poorly understood and relatively little research in New Zealand has tested assumptions that nocturnal floral visitors can act as pollinators (Buxton et al. 2018, but see Pattemore & Wilcove 2011). It is clear from this study that future work is required into the nature of all these nocturnal interactions before their role as nocturnal pollinators can be accurately gauged.

Avocado is characterised by high fruit and flower abscission, resulting in extremely low fruit set (approximately 0.3%), although artificial hand pollination has been shown to increase fruit set up to 5.2-8%, suggesting that pollination may be a limiting factor in avocado orchards (Evans et al. 2010; Ish-Am & Lahav 2011; Garner & Lovatt 2015). No study published to date has assessed the fruit set rate of nocturnal flowers, and it remains unknown if any of these flowers are retained throughout the fruit development process. Avocado often experience strong biennial bearing and increases in fruit set rates in typically low bearing years could help to ameliorate this pattern (Dixon et al 2006, Evans et al. 2010). This new information about the timing of the 'Hass' female-phase flowering and a potential new community of pollinators provides an opportunity to investigate new ways to improve avocado pollination and increase fruit set especially in low bearing years.

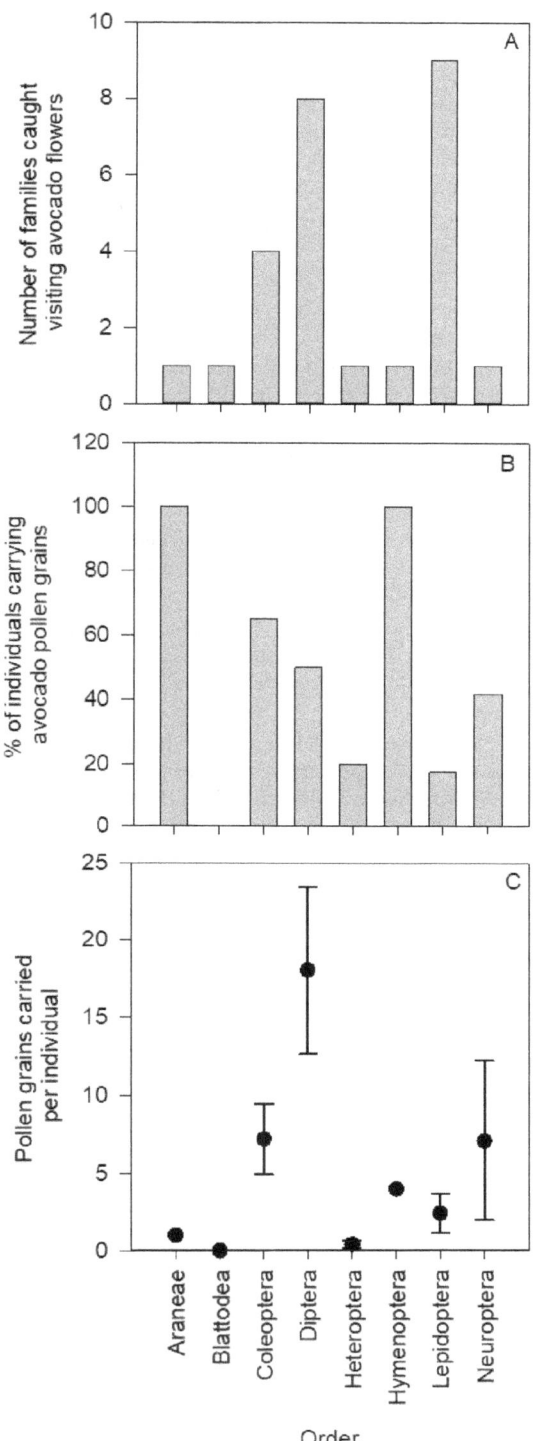

FIGURE 5:   Summary of arthropod flower visitors caught while visiting open 'Hass' avocado flowers at night, showing the number of families of arthropods caught in each order (A), the percentage of individuals of each order that were found to be carrying avocado pollen grains (B), and the mean (+/-SEM) number of pollen grains found on each individual of each order (C).

ACKNOWLEDGEMENTS

This project was funded through Plant & Food Research's internal research programme on avocado irregular bearing. We would like to thank Steve Noble, John Cottrell and Robert Lichtwark for letting us use their orchards for this study, Brian Patrick and Stephen Thorpe for their assistance with insect identification, Bill Snelgar for his assistance in monitoring temperature, Nick Gould for comments on this manuscript, and Melissa Broussard for advice and assistance with the data analysis.

APPENDICES

Additional supporting information may be found in the online version of this article:

APPENDIX I.   Graph showing daily maximum and minimum temperatures during the avocado flowering season in 2011, 2012 and 2013.

APPENDIX II.   Full list of nocturnal flower visitors caught on avocado flowers, including those not found to carry avocado pollen.

REFERENCES

Alcaraz ML, Montserrat M, Hormaza JI (2011) *In vitro* pollen germination in avocado (*Persea Americana* Mill.): Optimization of the method and effect of temperature. Scientia Horticulturae 130: 152-156.

Buxton MN, Anderson BJ, Lord JM 2018. The secret service – analysis of the available knowledge on moths as pollinators in New Zealand. New Zealand Journal of Ecology 42: 1-9.

Corbet SA, Fussell M, Ake R, Fraser A, Gunson C, Savage A, Smith K (1993) Temperature and the pollinating activity of social bees. Ecological Entomology 18:17-30.

Davenport TL (1986) Avocado flowering. Horticultural Reviews 8: 257-289.

Degani C, El-Batsri R, Gazit S (1997) Outcrossing rate, yield, and selective fruit abscission in 'Ettinger' and 'Ardith' avocado blocks. Journal of the American Society for Horticultural Science. 6: 813-817.

Dixon J, Smith DB, Greenwood AC, Elmsly TA (2006) Putative timing of irreversible commitment to flowering of Hass avocado trees in the Western Bay of Plenty. New Zealand Avocado Growers Association Annual Research Report 6: 13-20.

Evans LJ, Goodwin RM, McBrydie HM (2010) Factors affecting 'Hass" avocado (*Persea americana*) fruit set in New Zealand. New Zealand Plant Protection Society 63: 214-218.

Garner LC, Lovatt CJ (2015) Physiological factors affecting flower and fruit abscission of 'Hass' avocado. Scientia Horticulturae 199: 32-40.

Gazit S (1976) Pollination and fruit set of avocado. Proceedings of the First International Tropical Fruit Short Course: The Avocado. 88-92.

Gazit S, Gafni E (1986) Effect of hand pollination with different pollen donors on initial fruit set in avocado. Israel Agrisearch 1: 3-17.

Goldring A, Gazit S, Degani C (1987) Isozyme analysis of mature avocado embryos to determine outcrossing rate in a 'Hass' plot. Journal of the American Society for Horticultural Science. 112: 389-392Hedhly A, Hormaza JI, Herrero M (2005) The effect of temperature on pollen germination, pollen tube growth, and stigmatic receptivity in peach. *Plant Biology*, 7(5), 476-483.

Henshall WR, Snelgar WP (1989) A small unaspirated screen for air temperature measurement. New Zealand Journal of Crop and Horticultural Science 17: 103-107.

Ish-Am G, Eisikowitch D (1991) New insight into avocado flowering in relation to its pollination. California Avocado Society 1991 Yearbook, 75:125-137.

Ish-Am G, Eisikowitch D (1993) The behaviour of honey bees (*Apis melifera*) visiting avocado (*Persea americana*) flowers and their contribution to its pollination. Journal of Apicultural Research 32: 175-186.

Ish-Am G, Regev Y, Peterman Y, Lahav E, Degani C, Elbatzri R, Gazit S (1998) Improving avocado pollination with bumblebees: 3 seasons summary. California Avocado Society 1998 Yearbook 82: 119-135.

Ish-Am G, Lahav E (2011) Evidence for a major role of honeybees (*Apis mellifera*) rather than wind during avocado (*Persea americana* Mill.) pollination. The Journal of Horticultural Science and Biotechnology, 86:589-594.

King C, Ballantyne G, Willmer PG (2013) Why flower visitation is a poor proxy for pollination: measuring single-visit pollen deposition, with implications for pollination networks and conservation. Methods in Ecology and Evolution 4: 811-818.

Knight Jr RJ (2002) History, distribution and uses. The Avocado: Botany, Production and Uses, eds. AW Whiley, B Schaffer and BN Wolstenholme, CAB International.

Loupassaki M, Vasilakakis M, Androulakis I (1997) Effect of pre-incubation humidity and temperature treatment on the *in vitro* germination of avocado pollen grains. Euphytica 94: 247-251.

Milatović D, Nikolić D, Radović A (2016) The effect of temperature on pollen germination and pollen tube growth of apricot cultivars. Acta Horticulturae, 1139:359-362. DOI: 10.17660/ActaHortic.2016.1139.62

Pattemore DE, Wilcove DS (2012) Invasive rats and recent colonist birds partially compensate for the loss of endemic New Zealand pollinators. Proceedings of the Royal Society B: Biological Sciences. 279: 1597-1605.

Pham VT, Herrero M, Hormaza JI (2015) Effect of temperature on pollen germination and pollen tube growth in longan (Dimocarpus longan Lour.). Scientia Horticulturae 197, 470-475

Popic TJ, Wardle GM, Davila YC (2013) Flower-visitor networks only partially predict the function of pollen transport by bees. Austral Ecology 38: 76-86.

Read SF, Howlett BG, Jesson LK, Pattemore DE (2017) Insect visitors to avocado flowers in the Bay of Plenty, New Zealand. New Zealand Plant Protection. 70: 38-44.

Sedgley M (1977) The effect of temperature on floral behaviour, pollen tube growth and fruit set in the avocado. Journal of Horticultural Science 52: 135-141.

Sedgley M, Annells CM (1981) Flowering and fruit-set response to temperature in the avocado cultivar 'Hass". Scientia Horticulturae 14:27-33.

Sedgley M, Grant WJR (1983) Effect of low temperatures during flowering on floral cycle and pollen tube growth in nine avocado cultivars. Scientia Horticulture 18: 207-213.

Vázquez DP, Morris WF, Jordano P (2005) Interaction frequency as a surrogate for the total effect of animal mutualists on plants. Ecology Letters 8: 1088-1094.

Whiley AW, Winston EC (1987) Effect of temperature at flowering on varietal productivity in some avocado-growing areas in Australia. South African Avocado Growers' Association Yearbook 10:45-47.

Zinn KE, Tunc-Ozdemir M, Harper JF (2010) Temperature stress and plant sexual reproduction: uncovering the weakest links. Journal of Experimental Botany 61: 1959-1968.

24

Syrphine hoverflies are effective pollinators of commercial strawberry

Dylan Hodgkiss[1,2], Mark J.F. Brown[1], Michelle T. Fountain[2]

[1]*School of Biological Sciences, Royal Holloway University of London, Egham, Surrey, UK.*
[2]*NIAB EMR, New Road, East Malling, Kent, UK.*

Abstract—Recent declines in wild pollinators represent a significant threat to the sustained provision of pollination services. Insect pollinators are responsible for an estimated 45% of strawberry crop yields, which equates to a market value of approximately £99 million per year in the UK alone. As an aggregate flower with unconcealed nectaries, strawberries are attractive to a diverse array of flower-visiting insects. Syrphine hoverflies, which offer the added benefit of consuming aphids during their predatory larval stage, represent one such group of flower visitor, but the extent to which aphidophagous hoverflies are capable of pollinating strawberry flowers remains largely untested. In replicated cage experiments we tested the effectiveness of strawberry pollination by the aphidophagous hoverflies *Episyrphus balteatus* and *Eupeodes latifasciatus*, and a mix of four hoverfly taxa, when compared to hand pollination and insect pollinator exclusion. Hoverflies were released into cages, and the strawberry fruits that resulted from pollinated flowers were assessed for quality measures. Hoverfly visitation increased strawberry yields by over 70% and doubled the proportion of marketable fruit, highlighting the importance of hoverflies for strawberry pollination. A comparison between two hoverfly species showed that *Eupeodes latifasciatus* visits to flowers produced marketable fruit at nearly double the rate of *Episyrphus balteatus*, demonstrating that species may differ in their pollination efficacy even within a subfamily. Thus, this study offers compelling evidence that aphidophagous syrphine hoverflies are effective pollinators of commercial strawberry and, as such, may be capable of providing growers with the dual benefit of pollination and aphid control.

Keywords: Crop yield, ecosystem services, Fragaria, fruit quality, pollination, Syrphidae

Introduction

Compounding pressures from rising global food demand and recent declines in managed and wild pollinators pose a significant threat to the production of insect-dependent crops, which comprise 87 of the 115 leading crop species (Williams 1994; Klein et al. 2007; Ellis et al. 2010; Potts et al. 2010). Globally, the proportion of agricultural land devoted to pollinator-dependent crops has grown steadily over the last 50 years (Aizen et al. 2008), and animal-pollinated crops account for 35% of total crop yields worldwide (Klein et al. 2007). Thus, pollination represents a vital ecosystem service, contributing an estimated £121.8 billion to the global economy annually (Gallai et al. 2009).

Insect pollination not only boosts yields, but also enhances crop quality (Garibaldi et al. 2014). In commercial strawberry, *Fragaria* × *ananassa* Duch., open pollination by a range of wild bee species has been shown to result in fruit with fewer malformations, lower sugar-acid ratios, a more intense red colour, heavier berry weight and a longer shelf life than fruit from pollinator-excluded plants (Klatt et al. 2014). Thus, insect pollination can confer the dual economic benefits of larger yields and better-quality produce.

Research for the UK National Ecosystem Assessment has revealed that strawberry growers rely on insect pollination for 45% of crop yields (Smith et al. 2011), which equates to approximately £99 million/year in the UK alone (Defra 2015). With global strawberry production ballooning from 3.4 to 8.1 million tonnes/year between 1994 and 2014 (FAO 2017), the service provided by insect pollinators is becoming an increasingly vital natural resource. Therefore, gaining a clearer understanding of the species involved in this indispensable ecosystem service is paramount to ensuring that future strawberry harvests meet growing demands.

Strawberries are aggregate fruits with each flower receptacle containing multiple carpels (Free 1993). During fruit development the flesh around each achene, or seed, only expands once the achene has been fertilised with a pollen grain (Carew et al. 2003). Thus, poor pollination is one of the main reasons for malformations to occur. Carew et al. (2003) suggest that for fruit to develop properly, at least 70-80% of carpels must be pollinated. Due to their less specialised characteristics, such as radial symmetry, disc shape, easily accessible nectar and exposed anthers, strawberry flowers are visited by a wide range of pollinating insects (Nye & Anderson 1974; Albano et al. 2009a). Research into the effectiveness of various strawberry pollinators has shown that several insects are more or less equally important in the creation of high-quality fruit, and indeed that visits from pollinators with diverse morphologies and behavioural habits tend to produce fruit more frequently and with fewer malformations (Chagnon et al. 1993; Albano et al. 2009b). Therefore, multiple visits from insect pollinators are necessary in order to achieve full pollination (Free 1993).

To date most pollination research in agroecosystems has focused on bees, with comparatively few studies aimed at other insect pollinator taxa (Ssymank et al. 2008; Ssymank & Kearns 2009). Nevertheless, a growing body of research suggests that hoverflies, specifically honeybee-mimicking drone flies (*Eristalis* spp.), are among the most efficient pollinators of strawberry flowers (Nye & Anderson 1974; Albano et al. 2009b; Ssymank 2009; Gibson 2012). However, *Eristalis* hoverflies, which feed on decaying organic material as larvae, represent a tiny fraction of the Syrphidae family in Britain, and several other species may be equally, or indeed more, effective strawberry pollinators.

This study focused on the pollination effectiveness of a cohort of syrphine hoverflies, which possess aphid-eating larvae and are commonly found in strawberry fields. A series of cage trials was conducted to determine whether these syrphines are effective pollinators of strawberry flowers and if they differ between species in their pollination efficacy.

MATERIALS AND METHODS

Pollination effectiveness of a mix of hoverfly species on strawberry flowers

To determine the pollination effectiveness of a mixture of aphidophagous hoverfly species, 18 nylon mesh cages (47.5 × 47.5 × 93.0 cm; BugDorm, Taichung, Taiwan) were constructed and arranged on the ground in a 3 × 6 grid under a polytunnel at the NIAB EMR research institute, Kent, UK (51.286034° N, 0.449165° E, elevation: 35 m). The study site was surrounded by horticultural land which was comprised of other strawberry crops and arable fields, with mixed native hedgerows. Given that the cages were arranged in columns of six on each of three longitudinal drip irrigation lines, two sets of 3 × 3 randomly-generated Latin square designs were used to allocate treatments to the cages, with six cages, or replicates, per treatment. This method ensured that each treatment was represented in every row and twice in each column, reducing bias that may have resulted from distance from the drip irrigation source or from the sides of the tunnel. Ten cv. 'Finesse' strawberry plants in black plastic pots (11 × 11 × 12 cm; Soparco, Condé-sur-Huisne, France) were placed in each cage. All plants were watered and supplied with fertiliser (Ferticare 22-4-22, NutriAg Ltd., Toronto, Canada) at 06:00 and 18:00 daily for five minutes with individual drippers for each pot. The pollination period was started as soon as open flowers were present in each cage: 2 September – 9 October 2015.

The experiment had three treatments: (1) hand pollination (positive control, optimal pollination); (2) insect-exclusion (negative control); and (3) hoverfly visitation. For the hand pollination treatment, a size 12 paintbrush (Major Brushes Ltd., Cardiff, UK) was used to transfer pollen from dehisced strawberry anthers onto the entire receptacle of each open flower in the hand pollination cages. Hand-pollinated cages were visited ten times, approximately twice weekly, over the course of the pollination period and all open flowers were

brushed once with pollen on each visit. Pollinator-excluded cages were left undisturbed throughout the experiment to allow only self- or wind-pollination to occur.

A combination of four taxa of wild-caught aphidophagous hoverflies was used for the hoverfly visitation treatment. Nine hoverflies were released into each hoverfly-pollinated cage on 2 September, with at least one individual from each of the four groups. Subsequently, additional hoverflies were added to each cage on 17, 23 and 30 September once six individuals belonging to the same taxon were collected. This procedure ensured that the flower visitor assemblages remained consistent across the cages. Dead hoverflies were removed and frozen for identification to species level.

All four taxonomic groups had previously been observed visiting strawberry flowers in surveys at fruit farms in the southeast of England (unpublished data) and were released into cages in the following quantities: (1) five individuals of large-bodied (5.0 – 11.5 mm) species in the genera *Eupeodes* and *Syrphus*; (2) three individuals of large-bodied (6.0 – 10.3 mm) *Episyrphus balteatus* (De Geer); (3) five individuals of smaller (4.3 – 7.0 mm) species in the genus *Sphaerophoria*; and (4) eight individuals (4.5 – 8.0 mm) of the tribe Bacchini, which, in this study, were *Melanostoma* and *Platycheirus*. The first three hoverfly categories all belong to the tribe Syrphini, and all four groups include only species whose larvae predate aphids on herbaceous plants (Ball & Morris 2015). A species list can be found in Appendix I.

Comparison of pollination effectiveness of hand pollination and two hoverfly species

Because the hand-pollinated plants in the mixed-species experiment did not yield better-quality fruit than the hoverfly visitation treatment (see Results), we set up an experiment to determine the optimum frequency of hand pollinating strawberry flowers. Four nylon mesh cages were constructed and arranged on the ground in a single column under a small polytunnel at NIAB EMR to exclude insects from visiting the strawberry flowers. Ten 'Finesse' strawberry plants were arranged in each cage, following the procedure in the mixed-species experiment. Four pollination treatments were compared: (1) control, in which no flowers were pollinated by hand; (2) one brush, in which open flowers were brushed with a paintbrush once; (3) two brushes, in which flowers were brushed twice, with 24-48 hours between brushes; and (4) three brushes, in which flowers were brushed three times, again with 24-48 hours between brushes. Plants in each cage were assigned to the four treatments, so that each treatment was represented in every cage. When a flower was brushed, a felt-tipped marker was used to mark the peduncle so that the number of brushes could be tallied for each fruit.

The same general experimental design as the mixed-species experiment was then used to determine whether single species of hoverfly were effective 'Finesse' strawberry pollinators. Twenty cages were constructed to accommodate five replicates for each of four treatments: (1) *Episyrphus balteatus*; (2) *Eupeodes latifasciatus* (Macquart); (3) hand pollination (based on the results from the hand pollination experiment described above); and (4) pollinator-excluded. A randomised block design was employed, with the 20 cages split into five blocks of four cages, with each treatment represented in each block. Both *Episyrphus balteatus* and *Eupeodes latifasciatus* are common visitors to strawberry flowers, and are common in the southeast of England, where the study took place (Ball & Morris 2015).

The pollination period for the trial was 16 – 30 August 2016. Based on experience from the hand pollination study, the hand pollination procedure was modified so that each open flower was brushed with pollen on only two occasions. Each time an open flower was brushed with pollen, a mark was made on the peduncle with a felt-tipped marker.

Fruit quality assessments

At the end of the pollination period, all plants were transferred to a glasshouse to allow the fruit to ripen and to facilitate fruit collection. In the mixed-species experiment, berries from all cages were picked once at least 75% of the fruit surface was red (Klatt 2013). For the latter experiments, strawberries were picked when approximately 25-75% of the fruit surface area had turned pinkish-red to reduce losses to pests. As each berry was picked, a note was made of the cage it came from and its position on the fruit truss, hereafter referred to as "growth position:" primary, secondary or tertiary, following the nomenclature used in Darrow (1929). To compare fruit quality across the treatments, the following variables were recorded for each strawberry: fruit shape class, diameter, fresh weight, maximum firmness, dry weight, Brix (using soluble solids content as an index of Brix), number of fertilised achenes and marketability (Klatt et al. 2014).

Strawberries were given a shape score, ranging from 1-4 (1 = highly symmetrical fruit with no malformations; 2 = slightly asymmetrical fruit with minimal malformations; 3 = fruit with clear asymmetry and/or some malformations; 4 = fruit with major malformations). The diameter of each fruit was measured to the nearest tenth of a millimetre using calipers. Berries were then weighed on a scale (Sartorius, Göttingen, Germany) and the mass recorded to the nearest tenth of a gram. Firmness (maximum force in Newtons) was assessed for each fruit in the mixed-species experiment only using a texture analyser (Lloyds Instruments, Ametek, Berwyn, USA) with an 8 mm probe. Each berry was evenly sliced in half and one half was weighed again on the scale and reserved for drying overnight in an oven at 60°C. The following day the dried strawberry halves were weighed a second time and the dry weight recorded.

The other half of each berry was used for Brix measurement and counts of fertilised achenes. To measure the Brix, 1-2 drops of juice were squeezed onto a digital refractometer (Palette, Atago, Tokyo, Japan) and soluble solids concentration recorded to the nearest tenth of a percent. To separate achenes from the flesh of the fruit, each berry was placed in a blender (Minipro, Tefal, Rumilly, France) with 200 ml of water and blended for 20 seconds. The contents were then transferred to a 500 ml beaker and allowed to settle. All floating achenes were removed by gently pouring away the supernatant. The sunken achenes were collected by pouring the remaining contents through a sieve. These achenes were then transferred to a petri dish and dried overnight in an incubator at 20°C. The following day, the number of fertilised achenes per fruit half was counted and recorded for each strawberry. In the latter two experiments, rather than pouring out unfertilised seeds and drying the fertilised achenes in a petri dish, sunken fertilised seeds were simply counted by lifting the glass beaker and counting the achenes that had collected at the bottom. Lastly, strawberries with a minimum diameter of 18 mm and a shape score of 1 or 2 were classed as marketable (Conti et al. 2014; Klatt et al. 2014).

Data analysis

All analyses were carried out in R version 3.3.3 (R Core Team 2017). Average values were calculated for all fruit quality measurements and are presented as mean ± standard error. For fertilised seed counts from fruit halves, the mass of the fruit half divided by the mass of the whole fruit was calculated and used to weight the calculation of mean seed counts. Linear mixed models were then used on all normally-distributed fruit quality measurements in hoverfly experiments. Response variables were transformed where necessary. When transformations failed to produce normally-distributed data and in the case of fruit marketability, generalised linear mixed models were used instead. For continuous variables, a gamma distribution was used, and for marketability, a binomial distribution was chosen. Fruit shape score frequency distributions were analysed using cumulative link mixed models with a probit link function, as degree of misshapenness in strawberries is a latent continuous variable that was artificially separated into the four shape scores (Christensen 2015).

For all fruit quality measures apart from fruit yield, cage column, cage row, and the interaction between fruit growth position and pollination method were selected as fixed effects for the full model of the mixed- and single-species hoverfly pollination experiments. The optimal model was chosen by sequentially removing the least significant fixed effect from the full model and running the 'drop1' function on the reduced model to test the significance of the fixed effects (Ekstrom 2012). The optimal model was obtained once the reduced model contained only statistically significant fixed effect terms. The nested random effect for each model was growth position nested within cage, or when this term did not significantly influence the response variable, the random effect was simplified to cage. The significance of the random effect was tested by comparing the optimal model against an identical model that only contained fixed effects using the likelihood ratio test. To determine where the differences lay among levels of a fixed effect, least-square means were calculated with the 'lsmeans' function and Tukey-adjusted comparisons were made to reveal any significant differences among factor levels.

For the analysis of fruit yield per cage, general linear models were used in the mixed-species experiment, with cage column, cage row and pollination method as fixed effects. In the single-species hoverfly experiment, generalised linear models were chosen instead using a gamma distribution to account for non-normality in the fruit yield data. The fixed effects of the full model remained the same as those used in the mixed-species experiment. In both cases the 'drop1' function was used to select the optimal model.

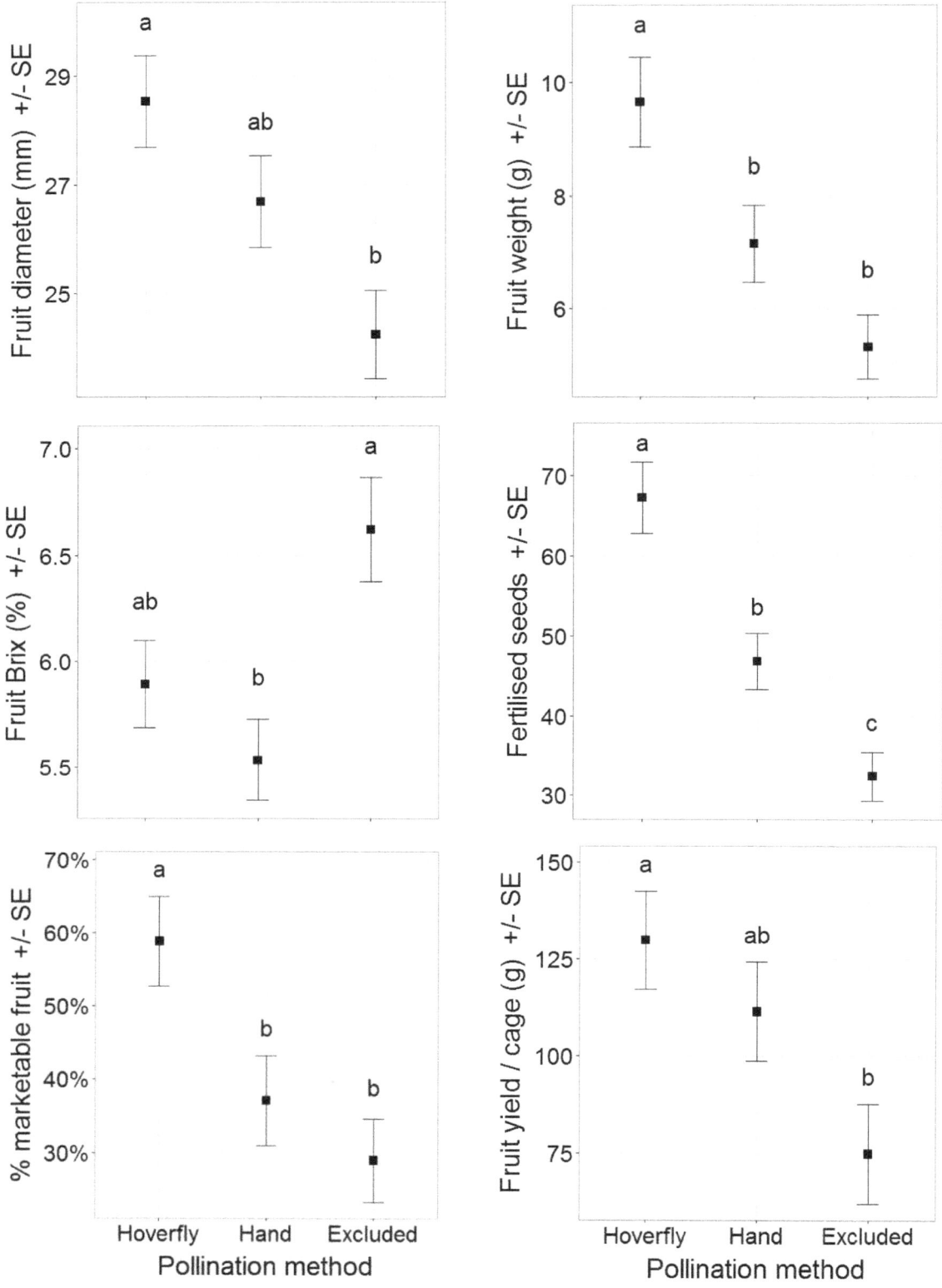

FIGURE I.   Mean fruit diameter, fruit weight, Brix, fertilised seeds per fruit half, proportion of marketable fruit and yield per cage by pollination method. Boxes indicate least square means ± standard error. Means sharing the same letter are not significantly different (Tukey-adjusted comparisons).

Finally, for the hand pollination efficacy experiment, generalised linear models were used to account for the unbalanced number of fruit per treatment. Unlike in the hoverfly pollination experiments, 'cage' was used as a blocking factor in the randomised block design of the hand pollination trial. Therefore, the fixed effects for this experiment were cage and pollination treatment. Response variables were transformed where necessary, and a binomial distribution was used for fruit marketability. Fruit shape score frequency distributions were compared using cumulative link models with a probit link function. Model selection was again performed using the 'drop1' function.

RESULTS

Pollination effectiveness of a mix of hoverfly species

Pollination by the mixed group of hoverflies had significant positive impacts on a range of strawberry quality measures. Across 215 strawberries, fruit diameter varied according to pollination treatment ($\chi^2(2) = 12.67$, $P = 0.0018$) and growth position ($\chi^2(2) = 21.55$, $P < 0.001$). Hoverfly-pollinated fruit had the largest mean diameter (28.5 ± 0.84 mm), compared to hand-pollinated fruit (26.7 ± 0.84 mm) and pollinator-excluded (24.2 ± 0.82 mm; Fig. 1). Primary fruit diameter averaged at 29.4 ± 0.64 mm, compared to 26.5 ± 0.60 mm for secondary fruit and 23.6 ± 1.24 mm for tertiary fruit. The interaction between pollination treatment and growth position was not significant.

Pollination method also had a significant effect on fruit weight ($\chi^2(2) = 17.08$, $P < 0.001$). Hoverfly-pollinated fruit weighed 9.7 ± 0.79 g, compared to 7.2 ± 0.68 g for hand-pollinated fruit and 5.3 ± 0.57 g for pollinator-excluded fruit (Fig. 1). Growth position similarly influenced fruit weight ($\chi^2(2) = 21.11$, $P < 0.001$), with primary fruit averaging at 9.7 ± 0.60 g, compared to 7.1 ± 0.48 g for secondary fruit and 5.3 ± 0.85 g for tertiary fruit. Again, the interaction between the two variables was not significant.

Fruit Brix was 6.2% ± 0.11% across the 194 berries that were assessed, but Brix varied according to pollination treatment ($\chi^2(2) = 16.61$, $P < 0.001$), cage column ($\chi^2(2) = 8.56$, $P = 0.014$) and cage row ($\chi^2(5) = 21.86$, $P < 0.001$). Pollinator-excluded fruit was higher in soluble solids (6.6% ± 0.24%) than hoverfly-pollinated fruit (5.9% ± 0.21%) and hand-pollinated fruit (5.5% ± 0.19%; Fig. 1). Fruit from columns 1 and 3 possessed a higher Brix (6.2% ± 0.21% and 6.2% ± 0.22%, respectively) compared to column 2 (5.6% ± 0.21%). Finally, Brix generally decreased as cage row number increased with the largest mean Brix of 6.7% ± 0.31% for row 2 and the smallest mean of 5.0% ± 0.23% for row 6.

The mean number of fertilised seeds per fruit half (215 berries) was 54.4 ± 2.37 seeds. Pollination method significantly influenced fertilised seed counts ($\chi^2(2) = 31.19$, $P < 0.001$). Hoverfly-pollinated fruit had the highest seed count (67.3 ± 4.43 seeds) followed by hand-pollinated (46.9 ± 3.56 seeds) and pollinator-excluded fruit (32.3 ± 3.09 seeds; Fig. 1). Cage row also affected the number of fertilised seeds ($\chi^2(5) = 17.82$, $P = 0.003$), which was lower as row number increased and ranged from 59.4 ± 4.72 to 37.4 ± 5.38 seeds.

A total of 215 strawberries were placed into one of four shape categories (ranging from 1-4). Pollination method was the only fixed effect to have a significant effect on the frequency distribution of shape scores ($\chi^2(2) = 14.60$, $P < 0.001$). Compared to the hand and insect-excluded

treatments, plants in the hoverfly-pollinated cages tended to produce the least-misshapen fruit (mean shape score = 2.38 ± 0.09), compared to hand-pollinated and pollinator-excluded fruit (mean shape score = 2.77 ± 0.10 and 3.07 ± 0.10, respectively). Moreover, the frequency distribution of shape scores for hoverfly-pollinated fruit was significantly different to the frequency distributions of both hand-pollinated ($Z = 2.63$, $P = 0.02$) and pollinator-excluded fruit ($Z = -4.62$, $P < 0.001$). The shape score frequency distributions of hand-pollinated and pollinator-excluded fruit did not differ significantly from each other ($Z = -2.16$, $P = 0.08$; Fig. 2).

Overall 41.4% ± 0.034% of 215 strawberries were deemed marketable. Plants in the hoverfly-pollinated cages tended to produce the highest proportion of marketable fruit at 58.8% ± 6.11%, compared to 37.1% ± 6.11% for hand-pollinated and 29.0% ± 5.61% for pollinator-excluded fruit ($\chi^2(2) = 10.48$, $P = 0.005$; Fig. 1).

Fruit yield per cage differed significantly according to pollination treatment and cage row. Pollination treatment significantly affected fruit yield per cage ($F_{2,10} = 4.84$, $P = 0.034$), with hoverfly-pollinated cages producing a mean of 129.8 ± 12.69 g, compared with 111.5 ± 12.69 g for hand-pollination and 75.0 ± 12.69 g for pollinator-excluded (Fig. 1). Cage row also affected the yield of strawberries per cage ($F_{5,10} = 4.74$, $P = 0.018$). Across rows, mean yields per cage ranged from 59.5 ± 17.95 g (row 4) to 171.9 ± 17.95 g (row 1).

The mean fruit firmness (64 strawberries) was 6.0 ± 0.20 Newtons (N) but varied among cage rows ($\chi^2(5) = 12.48$, $P = 0.029$), with means ranging from 5.3 ± 0.30 N (row 1) to

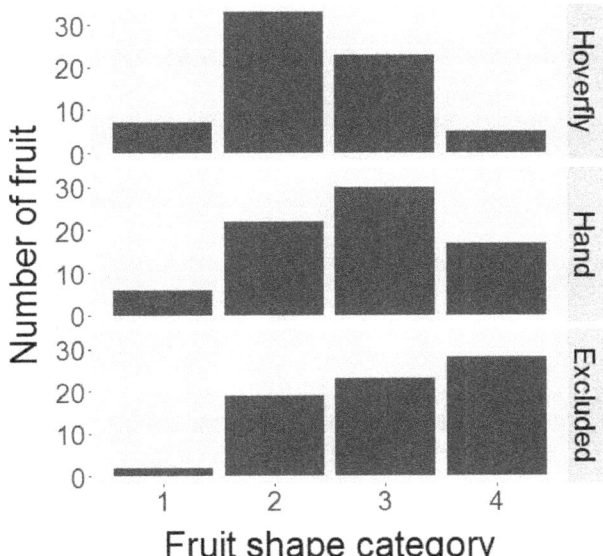

FIGURE 2. Fruit shape category frequency distributions by pollination treatment (1 = highly symmetrical fruit with no malformations; 2 = slightly asymmetrical fruit with minimal malformations; 3 = fruit with clear asymmetry and/or some malformations; 4 = fruit with major malformations). Fruit that fell into category 3 or 4 were deemed unmarketable.

6.6 ± 0.47 N (row 3). Pollination method had no effect on fruit firmness ($\chi^2(2) = 2.57$, $P = 0.28$). Lastly, none of the fixed effects affected percent dry matter. However, the random effect of cage significantly influenced fruit percent dry matter ($\chi^2(1) = 11.73$, $P < 0.001$).

Effect of varying brush pollination frequency on fruit quality

The frequency of brush pollinations had significant effects on berry weight, Brix and number of fertilised achenes. Mean fruit weight was influenced by the number of pollination events ($\chi^2(3) = 13.82$, $P = 0.003$). Fruit from flowers brushed twice were the heaviest (8.0 ± 0.49 g), compared to flowers brushed once (7.4 ± 0.61 g), unbrushed control strawberries (5.5 ± 0.71 g) or flowers brushed three times (5.0 ± 0.90 g; Fig. 3), suggesting that two hand pollination events with a paintbrush gave optimal pollination.

Pollination method also had a significant effect on fruit Brix ($\chi^2(3) = 19.92$, $P < 0.001$), with fruit brushed three times having the highest Brix levels (8.4% ± 0.43%) compared to fruit brushed once (7.0% ± 0.29%), fruit brushed twice (6.8% ± 0.24%) and unbrushed control fruit (6.0% ± 0.36%; Fig. 3). The fixed factor of cage had a significant effect on Brix ($\chi^2(3) = 24.71$, $P < 0.001$).

Pollination success, as measured by number of fertilised seeds, was significantly affected by the frequency of brushes used to hand-pollinate strawberry flowers ($\chi^2(3) = 14.27$, $P = 0.003$). Fruit brushed twice had more seeds (32.5 ± 2.99) than fruit brushed once (30.7 ± 3.54), unbrushed control fruit (18.8 ± 3.24) and fruit brushed three times (17.7 ± 4.09; Fig. 3). Thus, two brushes achieved the highest pollination success.

In contrast to the effects described above, the number of hand pollination events did not have a significant effect on fruit diameter (mean = 22.1 ± 0.43 mm, $N = 82$; $\chi^2(3) = 5.95$, $P = 0.11$), percent dry matter of strawberries (mean = 8.1% ± 0.23%, $N = 82$; $\chi^2(3) = 4.26$, $P = 0.24$), the frequency distribution of shape scores (mean shape score = 2.74 ± 0.11, $N = 82$; $\chi^2(3) = 1.62$, $P = 0.7$) or the proportion of marketable fruit (mean = 46.3% ± 0.055%, $N = 82$; $\chi^2(3) = 3.07$, $P = 0.4$).

Effect of hoverfly species flower visits on fruit quality

When compared to pollinator-excluded controls, pollination by *Episyrphus balteatus* and *Eupeodes latifasciatus* significantly improved strawberry yields and fruit shape score distributions, but only visits from *Eupeodes latifasciatus* enhanced additional fruit quality measures. Pollination treatment significantly influenced fruit weight ($\chi^2(3) = 9.52$, $P = 0.023$). Hand-pollinated fruit were the heaviest (4.5 ± 0.21 g), followed by fruit pollinated by *Eupeodes latifasciatus* (4.4 ± 0.22 g), *Episyrphus balteatus*-pollinated fruit (4.2 ± 0.21 g) and finally insect-excluded fruit (3.6 ± 0.21 g; Fig. 4). Growth position also had a significant effect on fruit weight ($N = 1083$; $\chi^2(2) = 231.67$, $P < 0.001$), with primary fruit larger (5.6 ± 0.19 g) than secondary fruit (4.4 ± 0.13 g) and tertiary fruit (2.8 ± 0.12 g). The random effect of cage

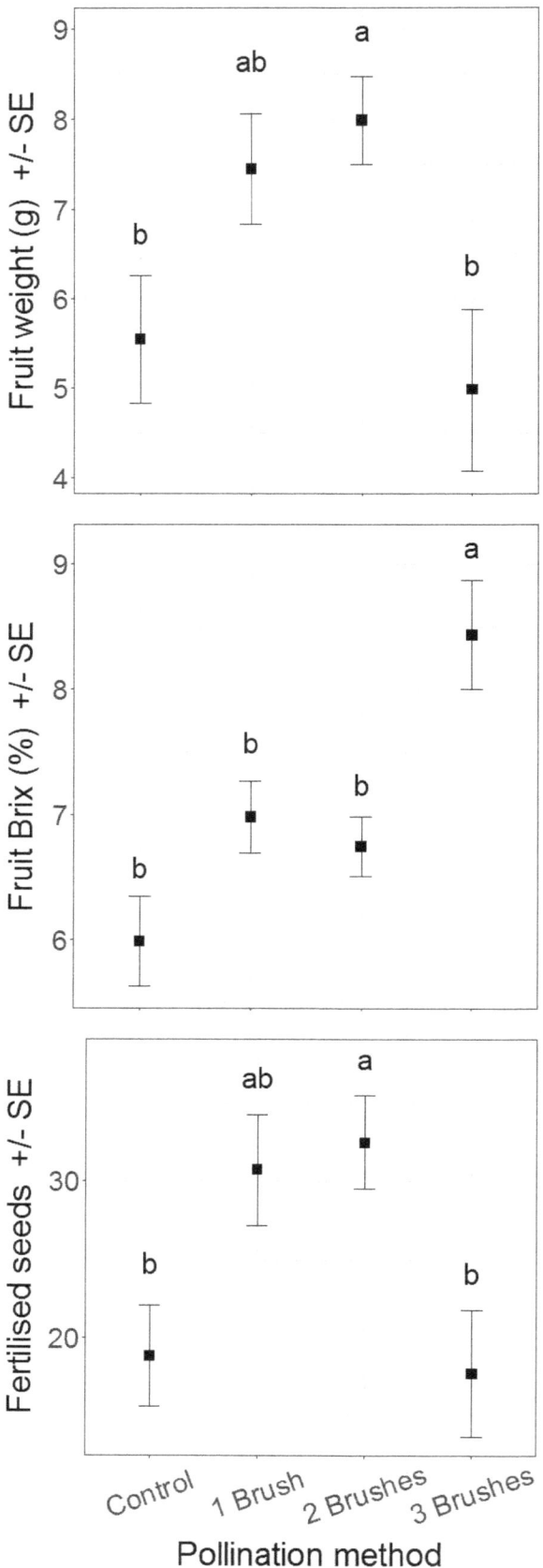

Figure 3. Mean fruit weight, Brix and number of fertilised seeds per fruit half by pollination method. Boxes indicate least square means ± standard error. Means sharing the same letter are not significantly different (Tukey-adjusted comparisons).

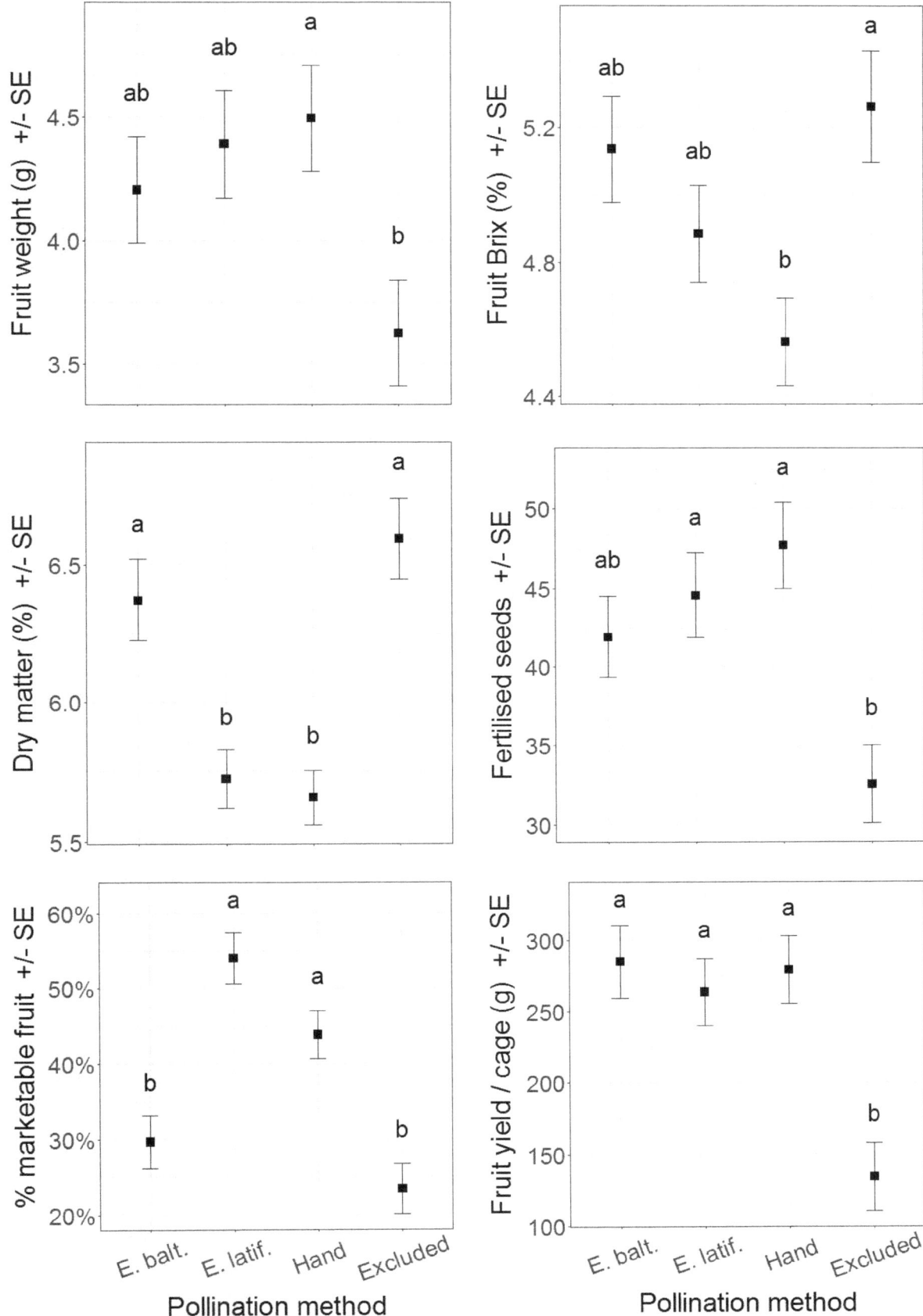

FIGURE 4.   Mean fruit weight, Brix, percent dry weight, fertilised seeds per fruit half, proportion of marketable fruit and yield per cage by pollination method. "*E. balt.*" is an abbreviation of *Episyrphus balteatus*. "*E. latif.*" is an abbreviation of *Eupeodes latifasciatus*. Boxes indicate least square means ± standard error. Means sharing the same letter are not significantly different (Tukey-adjusted comparisons).

also significantly influenced fruit weight ($\chi^2(1) = 6.95$, $P = 0.008$).

Pollination method also significantly affected Brix ($\chi^2(3) = 12.58$, $P = 0.006$). Pollinator-excluded fruit had higher Brix (5.3% ± 0.17%), compared with fruit pollinated by *Episyrphus balteatus* (5.1% ± 0.16%), *Eupeodes latifasciatus* (4.9% ± 0.14%) or hand-pollinated fruit (4.6% ± 0.13%; Fig. 4). Cage column affected fruit Brix ($N = 1076$; $\chi^2(2) = 7.70$, $P = 0.021$), with fruit from the central column of cages possessing the highest mean Brix (5.2% ± 0.13%), followed by fruit from the column nearest the irrigation source (4.9% ± 0.13%) and the column farthest from the irrigation source (4.8% ± 0.13%). Primary fruit tended to have higher sugar concentrations (5.3% ± 0.11%) than secondary (5.0% ± 0.083%) or tertiary fruit (4.7% ± 0.088%; $\chi^2(2) = 34.03$, $P < 0.001$). Lastly, the random effect of cage also significantly influenced fruit Brix ($\chi^2(1) = 9.52$, $P = 0.002$).

Pollinator-excluded fruit had the highest percent dry weight (6.6% ± 0.15%), followed by fruit pollinated by *Episyrphus balteatus* (6.4% ± 0.15%), *Eupeodes latifasciatus* (5.7% ± 0.11%) and hand-pollinated fruit (5.7% ± 0.10%; $\chi^2(3) = 16.68$, $P < 0.001$; Fig. 4). Fruit from the central column had the highest percent dry matter (6.3% ± 0.097%), followed by the column nearest the irrigation source (6.0% ± 0.091%) and the column farthest from the irrigation source (5.9% ± 0.11%; $N = 1075$; $\chi^2(2) = 7.72$, $P = 0.021$). Analysis of the influence of cage row revealed that percent dry weight generally decreased as distance from the irrigation source increased ($\chi^2(6) = 16.72$, $P = 0.010$).

Hand-pollinated fruit had the highest mean seed count (47.7 ± 2.70 seeds) compared with fruit pollinated by *Eupeodes latifasciatus* (44.6 ± 2.68 seeds), *Episyrphus balteatus* (41.9 ± 2.58 seeds) and pollinator-excluded fruit (32.5 ± 2.45 seeds; $\chi^2(3) = 15.90$, $P = 0.0012$; Fig. 4). Primary fruit had greatest number of fertilised seeds (51.5 ± 2.38), followed by secondary (44.7 ± 1.91) and tertiary fruit (29.8 ± 1.68; $N = 1141$; $\chi^2(2) = 45.74$, $P < 0.001$). However, the random effect of growth position nested within cage also significantly influenced fertilised seeds counts ($\chi^2(2) = 21.66$, $P < 0.001$).

Strawberries pollinated by *Eupeodes latifasciatus* had the best mean shape score (2.43 ± 0.054), compared to hand-pollinated fruit (2.46 ± 0.046), fruit pollinated by *Episyrphus balteatus* (2.63 ± 0.048) and pollinator-excluded fruit (2.99 ± 0.057). Moreover, the frequency distribution of shape scores for *Eupeodes latifasciatus*-pollinated fruit significantly differed from that of *Episyrphus balteatus*-pollinated fruit ($Z = 3.42$, $P = 0.004$). In contrast, the shape score distribution for hand-pollinated fruit was not significantly different from either of the hoverfly-pollinated treatments (hand-*Episyrphus balteatus* comparison: $Z = 2.06$, $P = 0.17$; hand-*Eupeodes latifasciatus*: $Z = -1.75$, $P = 0.30$). However, the shape score distribution for pollinator-excluded fruit differed significantly from all other treatments (excluded-*Episyrphus balteatus*: $Z = -4.25$, $P < 0.001$; excluded-*Eupeodes latifasciatus*: $Z = -8.30$, $P < 0.001$; excluded-hand: $Z = -7.31$, $P < 0.001$; Fig. 5). Cage row significantly influenced shape score ($\chi^2(6) = 24.69$, $P <$

0.001), with mean scores ranging from 2.47 – 2.80 across cage rows;. Primary fruit had the highest mean shape score (2.81 ± 0.060), followed by secondary (2.55 ± 0.037) and tertiary fruit (2.54 ± 0.045; $\chi^2(2) = 18.14$, $P < 0.001$).

Pollination method significantly affected the proportion of marketable fruit ($N = 1071$; $\chi^2(3) = 26.11$, $P < 0.001$). Plants in the *Eupeodes latifasciatus*-pollinated cages produced the highest proportion of marketable fruit (54.0% ± 3.42%), compared to hand-pollinated (43.8% ± 3.21%), *Episyrphus balteatus*-pollinated (29.7% ± 3.44%) and pollinator-excluded fruit (23.4% ± 3.31%; Fig. 4). Proportions of marketable fruit across cage rows varied from 19.1% - 48.7% ($\chi^2(6) = 20.65$, $P = 0.002$). Finally, secondary fruit possessed the highest proportion of marketable fruit (45.4% ± 2.37%), followed by primary (36.7% ± 3.28%) and tertiary fruit (29.8% ± 2.67%; $\chi^2(2) = 20.29$, $P < 0.001$).

Pollination treatment also significantly influenced fruit yield per cage ($F_{3,10} = 9.26$, $P = 0.003$), with *Episyrphus balteatus*-pollinated cages producing the highest yields (285.0 ± 25.73 g), compared with hand-pollinated (279.1 ± 23.76 g), *Eupeodes latifasciatus*-pollinated (263.4 ± 23.53 g) and pollinator-excluded cages (134.2 ± 23.58 g; Fig. 4). In addition, cage row had a significant effect on the yield of strawberries per cage ($F_{6,10} = 6.83$, $P = 0.004$).

Finally, growth position was the only fixed factor to have a significant effect on fruit diameter ($N = 1082$ strawberries; $\chi^2(2) = 252.53$, $P < 0.001$). Primary fruit were larger (22.9 ± 0.26 mm) compared to secondary fruit (20.7 ± 0.20 mm) and tertiary fruit (17.9 ± 0.24 mm). Pollination method did not affect fruit diameter ($\chi^2(3) = 5.90$, $P = 0.12$). In addition to growth position, the random effect of cage also influenced fruit diameter ($\chi^2(1) = 6.52$, $P = 0.011$).

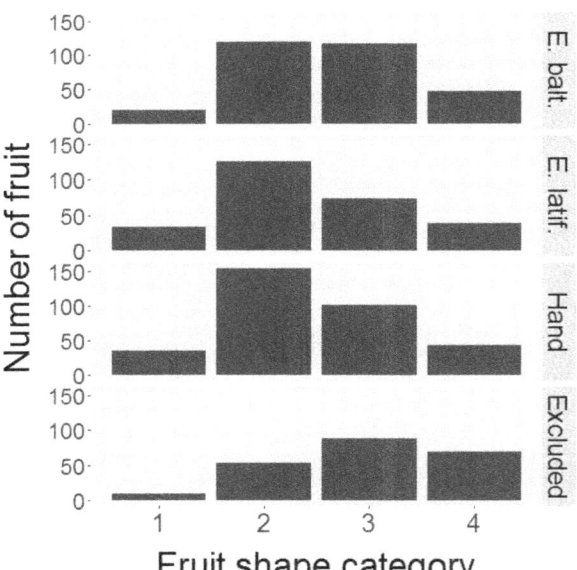

FIGURE 5. Fruit shape category frequency distributions by pollination treatment (1 = highly symmetrical fruit with no malformations; 2 = slightly asymmetrical fruit with minimal malformations; 3 = fruit with clear asymmetry and/or some malformations; 4 = fruit with major malformations). Fruit that fell into category 3 or 4 were deemed unmarketable. "E. balt." is an

abbreviation of *Episyrphus balteatus*. "E. latif." is an abbreviation of *Eupeodes latifasciatus*.

DISCUSSION

This study compared the effects of aphidophagous hoverfly flower visits on strawberry fruit quality and yield. Hoverfly pollination enhanced fruit quality and yield when compared to strawberry flowers that received no insect visits. Strawberry flowers visited by a mix of aphidophagous hoverfly species produced fruit with a greater diameter, weight, number of fertilised achenes and fewer malformations. These characteristics, in turn, meant that proportions of fruit that were marketable doubled from 29.0% in insect-excluded cages to 58.8% in hoverfly pollination cages. In addition to improving fruit quality, yields of strawberries increased by 73.1% when hoverflies were added to cages.

These improvements in fruit quality may be explained in part by the use of a mix of hoverfly species as flower visitors. Previous research has demonstrated that a diverse pollinator assemblage will more effectively pollinate crops (Blitzer et al. 2016), with several studies showing that diversity, rather than pollinator abundance *per se*, enhances seed set (Klein et al. 2003; Hoehn et al. 2008; Mallinger & Gratton 2015; Martins et al. 2015). These authors promote the concept of niche complementarity as an explanation for the positive relationship between pollinator diversity and crop quality. Different pollinator taxa tend to visit flowers at different heights and times of day. Furthermore, taxa with different body sizes carry varying pollen loads and behave differently on flower heads (Chagnon et al. 1993; Hoehn et al. 2008). All of these factors suggest that each pollinator functional group will deliver pollen grains in a unique manner. Moreover, when acting in concert, diverse pollinator guilds complement one another resulting in the provision of more complete pollination (Chagnon et al. 1993; Hoehn et al. 2008; Blitzer et al. 2016). In this study, hoverfly species varied in their average body size and typical behaviours on the strawberry flower receptacle, with larger species tending to feed while standing on the receptacle and smaller species touching the edge of the receptacle while standing on petals (personal obs.). Therefore, some degree of niche complementarity could have contributed to the improved pollination success and fruit quality observed in hoverfly-pollinated strawberries, and quantifying this should be the focus of future studies.

Despite these findings, fruit Brix, firmness and percent dry matter did not benefit from the introduction of a mix of hoverfly species. In each case, mean values for the hoverfly pollination treatment did not differ significantly from those of the insect-excluded treatment. One possible explanation is that any benefit from hoverfly pollination was mitigated by a subsequent increase in water concentration during the rapid cell expansion that occurs as a result of an influx of auxin and gibberellic acid when strawberries mature (Csukasi et al. 2011). This swelling of the fruit tissue may have lowered Brix, firmness and percent dry matter.

Although intended to serve as a positive control, the hand pollination treatment in the mixed-species experiment did not produce more marketable fruit. For most fruit quality measures, strawberries from the hand pollination treatment scored either significantly lower than hoverfly-pollinated fruit, or else not significantly different from either hoverfly-pollinated or insect-excluded berries. Overly vigorous brushing of the receptacle can result in poor pollination success (A. B. Whitehouse, pers. comm. 2017). Because all open strawberry flowers were brushed with pollen twice a week as long as they remained open, receptacles may have become damaged, thereby lowering the pollination success rate and causing the observed reductions in fruit quality.

The subsequent hand pollination experiment revealed that brush pollinating strawberry flowers twice only yielded better-quality fruit than either no brushing or three-brush treatments, both in terms of fruit weight and number of pollinated achenes. As with the hoverfly pollination experiment, better-pollinated fruit tended to have lower Brix, most likely due to the increased water content. The decrease in fruit quality observed in the three-brush treatment may represent the threshold at which the receptacles began to suffer damage from being brushed too often. This phenomenon may be analogous to the effect of having too many visits from insect pollinators, which has previously been shown to cause reduced pollination success (Gómez et al. 2007; Albrecht et al. 2012).

In the trial comparing the pollination effectiveness of two hoverfly species, strawberries visited by *Eupeodes latifasciatus* and hand-pollinated flowers yielded better-quality fruit than the insect-excluded treatment as evidenced by the 37.2% and 46.8% increases, respectively, in number of pollinated achenes, and the 130.8% and 87.2% increases in proportion of marketable fruit. Allowing *Episyrphus balteatus* to visit the strawberry flowers did not significantly improve fruit weight, pollination success or marketability. However, berries from both hoverfly pollination treatments and hand-pollinated fruit had lower frequencies of malformations than insect-excluded strawberries. Interestingly, the shape score distribution for *Eupeodes latifasciatus* differed significantly from that of *Episyrphus balteatus*, which possessed a smaller proportion of berries in the marketable fruit shape categories (45.9%) than the former species (58.6%). In both hoverfly species treatments and hand pollination cages, fruit yields per cage were enhanced by more than 90% when compared to pollinator-excluded cages. Thus, pollination by both hoverfly species would benefit strawberry growers by increasing yields and reducing rates of malformed fruit. However, based on its impacts on pollination success, fruit weight and marketability, *Eupeodes latifasciatus* appears to be a more effective pollinator of strawberry flowers than *Episyrphus balteatus*.

As in previous cage trials, Brix was higher for treatments that tended to have a lower pollination success rate. In this case, percent dry matter also followed Brix in having higher values for treatments with poorly-pollinated berries. In both instances, the smaller cells of poorly-pollinated fruit likely explain the observed differences in Brix and percent dry matter.

When the pollination efficacy of single species of hoverfly is compared against the results from the mixed-species experiment, several similarities emerge in the effect that the insects have on fruit quality parameters. Most notably, fruit yields were significantly augmented by both mixed-species assemblages of hoverflies and visits from only *Episyrphus balteatus* or *Eupeodes latifasciatus*. In the mixed-species experiment, fruit yields grew by 73.1% in hoverfly-pollinated cages when compared to controls, while the difference was even more pronounced in the single-species experiment. In that trial, introducing *Episyrphus balteatus* and *Eupeodes latifasciatus* to cages resulted in yield increases of 112.4% and 96.3%, respectively. The mean proportion of marketable fruit in mixed-species and in *Eupeodes latifasciatus* cages was over double that of pollinator-excluded cages in both experiments: mixed species of hoverflies increased proportions of marketable fruit by 102.8%, and *Eupeodes latifasciatus* enhanced rates of marketable fruit by 130.8%. By contrast, *Episyrphus balteatus* did not significantly improve fruit marketability when compared to the pollinator-excluded controls. In terms of pollination success rates, visitation from a mixed of hoverfly species led to a 108.4% increase in the number of fertilised seeds, while visits from *Eupeodes latifasciatus* improved pollination success rates by 37.2% over pollinator-excluded controls. Research by Klatt et al. (2014) documented a 61.7% rise in the number of fertilised achenes when bee-pollinated fruit were compared against self-pollinated controls using different strawberry cultivars; therefore, syrphine hoverflies may be as effective strawberry pollinators as bees.

Moreover, though *Eupeodes latifasciatus* outperformed mixed-species assemblages of hoverflies in enhancing yields and fruit marketability, visits from a group of hoverfly species resulted in a larger increase in numbers of fertilised achenes, when compared against fruit from control cages. Although these results seem to indicate slight differences in the pollination efficacy of *Eupeodes latifasciatus* as compared to a mixed group of hoverfly species, in order to uncover true differences, future research should compare single- and multiple-species assemblages in the same experiment.

The findings of this study provide the first evidence to suggest that hoverflies with aphidophagous larvae are effective pollinators of strawberry. Given that aphids are the primary prey of syrphine larvae (Rotheray & Gilbert 2011), these hoverflies may be capable of delivering both pollination and pest control ecosystem services for strawberry growers. Syrphine hoverflies have been shown to pollinate other crops, such as oilseed rape (Jauker & Wolters 2008; Jauker et al. 2012; Garratt et al. 2014) and apple (Garratt et al. 2016). Though these studies found that aphidophagous hoverflies were less effective pollinators than bees, syrphines may nonetheless supplement bee pollination and provide pest control services in these and other crops.

The main limitation of this study is that, as a cage trial, these results provide evidence that syrphines are capable of pollinating strawberry flowers; however, whether hoverflies pollinate strawberries effectively in the field remains to be demonstrated. Hoverflies may not visit strawberry flowers as frequently in the field and therefore their potential value as pollinators may not be as high as our findings imply (Albano et al. 2009b). Furthermore, although syrphine hoverflies are able to improve fruit quality and yields in cages, other flower-visiting taxa may prove to be even more effective pollinators of strawberry. Previous research has shown that honeybees,

bumblebees, halictid bees and eristaline hoverflies are also effective strawberry pollinators (Albano et al. 2009b; Gibson 2012). In order to assess the pollination efficacy of syrphines in relation to other taxa, one method that may prove useful is comparing the pollination success and fruit quality after a single visit from flower visitors (King et al. 2013). Such single visit deposition rates can then be coupled with flower visitation rates in the field to obtain a more complete picture of the pollination effectiveness of different species groups, as was done by Albano et al. (2009b) using honeybees, halictid bees and eristaline hoverflies as focal taxa.

To conclude, our findings demonstrate that aphidophagous syrphine hoverflies are effective pollinators of strawberry, boosting yields by over 70% and doubling proportions of marketable fruit. Moreover, even when strawberry flowers were only visited by a single species, both *Eupeodes latifasciatus* and *Episyrphus balteatus* were able to improve fruit yields by over 96% when compared to pollinator-excluded plants. These results suggest that syrphine hoverflies may provide the dual benefits of more complete pollination and aphid biocontrol in strawberry fields. Future studies could compare the pollination effectiveness of syrphine hoverflies with that of *Eristalis* hoverflies, the common strawberry-visiting hoverfly *Syritta pipiens* and bees in a field setting. Though our results suggest that syrphines are effective strawberry pollinators in cages, gaining a better understanding of how well these hoverflies pollinate in the field and how they perform relative to other flower visitors would improve our knowledge of their relative importance as strawberry pollinators.

ACKNOWLEDGEMENTS

This research was funded jointly by the East Malling Trust and Royal Holloway University of London. We would like to thank Dilly Rogers for providing glasshouse space and the polytunnel used in these experiments; Roger Payne for his assistance with the installation of the irrigation system; Alvaro Delgado for helping transfer plants between glasshouses and cages; and Phil Brain and Mark Jitlal for their advice on the statistical analyses.

APPENDICES

Additional supporting information may be found in the online version of this article:
    APPENDIX I.   List of hoverfly species used in mixed-species experiment

REFERENCES

Aizen MA, Garibaldi LA, Cunningham SA, Klein AM (2008) Long-term global trends in crop yield and production reveal no current pollination shortage but increasing pollinator dependency. Current Biology 18:1572-1575.

Albano S, Salvado E, Duarte S, Mexia A, Borges PAV (2009a) Floral visitors, their frequency, activity rate and Index of Visitation Rate in the strawberry fields of Ribatejo, Portugal: selection of potential pollinators. Part I. Advances in Horticultural Science 23:238-245.

Albano S, Salvado E, Duarte S, Mexia A, Borges PAV (2009b) Pollination effectiveness of different strawberry floral visitors in Ribatejo, Portugal: selection of potential pollinators. Part 2. Advances in Horticultural Science 23:246-253.

Albrecht M, Schmid B, Hautier Y, Müller CB (2012) Diverse pollinator communities enhance plant reproductive success. Proceedings of the Royal Society B: Biological Sciences 279:4845-4852.

Ball S, Morris R (2015) Britain's Hoverflies: A Field Guide, 2nd edn. Princeton UP, Princeton.

Blitzer EJ, Gibbs J, Park MG, Danforth BN (2016) Pollination services for apple are dependent on diverse wild bee communities. Agriculture, Ecosystems & Environment 221:1-7.

Carew JG, Morretini M, Battey NH (2003) Misshapen fruits in strawberry. Small Fruits Review 2:37-50.

Chagnon M, Ingras J, De Oliveira D (1993) Complementary aspects of strawberry pollination by honey and indigenous bees (Hymenoptera). Journal of Economic Entomology 86:416-420.

Christensen RHB (2015) Analysis of ordinal data with cumulative link models - estimation with the R-package 'ordinal'. [online] URL: https://cran.r-project.org/web/packages/ordinal/vignettes/clm_intro.pdf (accessed: September 2017).

Conti S, Villari G, Faugno S, Melchionna G, Somma S, Caruso G (2014) Effects of organic vs. conventional farming system on yield and quality of strawberry grown as an annual or biennial crop in southern Italy. Scientia Horticulturae 180:63-71.

Csukasi F, Osorio S, Gutierrez JR, Kitamura J, Giavalisco P, Nakajima M, Fernie AR, Rathjen JP, Botella MA, Valpuesta V (2011) Gibberellin biosynthesis and signalling during development of the strawberry receptacle. New Phytologist 191:376-390.

Darrow GM (1929) Inflorescence types of strawberry varieties. American Journal of Botany 16:571-585.

Defra (2015) Horticulture Statistics - 2014. Department for Environment, Food & Rural Affairs, London.

Ekstrom CT (2012) The R Primer. CRC Press, Boca Raton.

Ellis JD, Evans JD, Pettis J (2010) Colony losses, managed colony population decline, and Colony Collapse Disorder in the United States. Journal of Apicultural Research 49:134-136.

FAO (2017) FAOSTAT crops statistics. [online] URL: http://www.fao.org/faostat/en/#data/QC (accessed: September 2017).

Free JB (1993) Chapter 56: Rosaceae: *Fragaria*. In: Insect Pollination of Crops. Academic Press, London, pp 425-430.

Gallai N, Salles J-M, Settele J, Vaissière BE (2009) Economic valuation of the vulnerability of world agriculture confronted with pollinator decline. Ecological Economics 68:810-821.

Garibaldi LA, Carvalheiro LG, Leonhardt SD, Aizen MA, Blaauw BR, Isaacs R, Kuhlmann M, Kleijn D, Klein AM, Kremen C, Morandin L, Scheper J, Winfree R (2014) From research to action: enhancing crop yield through wild pollinators. Frontiers in Ecology and the Environment 12:439-447.

Garratt MPD, Breeze TD, Boreux V, Fountain MT, Mckerchar M, Webber SM, Coston DJ, Jenner N, Dean R, Westbury DB (2016) Apple pollination: demand depends on variety and supply depends on pollinator identity. PloS One 11:e0153889.

Garratt MPD, Coston DJ, Truslove CL, Lappage MG, Polce C, Dean R, Biesmeijer JC, Potts SG (2014) The identity of crop pollinators helps target conservation for improved ecosystem services. Biological Conservation 169:128-135.

Gibson RH (2012) Pollination Networks and Services in Agro-ecosystems. Biological Sciences, University of Bristol.

Gómez JM, Bosch J, Perfectti F, Fernández J, Abdelaziz M (2007) Pollinator diversity affects plant reproduction and recruitment: the tradeoffs of generalization. Oecologia 153:597-605.

Hoehn P, Tscharntke T, Tylianakis JM, Steffan-Dewenter I (2008) Functional group diversity of bee pollinators increases crop yield. Proceedings of the Royal Society B: Biological Sciences 275:2283-2291.

Jauker F, Bondarenko B, Becker HC, Steffan-Dewenter I (2012) Pollination efficiency of wild bees and hoverflies provided to oilseed rape. Agricultural and Forest Entomology 14:81-87.

Jauker F, Wolters V (2008) Hover flies are efficient pollinators of oilseed rape. Oecologia 156:819-823.

King C, Ballantyne G, Willmer PG (2013) Why flower visitation is a poor proxy for pollination: measuring single-visit pollen deposition, with implications for pollination networks and conservation. Methods in Ecology and Evolution 4:811-818.

Klatt BK (2013) Bee Pollination of Strawberries on Different Spatial Scales-from Crop Varieties and Fields to Landscapes. Niedersächsische Staats-und Universitätsbibliothek Göttingen.

Klatt BK, Holzschuh A, Westphal C, Clough Y, Smit I, Pawelzik E, Tscharntke T (2014) Bee pollination improves crop quality, shelf life and commercial value. Proceedings of the Royal Society B: Biological Sciences 281:20132440.

Klein A-M, Steffan-Dewenter I, Tscharntke T (2003) Fruit set of highland coffee increases with the diversity of pollinating bees. Proceedings of the Royal Society of London. Series B: Biological Sciences 270:955-961.

Klein A-M, Vaissiere BE, Cane JH, Steffan-Dewenter I, Cunningham SA, Kremen C, Tscharntke T (2007) Importance of pollinators in changing landscapes for world crops. Proceedings of the Royal Society B: Biological Sciences 274:303-313.

Mallinger RE, Gratton C (2015) Species richness of wild bees, but not the use of managed honeybees, increases fruit set of a pollinator-dependent crop. Journal of Applied Ecology 52:323-330.

Martins KT, Gonzalez A, Lechowicz MJ (2015) Pollination services are mediated by bee functional diversity and landscape context. Agriculture, Ecosystems & Environment 200:12-20.

Nye WP, Anderson J (1974) Insect pollinators frequenting strawberry blossoms and the effect of honey bees on yield and fruit quality. Journal of the American Society for Horticultural Science 99:40-44.

Potts SG, Biesmeijer JC, Kremen C, Neumann P, Schweiger O, Kunin WE (2010) Global pollinator declines: trends, impacts and drivers. Trends in Ecology & Evolution 25:345-353.

R Core Team (2017) R: A language and environment for statistical computing. R Foundation for Statistical Computing, Vienna, Austria. URL: https://www.R-project.org.

Rotheray GE, Gilbert F (2011) The Natural History of Hoverflies. Forrest Text, Cardigan, UK.

Smith P, Ashmore M, Black H, Burgess P, Evans C, Hails R, Potts SG, Quine T, Thomson A (2011) Chapter 14: Regulating Services. In: UK National Ecosystem Assessment. UNEP-WCMC, Cambridge, pp 535-596.

Ssymank A (2009) Flower Flies (Syrphidae). In: Ssymank A, Hamm A, Vischer-Leopold M (eds) Caring for Pollinators: Safeguarding Agro-biodiversity and Wild Plant Diversity. BfN, Skripten, pp 159-162.

Ssymank A, Kearns CA (2009) Flies – Pollinators On Two Wings. In: Ssymank A, Hamm A, Vischer-Leopold M (eds) Caring for Pollinators: Safeguarding Agro-biodiversity and Wild Plant Diversity. BfN, Skripten, pp 39-52.

Ssymank A, Kearns CA, Pape T, Thompson FC (2008) Pollinating flies (Diptera): a major contribution to plant diversity and agricultural production. Biodiversity 9:86-89.

PERMISSIONS

The contributors of this book come from diverse backgrounds, making this book a truly international effort. This book will bring forth new frontiers with its revolutionizing research information and detailed analysis of the nascent developments around the world.

We would like to thank all the contributing authors for lending their expertise to make the book truly unique. They have played a crucial role in the development of this book. Without their invaluable contributions this book wouldn't have been possible. They have made vital efforts to compile up to date information on the varied aspects of this subject to make this book a valuable addition to the collection of many professionals and students.

This book was conceptualized with the vision of imparting up-to-date information and advanced data in this field. To ensure the same, a matchless editorial board was set up. Every individual on the board went through rigorous rounds of assessment to prove their worth. After which they invested a large part of their time researching and compiling the most relevant data for our readers.

The editorial board has been involved in producing this book since its inception. They have spent rigorous hours researching and exploring the diverse topics which have resulted in the successful publishing of this book. They have passed on their knowledge of decades through this book. To expedite this challenging task, the publisher supported the team at every step. A small team of assistant editors was also appointed to further simplify the editing procedure and attain best results for the readers.

Apart from the editorial board, the designing team has also invested a significant amount of their time in understanding the subject and creating the most relevant covers. They scrutinized every image to scout for the most suitable representation of the subject and create an appropriate cover for the book.

The publishing team has been an ardent support to the editorial, designing and production team. Their endless efforts to recruit the best for this project, has resulted in the accomplishment of this book. They are a veteran in the field of academics and their pool of knowledge is as vast as their experience in printing. Their expertise and guidance has proved useful at every step. Their uncompromising quality standards have made this book an exceptional effort. Their encouragement from time to time has been an inspiration for everyone.

The publisher and the editorial board hope that this book will prove to be a valuable piece of knowledge for researchers, students, practitioners and scholars across the globe.

LIST OF CONTRIBUTORS

Marcos A. Caraballo-Ortiz and Eugenio Santiago-Valentín
Department of Biology, University of Puerto Rico, Río Piedras Campus, San Juan, Puerto Rico 00936–8377, USA
Herbarium, Botanical Garden of the University of Puerto Rico, 1187 Calle Flamboyán, San Juan, Puerto Rico 00926, USA
Center for Applied Tropical Ecology and Conservation, University of Puerto Rico, Río Piedras Campus, San Juan, Puerto Rico 00931–3341, USA

Fatih Kahrıman and Tuncay Aydın
Çanakkale Onsekiz Mart University, Faculty of Agriculture, Department of Field Crops, 17020, Çanakkale, Turkey

Cem Ömer Egesel and Selinnur Subaşı
Çanakkale Onsekiz Mart University, Faculty of Agriculture, Department of Agricultural Biotechnology, 17020, Çanakkale, Turkey

Barbara Gemmill-Herren
Food and Agriculture Organization of the United Nations, Viale delle terme di Caracalla, Rome 00153, Italy

Kwame Aidoo and Peter Kwapong
University of Cape Coast, Department of Entomology and Wildlife, School of Biological Sciences, Cape Coast, Ghana

Dino Martins
Turkana Basin Institute, Stony Brook University, Nairobi, Kenya

Wanja Kinuthia and Mary Gikungu
National Museums of Kenya, Centre for Bee Biology and Pollination, Nairobi, Kenya

Connal Eardley
Agricultural Research Council, Plant Protection Research Institute, Pretoria, 0001, South Africa / School of Biological and Conservation Sciences, University of Kwazulu Natal, Private Bag X01, Scottsville, Pietermaritzburg, 3209, South Africa

Carlos de Melo e Silva Neto, Flaviana Gomes Lima, Bruno Bastos Gonçalves, Leonardo Lima Bergamini, Barbara Araújo Ribeiro Bergamini, Marcos Antônio da Silva Elias and Edivani Villaron Franceschinelli
Departamento de Botânica, Instituto de Ciências Biológicas, Universidade Federal de Goiás, 74001-970, Goiânia, GO, Brazil

P. G. Willmer and K. Finlayson
Sir Harold Mitchell Building, School of Biology, University of St Andrews, St Andrews KY16 9TH

Eric A. Frimpong and Peter K. Kwapong
Department of Entomology and Wildlife, School of Biological Sciences, University of Cape Coast, Cape Coast, Ghana

Barbara Gemmill-Herren
AGPS – FAO, Viale Terme di Caracalla, Roma, Italy

Ian Gordon
International Centre of Insect Physiology and Ecology (ICIPE), Nairobi, Kenya

Saranya Arwen Carr and Priya Davidar
Department of Ecology and Environmental Sciences, Pondicherry University, Kalapet, Pondicherry 605014, India

Hannah R. Gaines-Day and Claudio Gratton
University of Wisconsin-Madison, Department of Entomology, 237 Russell Labs, 1630 Linden Dr., Madison, WI 53706, USA

David O. Chiawo
Strathmore University, Nairobi, Kenya,

Callistus K. P. O. Ogol
African Union Commission, Roosevelt Street, Addis Ababa, Ethiopia

Esther N. Kioko and Mary W. Gikungu
National Museums of Kenya, Nairobi, Kenya

Verrah A. Otiende
Pan African University, Nairobi, Kenya

Jeff Ollerton
School of Science and Technology, Newton Building, Avenue Campus, University of Northampton, Northampton, NN2 6JD, U. K

Clive Nuttman
The Tropical Biology Association, Department of Zoology, University of Cambridge, Downing Street, Cambridge, CB2 3EJ, U. K

V. J. Tepedino, W. R. Bowlin and T. L. Griswold
USDA ARS Bee Biology & Systematics Lab, Department of Biology, Utah State University, Logan UT 84322-5310

V. J. Tepedino
Department of Biology, Utah State University, Logan UT, USA 84322-5305

Laura C. Arneson
Department of Biology, Utah State University, Logan UT, USA 84322-5305
present address: 333 North Quince St., Salt Lake City UT, USA 84103

Susan L. Durham
Ecology Center, Utah State University, Logan UT, USA 84322-5205

Patrícia Nunes-Silva
Departamento de Biologia, Faculdade de Filosofia, Ciências e Letras de Ribeirão Preto, Universidade de São Paulo, Avenida Bandeirantes, 3900, 14040-901, Ribeirão Preto, Brazil

Michael Hnrcir and Vera Lucia Imperatriz-Fonseca
Departamento de Ciências Animais, Universidade Federal Rural do Semi-Árido, Avenida Francisco Mota, 572, 59.625-900, Mossoró, Brazil

Les Shipp
Agriculture and Agri-Food Canada, 2585 County Road #20, Harrow, Ontario, Canada, N0R 1G0

Peter G. Kevan
Canadian Pollination Initiative, School of Environmental Sciences, University of Guelph, ON N1G 2W1, Guelph, Canada

Lise Hansted, Brian W. W. Grout, Ivar B. Dencker and Torben B. Toldam-Andersen
Department of Agriculture and Ecology, University of Copenhagen, Højbakkegård Allé 13, 2630 Taastrup, Denmark

Jørgen Eilenberg
Department of Agriculture and Ecology, University of Copenhagen, Thorvaldsensvej 40, 1871 Frederiksberg C, Denmark

Diana B. Robson
The Manitoba Museum, 190 Rupert Avenue, Winnipeg, MB Canada R3B 0N2

Carlos H. Vergara and Paula Fonseca-Buendía
Laboratorio de Entomología, Departamento de Ciencias Químico-Biológicas, Universidad de las Américas Puebla. Ex-Hacienda Santa Catarina Mártir, 72820 Cholula, Puebla, México

Tuanjit Sritongchuay and Sara Bumrungsri
Department of Biology, Faculty of Science, Prince of Songkla University, Hat Yai, Thailand, 90122

Kristin Marie Lassen, Erik Dahl Kjær and Lene Rostgaard Nielsen
Department of Geosciences and Natural Resource Management, Faculty of Science, University of Copenhagen, Rolighedsvej 23, 1958 Frederiksberg C, Denmark

Moussa Ouédraogo
Centre National de Semences Forestières, Route de Kaya, 01 BP 2682 Ouagadougou, Burkina Faso

Yoko Luise Dupont
Department of Bioscience, Aarhus University, Vejlsøvej 25, 8600 Silkeborg, Denmark

Lina Herbertsson and Ida Gåvertsson
Lund University, Centre for Environmental and Climate Research, SE-223 62 Lund, Sweden

Björn Klatt and Henrik G. Smith
Lund University, Centre for Environmental and Climate Research, SE-223 62 Lund, Sweden
Lund University, Department of Biology, SE-223 62 Lund, Sweden

Victor H. Gonzalez
Undergraduate Biology Program and Department of Ecology and Evolutionary Biology, Haworth Hall, 1200 Sunnyside Ave., University of Kansas, Lawrence, Kansas, 66045, USA

Peter Cruz
Montclair State University, Montclair, New Jersey, 07043, USA

Nadiyah Folks
University of Texas at El Paso, El Paso, Texas, 79902

Sarah Anderson
University of Kansas, Lawrence, Kansas, 66045, USA

Dillon Travis
Boston University, Boston, Massachusetts, 02215, USA

John M. Hranitz
Biological and Allied Health Sciences, Bloomsburg University, Bloomsburg, PA, 17815, USA

John F. Barthell
Department of Biology and Office of Provost & Vice President for Academic Affairs, University of Central Oklahoma, Edmond, Oklahoma, 73034, USA

Jane C. Stout
School of Natural Sciences, Trinity College Dublin, Dublin, Republic of Ireland

Issa Nombre
Laboratoire de Biologie et Ecologie Végétales, Université Ouaga I Pr Joseph KI-ZERBO, Institut des sciences, 01 BP 1757 Ouagadougou 01, Burkina Faso

Bernd de Bruijn
Vogelbescherming Nederland - BirdLife in The Netherlands, 3700 AX Zeist, The Netherlands

Dzigbodi Adzo Doke
Faculty of Natural Resources and Environment, University for Development Studies, Tamale, Ghana

Thomas Gyimah
Ghana Wildlife Society, Accra, Ghana

Francois Kamano
BirdLife International, David Attenborough Building, Pembroke Street, Cambridge, CB2 3QZ, UK

Prudence Tankoano
Naturama, 01 B.P. 6133, 01, Ouagadougou, Burkina Faso

Peter Kwapong
Department of Conservation Biology and Entomology, University of Cape Coast and International Stingless Bee Centre, Cape Coast, Ghana

R Patchett and PG Willmer
Sir Harold Mitchell Building, School of Biology, University of St Andrews, St Andrews KY16 9TH

G Ballantyne
School of Applied Sciences, Edinburgh Napier University, Edinburgh, EH11 4BN

Max N. Buxton, Brian T. Cutting, Heather McBrydie and Mark Goodwin
The New Zealand Institute for Plant & Food Research Limited, Ruakura Research Centre, Hamilton 3210, New Zealand

David Pattemore
The New Zealand Institute for Plant & Food Research Limited, Ruakura Research Centre, Hamilton 3210, New Zealand
School of Biological Sciences, University of Auckland, Auckland, New Zealand

Arnon Dag
Gilat Research Center, Agricultural Research Organization, 85280, Israel

Mark J. F. Brown
School of Biological Sciences, Royal Holloway University of London, Egham, Surrey, UK

Dylan Hodgkiss
School of Biological Sciences, Royal Holloway University of London, Egham, Surrey, UK
NIAB EMR, New Road, East Malling, Kent, UK

Michelle T. Fountain
NIAB EMR, New Road, East Malling, Kent, UK

Index

www.ingramcontent.com/pod-product-compliance
Lightning Source LLC
Chambersburg PA
CBHW080649200326
41458CB00013B/4793